KU-198-441

THE URBAN SOCIOLOGY READER

The Urban Sociology Reader draws together seminal selections from the nineteenth to the twenty-first centuries. Contributions from Georg Simmel, Louis Wirth, Robert Ezra Park, Sharon Zukin, and Saskia Sassen are included among the 33 selections.

The urban world is a provocative terrain to contemplate central institutions, structures, and problems of society and how they have transformed over the past 200 years. The reader traverses this terrain with sections on urban social theory, social difference in the city, culture and the urban economy, globalization and urban change, and urban social movements. Selections are predominantly sociological but some readings go across disciplinary boundaries.

The reader provides an essential resource for students of urban studies, bringing together important and diverse writings. Editorial commentaries preceding the entries discuss their significance as well as related issues of relevance, while the bibliography allows deeper investigations.

Jan Lin is Associate Professor of Sociology at Occidental College, Los Angeles.

Christopher Mele is Associate Professor of Sociology at State University of New York at Buffalo.

THE ROUTLEDGE URBAN READER SERIES

Series editors

Richard T. LeGates

Professor of Urban Studies, San Francisco State University

Frederic Stout

Lecturer in Urban Studies, Stanford University

The Routledge Urban Reader Series responds to the need for comprehensive coverage of the classic and essential texts that form the basis of intellectual work in the various academic disciplines and professional fields concerned with cities.

The readers focus on the key topics encountered by undergraduates, graduates and scholars in urban studies and allied fields. They discuss the contributions of major theoreticians and practitioners and other individuals, groups, and organizations that study the city or practice in a field that directly affects the city.

As well as drawing together the best of classic and contemporary writings on the city, each reader features extensive general, section and selection introductions prepared by the volume editors to place the selections in context, illustrate relations among topics, provide information on the author and point readers toward additional related bibliographic material.

Each reader will contain:

- Approximately thirty-six *selections* divided into approximately six sections. Almost all of the selections will be previously published works that have appeared as journal articles or portions of books.
- A *general introduction* describing the nature and purpose of the reader.
- A two- to three-page *section introduction* for each section of the reader to place the readings in context.
- A one-page *selection introduction* for each selection describing the author, the intellectual background of the selection, competing views of the subject matter of the selection and bibliographic references to other readings by the same author and other readings related to the topic.
- A plate section with twelve to fifteen plates and illustrations at the beginning of each section.
- An index.

The types of readers and forthcoming titles are as follows:

THE CITY READER

The City Reader: third edition – an interdisciplinary urban reader aimed at urban studies, urban planning, urban geography and urban sociology courses – will be the *anchor urban reader.* Routledge published a first edition of *The City Reader* in 1996 and a second edition in 2000. *The City Reader* has become one of the most widely used anthologies in urban studies, urban geography, urban sociology and urban planning courses in the world.

URBAN DISCIPLINARY READERS

The series will contain *urban disciplinary readers* organized around social science disciplines. The urban disciplinary readers will include both classic writings and recent, cutting-edge contributions to the respective disciplines. They will be lively, high-quality, competitively priced readers which faculty can adopt as course texts and which will also appeal to a wide audience.

TOPICAL URBAN ANTHOLOGIES

The urban series will also include *topical urban readers* intended both as primary and supplemental course texts and for the trade and professional market.

INTERDISCIPLINARY ANCHOR TITLE

The City Reader: third edition
Richard T. LeGates and Frederic Stout (eds)

URBAN DISCIPLINARY READERS

The Urban Geography Reader
Nicholas R. Fyfe and Judith T. Kenny (eds)

The Urban Sociology Reader
Jan Lin and Christopher Mele (eds)

Forthcoming:

The Urban Politics Reader
Elizabeth Strom and John Mollenkopf (eds)

The Urban and Regional Planning Reader
Eugenie Birch (ed.)

The Urban Design Reader
Michael Larice and Elizabeth Macdonald (eds)

TOPICAL URBAN READERS

The City Cultures Reader: second edition
Malcolm Miles and Tim Hall, with Iain Borden (eds)

The Cybercities Reader
Stephen Graham (ed.)

The Sustainable Urban Development Reader
Stephen M. Wheeler and Timothy Beatley (eds)

Forthcoming:

The Global Cities Reader
Neil Brenner and Roger Keil (eds)

For further information on The Routledge Urban Reader Series
please visit our website:

www.geographyarena.com/geographyarena/urbanreaderseries

or contact:

Andrew Mould
Routledge
Haines House
21 John St
London WC1N 2BP
United Kingdom
andrew.mould@tandf.co.uk

Richard T. LeGates
Urban Studies Program
San Francisco State University
1600 Holloway Avenue
San Francisco, California 94132
(415) 338-2875
dlegates@sfsu.edu

Frederic Stout
Urban Studies Program
Stanford University
Stanford, California 94305-6050
(650) 725-6321
fstout@stanford.edu

The Urban Sociology Reader

Christopher Mele

LONDON AND NEW YORK

First published 2005
by Routledge
2 Park Square, Milton Park, Abingdon, Oxon OX14 4RN

Simultaneously published in the USA and Canada
by Routledge
270 Madison Ave, New York, NY 10016

Routledge is an imprint of the Taylor & Francis Group

Typeset in Amasis and Akzidenz Grotesk by
Graphicraft Limited, Hong Kong
Printed and bound in Great Britain by
Bell & Bain Ltd, Glasgow

British Library Cataloguing in Publication Data
A catalogue record for this book is available from the British Library

Library of Congress Cataloging in Publication Data
The urban sociology reader / edited by Jan Lin and Christopher Mele.
p. cm. — (The Routledge urban reader series)
Includes bibliographical references and index.
1. Sociology, Urban. I. Lin, Jan. II. Mele, Christopher. III. Series.
HT108.U733 2005
307.76—dc22 2004021893

ISBN 0–415–32342–8 (hbk)
ISBN 0–415–32343–6 (pbk)

Contents

Plates

Acknowledgments

This anthology is dedicated to Janet Abu-Lughod, currently Emeritus Professor of Sociology at the New School University in New York City. We were among the first cohort of sociologists that Janet trained at the Graduate Faculty of the New School upon her move from Northwestern University in 1987. We became keen urbanists under her guidance, stimulated by her deep comparative knowledge of communities, cities, and civilizations, and her passionate commitment to public intellectualism as an urban planner and a proponent of research with an action focus in the neighboring communities of Lower Manhattan. She supervised our dissertation research, Jan Lin working in Chinatown and Christopher Mele in the East Village. Lin and Mele have maintained a lively partnership through the years, in part through our association with the Community and Urban Sociology Section (CUSS) of the American Sociological Association (ASA). Our knowledge of the field has expanded through our work on the editorial board of the *Urban Affairs Review*, and as book review editors for the journals *Urban Affairs Review* (Christopher Mele) and *City and Community* (Jan Lin).

We gratefully acknowledge Richard T. LeGates, the pioneer and general editor (along with Frederic Stout) of The Routledge Urban Reader Series, for inviting us (with a kind referral from Janet Abu-Lughod) to undertake this volume. His active encouragement and valuable insights helped to improve our book proposal and manuscript immensely along every stage of its development. Andrew Mould, our editor at Routledge, was also of instrumental importance in shepherding our project through its various phases. His capable staff has been of critical assistance, including Melanie Attridge and Anna Somerville. We would also like to thank the anonymous reviewers for their very helpful critiques and comments during the proposal phase of our project.

We acknowledge the comraderie of other students associated with Research About Lower Manhattan (REALM), which Janet Abu-Lughod created at the New School University. This group includes: Dorien Greshof, Andy Van Kleunen, Abby Scher, John Dale, Umberto Blumati, Kumru Toktamisch, Juulia Kauste, Warren Goldstein, Joel Stillerman and Perry Chang. Other graduate students and New York-based faculty associated with REALM were Saskia Sassen, Neil Smith, Frank de Giovanni, Arthur Vidich, Bill Sites, Suzanne Wasserman, Marci Reaven, and Diana R. Gordon. Diane Davis and Sharon Zukin were very supportive.

We thank Dean Kenyon Chan of Occidental College for funding to support the purchase of visual images for our reader. We thank Urban Sociology students at the University of Houston, Amherst College, and Occidental College for being an academic testing ground over the years for many of the selections contained in this anthology. Mike Kerwin of Occidental College and Emilie Broderick at the University at Buffalo were helpful in scanning, digitizing, and editing the readings.

In preparing our reader, we found Philip Kasinitz's edited reader, *Metropolis: Center and Symbol of Our Times* (New York: New York University Press, 1995) to be very useful. Richard Sennett's edited reader, *Classic Essays on the Culture of Cities* (Englewood Cliffs, NJ: Prentice-Hall, Inc., 1969) was also very enlightening.

GENERAL INTRODUCTION

Cities are focal arenas for the contemplation of the human condition and man's struggle for self-expression. Cities are landscapes of cultural diversity and subcultural differentiation, what Robert Park called a "mosaic of social worlds." The bohemias, bright light districts, and red light districts of the city are crucibles for the exploration of artistic, cultural, and sexual expression. The city contains our workplaces, our residences, and our commercial marketplaces. The metropolis is a terrain of social inequality, from the decline and deterioration of disadvantaged and marginalized places like the Southside of Chicago and New York's Bowery to the affluence of prime spaces like Midtown Manhattan and Rodeo Drive in Los Angeles. Cities are landscapes of gender inequality and social differences in sexuality. Cities are key sites in the transformation of the global economy. There is a new cultural economy of cities that gives us an analytic window on the character of our post-industrial society. Rising inequality has led to a climate of fear in cities, which have become high-security fortresses. Urban social movements have arisen to articulate the demands of the socially and economically disenfranchised in our cities.

The Urban Sociology Reader offers seminal selections that span the sub-field from the nineteenth to the twenty-first century. We offer the kinds of readings that will spark broad, humanistic debate and discourse in the classic liberal studies tradition as well as advanced disciplinary debates within urban sociology. *The Urban Sociology Reader* is aimed at graduate urban sociologists as well as the general liberal studies readers. We target the kinds of readers drawn to urban sociology while contemplating or involved in careers such as education, urban policy and planning, public administration and government, community organizing, and even the arts. We promote the exploration of urban sociological practice as well as theory.

The urban world is a provocative terrain to contemplate central experiences, structures, and problems of the social world, and how they have transformed over the past two hundred years. Our reader traverses that terrain through central themes of urban social theory (classic and contemporary), inequality and social difference (by class, race/ethnicity, gender and sexuality), globalization and the world-system, culture and the urban economy, and urban social movements. Our selections are predominantly sociological, but some readings cut across disciplinary boundaries, reflecting underlying movements in the social sciences and social changes in the real world since the 1960s. These movements include those under the broad rubrics of "multiculturalism," "globalization," "postmodernism," and "cultural studies." Our selections are primarily American, but we give some attention to European writings (both classic and contemporary) as well cities in the developing world. We have attempted to offer a balanced mix of political–economic as well as cultural perspectives.

Ethnographic studies represent a strong tradition in urban sociology, beginning with the many community studies published in the early years of the Chicago School. Undergraduates find ethnographies enjoyable to read; they are drawn to their clear, jargon-free presentation and story-telling plotlines. The readings we have chosen represent the tradition of "thick description," and are written in a lively style. These include readings by Loïc Wacquant, James Duncan, Donald Donham, Paul Stoller and Jasmin Tahmaseb McConatha, and Teresa Caldeira.

Our reader is aimed primarily at a North American audience, but we have included a variety of non-North American selections to expose readers to important international and comparative urban issues and perspectives. We have included selections from classic and contemporary European sociology, including Ferdinand Tönnies, Georg Simmel, Loïc Wacquant, Pierre Hamel, Henri Lustiger-Thaler and Margit Mayer, and Bettina Kohler and Markus Wissen. We have also included selections from other world regions, such as Africa (Donald Donham) and Latin America (Teresa Caldeira).

Our anthology begins with Part One, Urbanism and Community, drawing from classic European as well as American texts in urban sociology spanning the nineteenth and twentieth centuries. The selections by Ferdinand Tönnies and Georg Simmel consider the major social and psychological changes that accompanied urbanization and the development of capitalism in Europe. Tönnies was comparable to Emile Durkheim in linking the decline of primary ties and community life with increasing specialization in the division of labor and social life. Simmel saw the capitalist city as a sensorium that assaulted the urbanite with a cacophony of sights and sounds, including advertising, commodities, pedestrians, and vehicular traffic. The decline of traditional mores and small town prejudices had fostered greater freedom and cosmopolitanism for the individual. Simmel believed the experience of modernity was somewhat paradoxical; the urban commercial sensorium fed the self while starving the spirit. The liberated individual was also a restless one. There was a rootlessness that came with participation in urban society and the modern marketplace.

Louis Wirth carried the perspectives of the nineteenth century European theorists into the American city of the early twentieth century. He drew greater attention than Simmel to the negative consequences of modernity, especially the status of Durkheimian *anomie* and the onset of urban social problems and personality disorders such as crime, delinquency, and mental illness. He felt that differentiation and the "mosaic of social worlds" fostered social pathology and social distance between people. Herbert Gans repudiated Wirthian urbanism and the idea of urban effects, arguing that suburbanism was just as legitimate a concept as urbanism. His concept of quasi-primary relationships in the suburbs provides a different view of the decline of Tönnesian *Gemeinschaft*. Fischer accepted the concept of general urban effects, but interpreted the "mosaic of social worlds" as a more positive phenomenon of subcultural differentiation. Fischer saw social differentiation as a creative process rather than a symptom of moral drift or social decline. Traditional communities gave way to subcultural communities with the growth of the metropolis.

Part Two examines classic and contemporary theories of the Form and Function of Cities from the Chicago School of human ecology to the rise of the urban political economy perspective and the L.A. School. The founder of the Chicago School, Robert Park, applied the ideas of Charles Darwin to justify the presence of urban social inequality, though he ultimately diverged from Social Darwinists like Herbert Spencer, who were advocates of imperialism and racial eugenics. Ernest Burgess promulgated the Chicago School's famous concentric zone theory of urban development, which was based on the city of Chicago during the railroad phase of its development, and Harvey Zorbaugh codified their understanding of natural areas. Burgess, like Zorbaugh, believed that increasingly geographic and social mobility led to urban social disorganization. Zorbaugh felt that slums attracted unstable people who further undermined the social order and group identity of the city. The Chicago School had a totally negative view of the consequences of human mobility and cultural change. Scholars like Simmel and Fischer had more forgiving views on mobility, identifying positive outcomes such as cultural freedom, cosmopolitanism, and subcultural creativity.

Walter Firey offered a sociocultural critique of human ecology that debunked the notion that market pressures always triumphed over cultural factors in the form and function of cities. Socioculturalists such as Firey presaged the growth of historic preservation interests in U.S. cities, and were a harbinger of the "cultural turn" in urban sociology associated with urban revitalization in the 1980s. The socioculturalists also gave way to the urban political economy perspective that emerged in the 1960s in conjunction with a variety of urban social movements. John Logan and Harvey Molotch are representative of this perspective, which is highly critical of social inequality and urban elites. They promote a political economy of place that renunciates the human ecology perspective. Their concept of the urban power elite as a "growth machine" has influenced a new generation of urban scholars. Michael Dear completes the critique of human ecology with his discussion of the L.A. School of urban studies. Dear takes the sociocultural critique into the terrain of postmodernism, suggesting that the modernist hegemony of urban elites has given way to a polycentric, polyglot, and polycultural pastiche of urban development. Los Angeles has superseded Chicago as the paradigm of urban growth in the twenty-first century.

Part Three examines issues of Inequality and Social Difference with reference to the poor and homeless, who inhabit the ghettoes, barrios, skid rows and other "badlands" of the metropolis. Cultural issues come to the fore again as we consider how the African American and Latino underclass, along with the homeless are stigmatized as the dangerous "other" by the media and urban power elite. This cultural marginalization reinforces their spatial segregation in the marginal spaces of the metropolis. Loïc Wacquant and William Wilson begin with a chapter that considers the "hyperghettoization" of the ghetto underclass in the U.S. despite the passage of Civil Rights legislation in the 1960s. Suburbanization and economic change has left the underclass socially isolated in the inner city, bereft of access to good jobs, schools, and housing. Douglas Massey and Nancy Denton place greater emphasis on racial issues with their outrage at the "hypersegregation" that continues to plague the black ghetto with the failure of civil rights legislation in America. They decry the condition of "American Apartheid" that is the result of discriminatory lending, racial steering by realtors, and redlining practices by banks and mortgage companies. White flight leads to the disenfranchisement of the white suburbs from the needs and concerns of the black inner cities. The spatial isolation of the underclass intensifies the presence of an "oppositional culture" in the ghetto that further marginalizes the residents from mainstream society. Massey and Denton point to a very clear dialectic between spatial segregation and social exclusion. Race trumps class as the leading problem confronting the underclass, an argument at odds with the early writings of William Wilson.

Loïc Wacquant engages in comparative research on the U.S. and French underclass, and finds that African Americans feel a more profound sense of social and political exclusion than French North Africans. James S. Duncan extends our understanding of segregation to the analysis of the homeless. He offers a symbolic interactionist perspective on the ways that the police and general public tolerate homeless people in the prime public spaces of the city as long as they follow a particular etiquette. Alejandro Portes and Robert Manning examine the phenomenon of immigrant enclaves. Immigrant enclaves are significant phenomena, being alternative sub-economies separate from the U.S. mainstream economy and the dead-end jobs sector in which poor minorities proliferate. They are a kind of protected sector offering certain opportunities in exchange for exploitative working conditions. While some immigrant enterprises primarily serve co-ethnic clientele, other immigrant businesses serve as "middlemen" between white elites and poor minorities. These middleman minorities act as a social buffer between the rich and the poor and may sometimes bear the brunt of underclass anger against the broader system that marginalizes them.

Part Four considers issues of Gender and Sexuality. The concept that women may be segregated is considered with reference to the suburbs. The spatial entrapment of women in suburban housing is culturally reinforced, and this segregation impedes their job opportunities and social mobility so long as employment is based in the central city. Ann Markusen opines that the suburbs were a creation of patriarchy, more than being a benign outcome of consumer choice, highway lobbyists, suburban developers or the postwar Federal Housing Administration. She believes that the widespread separation of residential suburbs from central city workplaces leads to gross inefficiencies due to commuting times and energy consumption, as well as the alienation of many individuals and families. The privatizing of family life in the suburbs fosters the decline of extended family networks and community social capital. Markusen calls for urban social policies that re-collectivize child-care facilities, housing, job opportunities, and recreational activities for families in America. We invite speculation on the impact of the Internet on the economic and cultural life of women in the suburbs.

Melissa Gilbert considers a different kind of spatial entrapment, that of poor women of color in the inner city. She finds that poor women can adapt to this spatial segregation, effectively generating opportunity out of constraint, by exploiting the social capital inherent in dense networks of kin and friends to procure child-care assistance, housing, education, and other resources. Sy Adler and Johanna Brenner similarly explore the interpersonal social capital networks that operate among lesbian women in the city. Lesbian networks are often invisible relative to the high visibility of gay men in the city. Donald Donham explores the new visibility and more public sexual culture available to gay men in South Africa in the post-apartheid period.

Part Five moves back to the macro-sociological terrain to explore the phenomenon of Globalization and Urban Change. John Friedmann is the seminal writer on this subject, and his chapter considers the rise of world cities as command centers in the new international division of labor since the 1960s. Saskia Sassen considers how global cities are not the outcome of a benign economic logic, but emerge through the strategic machinations of transnational corporations working with governmental actors. Michael Peter Smith in a similar fashion denies the inevitability of globalization, and explores in greater detail the interplay between global and local processes in economic as well as cultural dimensions. Paul Stoller and Jasmin Tahmaseb McConatha take a closer look at the local dimensions of globalization through their study of West African street vendors in New York City as a transnational community network. Jan Lin examines the role of Asian and Latino enclaves as purveyors of global processes in a number of U.S. immigration gateway cities. He discusses both the cultural and economic activities of immigrants in leading some cities out of deindustrialization and urban decline, attracting the interests of planners and public officials, but leading to conflicts and contradiction among sub-ethnic interest groups.

Part Six examines the growth of Culture and the Urban Economy, a phenomenon that has balanced the decline in manufacturing employment in many cities. The cultural economy is related to the sectors known as the "information economy" or "high technology" but includes a wider array of activities related to the consumption of culture and cultural products. Sharon Zukin uses the term "symbolic economy" to describe the ensemble of activities that includes the arts, entertainment, sports, fashion, restaurants, and tourism. Public officials and other growth machine interests have begun to perceive the advantages of linking these kinds of activities to the redevelopment and gentrification process in central cities, provoking the displacement of marginal players such as the homeless, the poor, and mom-and-pop businesses. As the inner city is restructured and "revitalized," the "other" is simultaneously evicted in a physical sense, and appropriated in a symbolic sense for middle-class consumers. A series of conflicts and controversies are provoked over the ownership and control of culture, in its aesthetic, historical, and ethnic dimensions.

Richard Florida offers a more positive perspective on the cultural economy of cities with his concept of "creative capital," which is a variant of the "social capital" concept of Robert Putnam and others. He has provoked great interest from scholars, planners, and public officials through his attention to creative capital as a strategy for promoting local economic innovation and regional growth. Mark Gottdiener follows with a chapter on the growing production and consumption of themed environments in urban life. Applying the ideas of Jean Baudrillard, he observes that these themes and motifs are often floating signifiers completely disarticulated from local cultural contexts. Las Vegas is perhaps the paradigm of the American themed metropolis, with a multiplicity of symbols, motifs, and themed environments superimposed upon the barren desert. Christopher Mele returns to the issue of conflict and contradiction as he examines the displacement of a low-income Latino community and struggling artistic bohemians in the redevelopment and gentrification of the Lower East Side of Manhattan in the 1980s and 1990s. In a process of neighborhood invention, the local bohemian artistic culture was appropriated and repackaged as the district was marketed as a less gritty and sanitized "East Village" for an incoming population of well-heeled gentrifiers.

Part Seven explores the forces of Urban Exclusion and Social Resistance. Mike Davis makes cinematic references to both Roman Polanski's *Chinatown* (1974) and Ridley Scott's *Bladerunner* (1982) as he portrays Los Angeles as a metropolis of socioeconomic polarization and elite domination in the late twentieth century. The Community Redevelopment Agency (CRA) and transnational corporations (especially from East Asia) are the main growth machine interests behind the redevelopment of downtown as a command and headquarters center for the world-economy. Downtown Los Angeles is being restructured as a citadel for the Pacific Rim, with the homeless and poor displaced to the margins. Teresa Caldeira similarly paints a portrait of social inequality and spatial segregation in São Paulo, where middle- and upper-class residents wall themselves in gated communities for fear of the dangerous "other" beyond high-security gates.

These ominous portrayals of global cities as fortresses of social polarization and exclusion give way to our last two chapters, which examine the efforts of grassroots community-based movements to promote the interests of those marginalized by urban restructuring in the age of globalization. Pierre Hamel, Henri Lustiger-Thaler, and Margit Mayer analyze a variety of struggles to advocate for the urban poor, refugees and immigrants, and the homeless. Finally, Bettina Kohler and Markus Wissen examine the ensemble of environmental, labor, and citizen movements that have emerged in opposition to corporate globalization and neoliberal policies of free trade and government privatization around the globe.

PART ONE

Urbanism and Community

Plate 1 A transatlantic arriving in the harbor, New York City, 1959 by Henri Cartier-Bresson (reproduced by permission of Magnum Photos).

Plate 2 Skateboarder and biker (reproduced by permission of Corbis).

Plate 3 Breakdancer on Pier 39, Fisherman's Wharf, San Francisco by Lindsay Hagan (reproduced by permission of the artist).

Plate 4 Woman with tattoos, Haight Street, San Francisco by Lindsay Hagan (reproduced by permission of the artist).

INTRODUCTION TO PART ONE

Modern cities are creations of the capitalist system. As the industrial revolution and the market economy began to transform rural society beginning in the eighteenth and nineteenth centuries in countries such as England, France, Germany, and the United States, the emerging field of sociology began to consider the links between urbanization and economic and social change. As crucibles of modernity, cities were becoming workplaces as well as showplaces of capitalism, with the emergence of manufacturing districts as well as commercial shopping boulevards. Rural-to-urban migrants filled the growing cities, making the adjustment from traditional to modern ways of life. Feudal and communal traditions gave way to the modern capitalist city of factories, banks, and marketplaces. The increase in the mobility of labor, capital, commodities, and culture also bred social inequality, crime, and related social problems. The decline of traditional norms and mores accompanied the growing mobility of ideas and culture, stimulating spiritual restlessness, artistic creativity, and subcultural expression. The city emerged as a focal site of modernity that promoted contradictory forces of freedom and rootlessness. The metropolis became a crucial arena in man's modern struggle with self-definition.

Karl Polanyi has elucidated how the "great transformation" of the nineteenth century was the "grand idea" that stimulated a host of intellectuals, such as Karl Marx, Max Weber, and Emile Durkheim, as they struggled to understand class struggle, rational bureaucracy, and the division of labor. Thinkers began to speculate on the effects of urbanization and modernization on society as well as the self. The idea of urbanism as a way of life became a central subject of urban sociological study. Ferdinand Tönnies was one of the first sociologists to elucidate the rural–urban shift through the conceptual categories, *Gemeinschaft* (community) and *Gesellschaft* (urban society). These concepts are good illustrations of the use of *ideal types* in sociological analysis. The ideal type functions as an analytical paradigm or model that can be analyzed and tested for its validity through comparison. Tönnies did not consider these societal types as mutually exclusive polar opposites, but as two categories in a continuum of societies undergoing social change. The shift from *Gemeinschaft* to *Gesellschaft* may be compared with Emile Durkheim's conception of society undergoing a transition from mechanical to organic solidarity. What Tönnies described as societal *will* is similar to what Durkheim described as collective consciousness, a collective soul or conscience that guides the behavior of individuals.

Both Tönnies and Durkheim tied social change to increasing specialization in the division of labor and differentiation in the body social. They recognized the fading of primary bonds of kinship, ritual, and community life, superseded by the growth of secondary bonds of association linked with occupation, law, and politics with the expansion of capitalism and urbanization. Tönnies has traditionally been viewed as having a romantic view of the loss of *Gemeinschaft* but in fact he saw *Gesellschaft* as a rational and necessary vehicle for guiding a more specialized and diverse society. The transition from village to city could be likened to the shift from a simple to a complex social organism, from infancy to maturity. Tönnies' outlook on an urban society of rational specialization and interdependency is a contrast from Marx's view that class conflict and worker revolt would arise with the growth of capitalism. Tönnies' concern was that *Gesellschaft* not be sabotaged or kidnapped by corrupt or totalitarian political interests such as Fascism. Durkheim, in contrast, was more concerned with the moral consequences of the rise of a *Gesellschaft* society.

Georg Simmel also believed in the contradictory character of urbanism and modernity. He recognized the negative effects of urban intensification on psychic and social life in the city, which led to growing anonymity and impersonality. Simmel provided an intriguing outlook on the importance of commerce and money, which he perceived as permeating modern urban life with a calculating and discriminating nature. The blasé urban personality is jaded, materialistic, and cosmopolitan. This cosmopolitanism provides some positive returns to the individual in fostering greater social tolerance for unconventional behaviors and freedom from traditionalism, provinciality, and prejudice. The oversaturation of our social life with materialism, superficiality, and objective values, however, has suppressed our subjectivity, spirituality, and social life. His quote on the *atrophy of individual culture through the hypertrophy of objective culture* may be understood as the idea that the capitalist city starves the spirit while overfeeding the self. The urban personality is both bombarded and liberated by the sensory commercial marketplaces of modern capitalism. For Simmel, the experience of modern urban life is equivalent to the experience of a money economy where quality has been reduced to quantity and consumers are materially rewarded but spiritually deprived. There is a loneliness that is brought about by a society of material plenty and liberation that has freed people to explore their individualism but left their souls in a state of restlessness, flux, and even anxiety.

German expressionist artists who were contemporaneous with Simmel depicted this jarring sense of sensory intensity, commercial spectacle, and social change in their paintings and drawings. Ernst Ludwig Kirchner devoted a set of paintings (such as "Two Women in the Street" and "Street: Dresden") to depictions of urban street scenes in bold, electric colors, conveying images of pedestrians attired in cosmopolitan fashion but also confronted by the cacophony of traffic and the claustrophobia of the crowd. Their faces are expressionless and even mask-like, conveying the anonymity and loneliness of the modern urban hustle and bustle. George Grosz, in his drawings such as "Berlin Alexanderplatz" (1918) depicts a similar atmosphere of cacophony and flux with pedestrians and urban landmarks drawn in a montage of jarring juxtapositions rather than linear perspective. Military men mingle with pedestrians in his composition, reflecting the growing militarization of German society and the creeping Fascism that was organizing during the late Weimar period in cities like Berlin and Dresden.

The anxiety and disquiet that was reflected in Simmel and the German expressionists contrasts with the more dreamy and romantic depictions of Paris as a capital of culture and commodity spectacle, through the writing of Walter Benjamin (see his essay "Paris, Capital of the Nineteenth Century," in *The Arcades Project* (Cambridge, MA: Harvard University Press, 1999 [1939]) and writers and poets such as Charles Baudelaire. Their pedestrian protagonist is a bourgeois dandy, or "flaneur," who wanders in a drunken, dream-like reverie through the shopping arcades of Paris with less of a sense of neurosis, but in a similar state of restlessness. These comparisons suggest both similarities and differences between the French and German experience of urbanism and modernity.

Louis Wirth provides a seminal application of a Durkheimian viewpoint on urbanism. Durkheim offered an interpretation of the contradictory force in modernity in that the liberating of the individual from traditional religious and social rules simultaneously reinforced the need for new legal codes and universal rules of behavior. Wirth perceived that factors of size, density, and heterogeneity, which accompanied urbanism, fostered role segmentation through the emancipation of the individual from traditional rules and morals. This status of *anomie*, the social void, resulted in the onset of urban social problems, such as crime, delinquency, mental breakdown, and other forms of psychological and social disorder. He adapted Robert Park's famous quote of the city (see "The City: Suggestions for the Investigation of Human Behavior in the City Environment," *American Journal of Sociology* 20, 5 (March 1915): 577–612) as "a mosaic of social worlds which touch but do not interpenetrate" to the concept of the city as a "mosaic of social worlds in which the transition from one to the other is abrupt." Geographic mobility, the growing decline of traditional norms and mores, and social heterogeneity were breeding social and personality disorders in the city. Wirth drew more attention than Simmel to the negative consequences of urbanism, and suggested that sociologists had a mission to analyze and ameliorate urban social problems.

Herbert Gans rebuked the Wirthian theory of urbanism for its overemphasis on the urban core and neglect of suburban life. Gans posited the presence of suburbanism as a new paradigm of neighborliness through quasi-primary relationships that emerge as a surrogate for kinship and communal ties. Gans downplayed the importance of "urban effects," saying urban social behavior is determined more by social class, ethnicity, and life-cycle characteristics than size, density, and heterogeneity. Claude Fischer reformulated rather than repudiating Wirthian urbanism, accepting the concrete urban effects of size, density, and heterogeneity in more positive than negative terms. He suggested that size and density of population in cities created the "critical mass" necessary to give birth to new subcultural communities. The increasingly heterogeneous "mosaic of social worlds" further intensified subcultures through his concept that they touch, but then "recoil, with sparks flying." His concept of subculture includes an eclectic assortment of special hobbyists, interest groups, artists, innovative thinkers, ethnic groups, religious subcultures, homosexuals and others commonly classified as "deviant." He could include breakdancers, skateboarders, and BMX bikers. That they congregate socially and spatially as communities reverses the traditional thinking that urbanism leads to the decline of community and the growth of social disorganization. Fischer sees cities as crucibles of subcultural diversity and social difference.

Simmel's jaded metropolitan man of the early twentieth century may be compared with the subcultural urbanists of the contemporary metropolis. Consider also the restlessness and emotional turmoil communicated by modern hip-hop music. The song "The Message," released by Grand Master Flash and the Furious Five, in 1982, considered by rapper aficionados as an "old school" form of hip-hop, expresses many of the themes enunciated by Simmel many decades earlier. The protagonist in "The Message" expresses a similar sense of urban schizophrenia, but there is more of a sense of angst, violence, and social marginalization expressed by these subcultural protagonists than Simmel's cosmopolitans of the boulevards and theaters. New school hip-hop has somewhat departed from its roots in an identity politics of street credibility to impersonate an image of subcultural cool as "bling-bling" and material success. Many gangsta rappers now emulate the conspicuous consumption of the elites, impersonating success and fashionability, rather than projecting a sense of deprivation and schizophrenia in street life.

Examine the phenomenon of breakdancing and rap music sampling, as illustrated in films such as Charlie Ahearn's *Wild Style* (1997). The breakdancers move in mechanical fashion, much like robots, expressing the experience of living in a society increasingly permeated by technology, mechanization, and dehumanization. Sometimes the dancers sport gloves, and pantomine the experience of being trapped in invisible boxes, somewhat expressing the condition of compartmentalization and bureaucratization in our rationalized society. The cut-and-mix sampling techniques of rapper disc jockeys and emcees are additionally significant in that they use the technique of montage in recycling, transforming, and juxtaposing different elements from a diverse repertoire of sounds.

"Community and Society"

from C. P. Loomis (ed.), *Community and Society* (1963)[1887]

Ferdinand Tönnies

Editors' Introduction

Ferdinand Tönnies (1855–1936) was born into a wealthy farming family in Schleswig-Holstein, Germany, in an era in which the peasant culture of the rural province was being transformed by mechanization and the money-economy. His oldest brother was engaged in a thriving trade with English merchants, exposing Tönnies first hand to the world of English capitalism. In 1881 he became a lecturer at the University of Kiel, where he remained until ousted by the Nazis in 1933 because of his social democratic political associations. Though less influential than his contemporaries Max Weber and Emile Durkheim, Tönnies may be recognized as a founding father of sociology.

His enduring contribution to urban sociology is the distinction between two basic types of social formations, *Gemeinschaft* (community) and *Gesellschaft* (society), with a general historical trend from the former to the latter. Societies of the earlier form are organized around family, village, and town, with a mainly agricultural economy and local political culture. The latter form of society, by contrast, is exemplified by larger level social units of metropolis and nation-state, and based on complex trade and industry. Primary sentimental relationships predominate in *Gemeinschaft*, while secondary associational relationships proliferate in *Gesellschaft*. While some of his interpreters proliferated the impression that Tönnies sentimentalized *Gemeinschaft* while criticizing *Gesellschaft*, he disclaimed such intention. For him, the shift was a normal developmental process of the body social, comparable to the transition from youth to adulthood.

Tönnies was strongly influenced by English thinkers, including the political philosopher Thomas Hobbes, Sir Henry Maine, and the Social Darwinist Herbert Spencer. The concept of will was central to his theory. Tönnies argued that there are two basic forms of human volition, or will. *Gemeinschaft* is formed around *Wesenwille*, or essential will, which is the underlying, organic, self-fulfilling or instinctive driving force, while *Gesellschaft* is characterized by *Kurwille*, or arbitrary will, which is deliberative, purposive, instrumental, and future (goal) oriented. *Wesenwille* is that which springs intrinsically from a person's temper and character. *Kurwille* is the capacity to distinguish means from ends and to act practically out of rational self-interest.

Tönnies decried totalitarianism (including the Nazism that emerged in Germany), but he was intrigued by the force of "public opinion" that enforces the communal will of society and may involve the use of sanctions against dissidents. He dealt with these ideas in other publications, including *Die Sitte* (1909) and *Critique of Public Opinion* (Lanham: Rowman & Littlefield, 2002, edited and translated by Hanno Hardt and Slavko Splichal from *Kritik der Offentlichen Meinung*, 1922). His concept of *Kurwille* can thus be related to the Hobbesian social contract, whereby citizens control the state through deliberation and reasoned discussion to counter tyrannical authority and avaricious despotism.

Tönnies developed his concepts *Gemeinschaft/Gesellschaft* as "ideal types" which are paradigms or models that may not fully conform to social reality, but are useful for purposes of analytical comparison. Rather than being polar extremes, the two ideal types can be seen as being on opposite ends of a continuum. Tönnies conceived of any society as always to some degree possessing characteristics of both ideal types. The original concept of ideal types may be credited to the German sociologist Max Weber. *Gemeinschaft* may be compared with the traditional society conceived by the French sociologist Emile Durkheim (*The Division of Labor in Society* [1893], translated by George Simpson. New York: Free Press, 1933) through his notion of mechanical solidarity, characterized by a simple division of labor and a morally homogeneous population bound by similar values and beliefs. *Gesellschaft* corresponds with Durkheim's notion of organic solidarity, found in the modern society that has a complex division of labor and a heterogeneous population held together by interdependency, laws, and contracts. The American sociologist Robert Redfield, on the basis of fieldwork in rural Mexico, later characterized the traditional society as the "folk society" ("The Folk Society," *American Journal of Sociology* 52 (1947), 293–308).

Gemeinschaft und Gesellschaft is also available in an earlier edition, which also contained some of Tönnies' later essays, as *Fundamental Concepts of Sociology* (1940). Tönnies' ten other books, of which the major work dealing with sociology is his 1931 *Einführung in die Soziologie* (*An Introduction to Sociology*), plus most of his essays, still await English translations. A full bibliography of Tönnies' work can be found in: *American Journal of Sociology*, 42 (1937), 100–101. A brief critique of Tönnies' works can be found in Louis Wirth, "The Sociology of Ferdinand Tönnies," *American Journal of Sociology*, 32 (1927), 412–422.

ORDER – LAW – MORES

There is a contrast between a social order which – being based upon consensus of wills – rests on harmony and is developed and ennobled by folkways, mores, and religion, and an order which – being based upon a union of rational wills – rests on convention and agreement, is safeguarded by political legislation, and finds its ideological justification in public opinion.

There is, further, in the first instance a common and binding system of positive law, of enforcible norms regulating the interrelation of wills. It has its roots in family life and is based on land ownership. Its forms are in the main determined by the code of the folkways and mores. Religion consecrates and glorifies these forms of the divine will, i.e., as interpreted by the will of wise and ruling men. This system of norms is in direct contrast to a similar positive law which upholds the separate identity of the individual rational wills in all their interrelations and entanglements. The latter derives from the conventional order of trade and similar relations but attains validity and binding force only through the sovereign will and power of the state. Thus, it becomes one of the most important instruments of policy; it sustains, impedes, or furthers social trends; it is defended or contested publicly by doctrines and opinions and thus is changed, becoming more strict or more lenient.

There is, further, the dual concept of morality as a purely ideal or mental system of norms for community life. In the first case, it is mainly an expression and organ of religious beliefs and forces, by necessity intertwined with the conditions and realities of family spirit and the folkways and mores. In the second case, it is entirely a product and instrument of public opinion, which encompasses all relations arising out of contractual sociableness, contacts, and political intentions.

Order is natural law, law as such = positive law, mores = ideal law. Law as the meaning of what may or ought to be, of what is ordained or permitted, constitutes an object of social will. Even the natural law, in order to attain validity and reality, has to be recognized as positive and binding. But it is positive in a more general or less definite way. It is general in comparison with special laws. It is simple compared to complex and developed law.

DISSOLUTION

The substance of the body social and the social will consists of concord, folkways, mores, and religion, the manifold forms of which develop under favorable conditions during its lifetime. Thus, each individual receives his share from this common center, which is manifest in his own sphere, i.e., in his sentiment, in his mind and heart, and in his conscience as well as in his environment, his possessions, and his activities. This is also true of each group. It is in this center that the individual's strength is rooted, and his rights derive, in the last instance, from the one original law which, in its divine and natural character, encompasses and sustains him, just as it made him and will carry him away. But under certain conditions and in some relationships, man appears as a free agent (person) in his self-determined activities and has to be conceived of as an independent person. The substance of the common spirit has become so weak or the link connecting him with the others worn so thin that it has to be excluded from consideration. In contrast to the family and co-operative relationship, this is true of all relations among separate individuals where there is no common understanding, and no time-honored custom or belief creates a common bond. This means war and the unrestricted freedom of all to destroy and subjugate one another, or, being aware of possible greater advantage, to conclude agreements and foster new ties. To the extent that such a relationship exists between closed groups or communities or between their individuals or between members and non members of a community, it does not come within the scope of this study. In this connection we see a community organization and social conditions in which the individuals remain in isolation and veiled hostility toward each other so that only fear of clever retaliation restrains them from attacking one another, and, therefore, even peaceful and neighborly relations are in reality based upon a warlike situation. This is, according to our concepts, the condition of Gesellschaft-like civilization, in which peace and commerce are maintained through conventions and the underlying mutual fear. The state protects this civilization through legislation and politics. To a certain extent science and public opinion, attempting to conceive it as necessary and eternal, glorify it as progress toward perfection.

But it is in the organization and order of the Gemeinschaft that folk life and folk culture persist. The state, which represents and embodies Gesellschaft, is opposed to these in veiled hatred and contempt, the more so the further the state has moved away from and become estranged from these forms of community life. Thus, also in the social and historical life of mankind there is partly close interrelation, partly juxtaposition and opposition of natural and rational will.

THE PEOPLE (VOLKSTUM) AND THE STATE (STAATSTUM)

In the same way as the individual natural will evolves into pure thinking and rational will, which tends to dissolve and subjugate its predecessors, the original collective forms of Gemeinschaft have developed into Gesellschaft and the rational will of the Gesellschaft. In the course of history, folk culture has given rise to the civilization of the state.

The main features of this process can be described in the following way. The anonymous mass of the people is the original and dominating power which creates the houses, the villages, and the towns of the country. From it, too, spring the powerful and self-determined individuals of many different kinds: princes, feudal lords, knights, as well as priests, artists, scholars. As long as their economic condition is determined by the people as a whole, all their social control is conditioned by the will and power of the people. Their union on a national scale, which alone could make them dominant as a group, is dependent on economic conditions. And their real and essential control is economic control, which before them and with them and partly against them the merchants attain by harnessing the labor force of the nation. Such economic control is achieved in many forms, the highest of which is planned capitalist production or large-scale industry. It is through the merchants that the technical conditions for the national union of independent individuals and for capitalistic production are created. This merchant class is by nature, and mostly also by origin, international as well as national and urban, i.e., it belongs to Gesellschaft, not Gemeinschaft. Later all social groups and dignitaries and, at least in tendency, the whole people acquire the characteristics of the Gesellschaft.

Men change their temperaments with the place and conditions of their daily life, which becomes hasty and changeable through restless striving. Simultaneously, along with this revolution in the social order, there takes place a gradual change of the law, in meaning as well as in form. The contract as such becomes the basis of the entire system, and rational will of Gesellschaft, formed by its interests, combines with authoritative will of the state to create, maintain and change the legal system. According to this conception, the law can and may completely change the Gesellschaft in line with its own discrimination and purpose; changes which, however, will be in the interest of the Gesellschaft, making for usefulness and efficiency. The state frees itself more and more from the traditions and customs of the past and the belief in their importance. Thus, the forms of law change from a product of the folkways and mores and the law of custom into a purely legalistic law, a product of policy. The state and its departments and the individuals are the only remaining agents, instead of numerous and manifold fellowships, communities, and commonwealths which have grown up organically. The characters of the people, which were influenced and determined by these previously existing institutions, undergo new changes in adaptation to new and arbitrary legal constructions. These earlier institutions lose the firm hold which folkways, mores, and the conviction of their infallibility gave to them.

Finally, as a consequence of these changes and in turn reacting upon them, a complete reversal of intellectual life takes place. While originally rooted entirely in the imagination, it now becomes dependent upon thinking. Previously, all was centered around the belief in invisible beings, spirits and gods; now it is focalized on the insight into visible nature. Religion, which is rooted in folklife or at least closely related to it, must cede supremacy to science, which derives from and corresponds to consciousness. Such consciousness is a product of leaning and culture and, therefore, remote from the people. Religion has an immediate contact and is moral in its nature because it is most deeply related to the physical–spiritual link which connects the generations of men. Science receives its moral meaning only from an observation of the laws of social life, which leads it to derive rules for an arbitrary and reasonable order of social organization. The intellectual attitude of the individual becomes gradually less and less influenced by religion and more and more influenced by science. Utilizing the research findings accumulated by the preceding industrious generation, we shall investigate the tremendous contrasts which the opposite poles of this dichotomy and these fluctuations entail. For this presentation, however, the following few remarks may suffice to outline the underlying principles.

TYPES OF REAL COMMUNITY LIFE

The exterior forms of community life as represented by natural will and Gemeinschaft were distinguished as house, village, and town. These are the lasting types of real and historical life. In a developed Gesellschaft, as in the earlier and middle stages, people live together in these different ways. The town is the highest, viz., the most complex, form of social life. Its local character, in common with that of the village, contrasts with the family character of the house. Both village and town retain many characteristics of the family; the village retains more, the town less. Only when the town develops into the city are these characteristics almost entirely lost. Individuals or families are separate identities, and their common locale is only an accidental or deliberately chosen place in which to live. But as the town lives on within the city, elements of life in the Gemeinschaft, as the only real form of life, persist within the Gesellschaft, although lingering and decaying. On the other hand, the more general the condition of Gesellschaft becomes in the nation or a group of nations, the more this entire "country" or the entire "world" begins to resemble one large city. However, in the city and therefore where general conditions characteristic of the Gesellschaft prevail, only the upper strata, the rich and the cultured, are really active and alive. They set up the standards to which the lower strata have to conform. These lower classes conform partly to supersede the others, partly in imitation of them in order to attain for themselves social power and independence. The city consists, for both groups (just as in the case of the "nation" and the "world"), of free persons who stand in contact with each other, exchange with each other and cooperate without any Gemeinschaft or will thereto developing among them except as such might develop sporadically or as a leftover from former conditions. On the contrary, these numerous external contacts, contracts, and contractual relations only cover up as many inner hostilities and antagonistic interests. This is especially true of the antagonism between the rich or the so-called cultured class and the poor or the servant class, which try to obstruct and destroy each other. It is this contrast which, according to Plato, gives the "city" its dual character and makes it divide in itself. This itself, according to our concept, constitutes the city, but the same contrast is also manifest in every large-scale relationship between capital and labor. The common town life remains within the Gemeinschaft of family and rural life; it is devoted to some agricultural pursuits but concerns itself especially with art and handicraft which evolve from these natural needs and habits. City life, however, is sharply distinguished from that; these basis activities are used only as means and tools for the special purposes of the city.

The city is typical of Gesellschaft in general. It is essentially a commercial town and, in so far as commerce dominates its productive labor, a factory town. Its wealth is capital wealth which, in the form of trade, usury, or industrial capital, is used and multiplies. Capital is the means for the appropriation of products of labor or for the exploitation of workers. The city is also the center of science and culture, which always go hand in hand with commerce and industry. Here the arts must make a living; they are exploited in a capitalistic way. Thoughts spread and change with astonishing rapidity. Speeches and books through mass distribution become stimuli of far-reaching importance.

The city is to be distinguished from the national capital, which, as residence of the court or center of government, manifests the features of the city in many respects although its population and other conditions have not yet reached that level. In the synthesis of city and capital, the highest form of this kind is achieved: the metropolis. It is the essence not only of a national Gesellschaft, but contains representatives from a whole group of nations, i.e., of the world. In the metropolis, money and capital are unlimited and almighty. It is able to produce and supply goods and science for the entire earth as well as laws and public opinion for all nations. It represents the world market and world traffic; in it world industries are concentrated. Its newspapers are world papers, its people come from all corners of the earth, being curious and hungry for money and pleasure.

COUNTERPART OF GEMEINSCHAFT

Family life is the general basis of life in the Gemeinschaft. It subsists in village and town life. The village community and the town themselves can be considered as large families, the various clans and houses representing the elementary organisms of its body; guilds, corporations, and offices, the tissues and organs of the town. Here original kinship and inherited status remain an essential, or at least the most important, condition of participating fully in common property and other rights. Strangers may be accepted and protected as serving-members or guests either temporarily or permanently. Thus, they can belong to the Gemeinschaft as objects, but not easily as agents and representatives of the Gemeinschaft. Children are, during minority, dependent members of the family, but according to Roman custom they are called free because it is anticipated that under possible and normal conditions they will certainly be masters, their own heirs. This is true neither of guests nor of servants, either in the house or in the community. But honored guests can approach the position of children. If they are adopted or civic rights are granted to them, they fully acquire this position with the right to inherit. Servants can be esteemed or treated as guests or even, because of the value of their functions, take part as members in the activities of the group. It also happens sometimes that they become natural or appointed heirs. In reality there are many gradations, lower or higher, which are not exactly met by legal formulas. All these relationships can, under special circumstances, be transformed into merely interested and dissolvable interchange between independent contracting parties. In the city such change, at least with regard to all relations of servitude, is only natural and becomes more and more widespread with its development. The difference between natives and strangers becomes irrelevant. Everyone is what he is, through his personal freedom, through his wealth and his contracts. He is a servant only in so far as he has granted certain services to someone else, master in so far as he receives such services. Wealth is, indeed, the only effective and original differentiating characteristic; whereas in Gemeinschaften property it is considered as participation in the common ownership and as a specific legal concept is entirely the consequence and result of freedom or ingenuity, either original or acquired. Therefore, wealth, to the extent that this is possible, corresponds to the degree of freedom possessed.

In the city as well as in the capital, and especially in the metropolis family life is decaying. The more and the longer their influence prevails the more the residuals of family life acquire a purely accidental character. For there are only few who will confine their energies within such a narrow circle; all are attracted outside by business, interests, and pleasures, and thus separated from one another. The great and mighty, feeling free and independent, have always felt a strong inclination to break through the barriers of the folkways and mores. They know that they can do as they please. They have the power to bring about changes in their favor, and this is positive proof of individual arbitrary power. The mechanism of money, under usual conditions and if working under high pressure, is means to overcome all resistance, to obtain everything wanted and desired, to eliminate all dangers and to cure all evil. This does not hold always. Even if all controls of the Gemeinschaft are eliminated, there are nevertheless controls in the Gesellschaft to which the free and independent individuals are subject. For Gesellschaft (in the narrower sense), convention takes to a large degree the place of the folkways, mores, and religion. It forbids much as detrimental to the common interest which the folkways, mores, and religion had condemned as evil in and of itself.

The will of the state plays the same role through law courts and police, although within narrower limits. The laws of the state apply equally to everyone; only children and lunatics are not held responsible to them. Convention maintains at least the appearance of morality; it is still related to the folkways, mores, and religious and aesthetic feeling, although this feeling tends to become arbitrary and formal. The state is hardly directly concerned with morality. It has only to suppress and punish hostile actions which are detrimental to the common weal or seemingly dangerous for itself and society. For as the state has to administer the common weal, it must be able to define this as it pleases. In the end it will probably realize that no increase in knowledge and culture alone will make people kinder, less egotistic, and more content and that dead folkways, mores, and religions cannot be revived by coercion and teaching. The state will then arrive at the conclusion that in order to create moral forces and moral beings it must prepare the ground and fulfill the necessary conditions, or at least it must eliminate counteracting forces. The state, as the reason of Gesellschaft, should decide to destroy Gesellschaft or at least to reform or renew it. The success of such attempts is highly improbable.

THE REAL STATE

Public opinion, which brings the morality of Gesellschaft into rules and formulas and can rise above the state, has nevertheless decided tendencies to urge the state to use its irresistible power to force everyone to do what is useful and to leave undone what is damaging. Extension of the penal code and the police power seems the right means to curb the evil impulses of the masses. Public opinion passes easily from the demand for freedom (for the upper classes) to that of despotism (against the lower classes). The makeshift convention has but little influence over the masses. In their striving for pleasure and entertainment they are limited only by the scarcity of the means which the capitalists furnish them as price for their labor, which condition is as general as it is natural in a world where the interests of the capitalists and merchants anticipated all possible needs and in mutual competition incite to the most varied expenditures of money. Only through fear of discovery and punishments, that is, through fear of the state, is a special and large group, which encompasses far more people than the professional criminals, restrained in its desire to obtain the key to all necessary and unnecessary pleasures. The state is their enemy. The state, to them, is an alien and unfriendly power; although seemingly authorized by them and embodying their own will, it is nevertheless opposed to all their needs and desires, protecting property which they do not possess, forcing them into military service for a country which offers them hearth and altar only in the form of a heated room on the upper floor or gives them, for native soil, city streets where they may stare at the glitter and luxury in lighted windows forever beyond their reach! Their own life is nothing but a constant alternative between work and leisure, which are both distorted into factory routine and the low pleasure of the saloons. City life and Gesellschaft down the common people to decay and death; in vain they struggle to attain power through their own multitude, and it seems to them that they can use their power only for a revolution if they want to free themselves from their fate. The masses become conscious of this social position through the education in schools and through newspapers. They proceed from class consciousness to class struggle. This class struggle may destroy society and the state which it is its purpose to reform. The entire culture has been transformed into a civilization of state and Gesellschaft, and this transformation means the doom of culture itself if none of its scattered seeds remain alive and again bring forth the essence and idea of Gemeinschaft, thus secretly fostering a new culture amidst the decaying one.

"The Metropolis and Mental Life"

from Kurt H. Wolff (ed.), *The Sociology of Georg Simmel* (1950)[1903]

Georg Simmel

Editors' Introduction

Georg Simmel (1858–1918) was born to a prosperous Jewish family, at the intersection of Friedrichstrasse and Leipzigerstrasse, the very heart of Berlin's commercial and theatrical bright light district, the equivalent of New York's Times Square or London's Piccadilly Circus. Simmel obtained his doctorate in philosophy in 1881 at the University of Berlin. Marginalized by the German academic system because of Jewish ancestry and intellectual radicalism, Simmel did not obtain a regular academic appointment until the last four years of his life. For most of his career, he maintained a recurring lecturing position at the University of Berlin, where his lectures influenced an extraordinary legacy of students, including Georg Lukacs, Ernst Bloch, Karl Mannheim and Robert Park. Despite being somewhat an academic outsider, he was nevertheless an engaged public intellectual who frequented fashionable salons and enjoyed the friendship of eminent sociologists like Max Weber and the German poet Rainer Maria Rilke. As a nonobservant Jew in Weimar Berlin, he was a rootless cosmopolitan while being a public figure.

Simmel's seminal essay ("Die Grossstätde und das Geistesleben") was originally delivered as a lecture within a series during the winter of 1902–03 connected to an exhibition held in Dresden by the Gehe Foundation on the emergence of the modern metropolis. This First German City Exhibit (*Erste Deutsche Städte-Ausstellung*) was following upon the 1896 Berlin Trade Exhibition, part of a historical vogue in world city expositions, such as those held in Paris in 1886 and Chicago in 1893. The lectures and exhibits examined the intellectual, economic, and political dimensions of German urbanism, and addressed planning problems and social issues related to public transportation, housing, employment, health, welfare, and cultural institutions. Simmel's essay focused more upon the philosophical and psychological implications of these transitions. Simmel was interested in the social construction of the modern urban self.

The commercial emporium of the world city expositions framed Simmel's view of the metropolis as the nexus point for the circulation of capital, commodities, and people. That commerce was central to the great transformations of modernity was not lost upon Marx in his writings on the political economy of capital, but what Simmel explored in this essay as well as his magnum opus, *The Philosophy of Money* (1909), were the philosophical and psychological dimensions of money in modern culture. Simmel discussed the triumph of the money-economy over traditional society, the rise of objectification and quantification. He saw the capitalist city as a sensorium of psychic overstimulations and commodity temptations. The decline of fixed, ancient, and venerable traditions with the rise of flux, transitoriness, and arbitrary value is what Marx described with his famed adage: "all that is solid melts into air" (see "The Communist Manifesto").

Simmel's detached and capricious urban cosmopolitan is much similar to the "flaneur" of philosopher Walter Benjamin ("Paris, Capital of the Nineteenth Century,") and poet Charles Baudelaire, the Parisian pedestrian who sumptuously wandered the shopping arcades and boulevards, intoxicated by the spectacle of commerce and the anonymity of the urban crowd. The notion of the loneliness of life in the crowd reflects the particular reserve and impersonality that is displayed by the pedestrian dandy, the bourgeois shopper, or the urban commuter. The barrage of lures, stimulations, and choices in the modern city of commerce has induced a kind of monkish self-reflection that can be seen as transcendence as much as retreat. Freed from the prejudices and obligations of family and community, the bourgeois urbanist experiences the restlessness of liberation, a new condition of self-consciousness and inner emotional development. For all his liberation from the communal society, the urban modernist is now embedded in the iron cage of a world of work and bureaucracy, and the consumer's dilemma of a search for identity in a soulless mass society.

Simmel's seminal essay on "The Stranger" further elucidates his interpretation of the soul of the metropolitan man who is more marginal to the axes of power. The short but powerful essay expresses some of the outsider status Simmel experienced in the academy, while communicating some general properties of Simmel's thought regarding the dialectic between the individual and the society. Simmel's wandering Jewish trader bears the stigmata of the quintessential outsider who is not regarded as an individual, but as a type of social type or category. This status may be extended to other social types such as the black underclass, immigrant foreigner, the social or sexual deviant as Simmel suggests: "the stranger, like the poor and like sundry 'inner enemies,' is an element of the group itself." The presence of the stranger establishes spatial rules and social etiquettes of social distance. Differentiation of the "other" as well as the "deviant" establishes rules of conduct in a secular society, sustaining the solidarity of the in-group.

Simmel's interest in micro-sociological realms, the minutiae of everyday life, has attracted sociologists associated with the "cultural turn" in sociology since the 1960s. The British sociologist David Frisby has recently done much in the translating and popularizing of Simmel in this way. His writings include *Georg Simmel* (London and Chichester: Tavistock/Ellis Horwood, 1984), *Simmel and Since: Essays on Georg Simmel's Social Theory* (London: Routledge, 1992), *Sociological Impressionism: A Reassessment of Georg Simmel's Social Theory* (London: Routledge, 1992), and a reader of original translated writings titled *Simmel on Culture: Selected Writings*, edited by David Frisby and Mike Featherstone (Thousand Oaks, CA: Sage Publications, 1997). An excellent biography of Simmel can be found in Lewis A. Coser, *Masters of Sociological Thought: Ideas in Historical and Social Context*, Second Edition (New York: Harcourt Brace Jovanovich, 1977).

The deepest problems of modern life derive from the claim of the individual to preserve the autonomy and individuality of his existence in the face of overwhelming social forces, of historical heritage, of external culture, and of the technique of life. The fight with nature which primitive man has to wage for his bodily existence attains in this modern form its latest transformation. The eighteenth century called upon man to free him of all the historical bonds in the state and in religion, in morals and in economics. Man's nature, originally good and common to all, should develop unhampered. In addition to more liberty, the nineteenth century demanded the functional specialization of man and his work; this specialization makes one individual incomparable to another, and each of them indispensable to the highest possible extent. However, this specialization makes each man the more directly dependent upon the supplementary activities of all others. Nietzsche sees the full development of the individual conditioned by the most ruthless struggle of individuals; socialism believes in the suppression of all competition for the same reason. Be that as it may, in all these positions the same basic motive is at work: the person resists to being leveled down and worn out by a social–technological mechanism. An inquiry into the inner meaning of specifically modern life and its products, into the soul of the cultural body, so to speak, must seek to solve the equation that structures like the metropolis set up between the individual and the super-individual contents of life. Such an inquiry must answer the question of how the personality accommodates itself in the adjustments to external forces. This will be my task today.

The psychological basis of the metropolitan type of individuality consists in the *intensification of nervous stimulation* that results from the swift and uninterrupted change of outer and inner stimuli. Man is a differentiating creature. His mind is stimulated by the difference between a momentary impression and the one that preceded it. Lasting impressions, impressions which differ only slightly from one another, impressions which take a regular and habitual course and show regular and habitual contrasts – all these use up, so to speak, less consciousness than does the rapid crowding of changing images, the sharp discontinuity in the grasp of a single glance, and the unexpectedness of onrushing impressions. These are the psychological conditions that the metropolis creates. With each crossing of the street, with the tempo and multiplicity of economic, occupational and social life, the city sets up a deep contrast with small town and rural life with reference to the sensory foundations of psychic life. The metropolis exacts from man as a discriminating creature a different amount of consciousness than does rural life. Here the rhythm of life and sensory mental imagery flows more slowly, more habitually, and more evenly. Precisely in this connection the sophisticated character of metropolitan psychic life becomes understandable – as over against small town life which rests more upon deeply felt and emotional relationships. These latter are rooted in the more unconscious layers of the psyche and grow most readily in the steady rhythm of uninterrupted habituations. The intellect, however, has its locus in the transparent, conscious, higher layers of the psyche; it is the most adaptable of our inner forces. In order to accommodate to change and to the contrast of phenomena, the intellect does not require any shocks and inner upheavals; it is only through such upheavals that the more conservative mind could accommodate to the metropolitan rhythm of events. Thus the metropolitan type of man – which, of course, exists in a thousand individual variants – develops an organ protecting him against the threatening currents and discrepancies of his external environment which would uproot him. He reacts with his head instead of his heart. In this an increased awareness assumes the psychic prerogative. Metropolitan life, thus, underlies a heightened awareness and a predominance of intelligence in metropolitan man. The reaction to metropolitan phenomena is shifted to that organ which is least sensitive and quite remote from the depth of the personality. Intellectuality is thus seen to preserve subjective life against the overwhelming power of metropolitan life, and intellectuality branches out in many directions and is integrated with numerous discrete phenomena.

The metropolis has always been the seat of the money economy. Here the multiplicity and concentration of economic exchange gives an importance to the means of exchange that the scantiness of rural commerce would not have allowed. Money economy and the dominance of the intellect are intrinsically connected. They share a matter-of-fact attitude in dealing with men and with things; and, in this attitude, a formal justice is often coupled with an inconsiderate hardness. The intellectually sophisticated person is indifferent to all genuine individuality, because relationships and reactions result from it that cannot be exhausted with logical operations. In the same manner, the individuality of phenomena is not commensurate with the pecuniary principle. Money is concerned only with what is common to all: it asks for the exchange value, it reduces all quality and individuality to the question: How much? All intimate emotional relations between persons are founded in their individuality, whereas in rational relations man is reckoned with like a number, like an element that is in itself indifferent. Only the objective measurable achievement is of interest. Thus metropolitan man reckons with his merchants and customers, his domestic servants and often even with persons with whom he is obliged to have social intercourse. These features of intellectuality contrast with the nature of the small circle in which the inevitable knowledge of individuality as inevitably produces a warmer tone of behavior, a behavior which is beyond a mere objective balancing of service and return. In the sphere of the economic psychology of the small group it is of importance that under primitive conditions production serves the customer who orders the good, so that the producer and the consumer are acquainted. The modern metropolis, however, is supplied almost entirely by production for the market, that is, for entirely unknown purchasers who never personally enter the producer's actual field of vision. Through this anonymity the interests of each party acquire an unmerciful matter-of-factness; and the intellectually calculating economic egoisms of both parties need not fear any deflection because of the imponderables of personal relationships. The money economy dominates the metropolis; it has displaced the last survivals of domestic production and the direct barter of goods; it minimizes, from day to day, the amount of work ordered by customers. The matter-of-fact attitude is obviously so intimately interrelated with the money economy, which is dominant in the metropolis, that nobody can say whether the intellectualistic mentality first promoted the money economy or whether the latter determined the former. The metropolitan way of life is certainly the most fertile soil for this reciprocity, a point which I shall document merely by citing the dictum of the most eminent English constitutional historian: throughout the whole course of English history, London has never acted as England's heart but often as England's intellect and always as her moneybag!

In certain seemingly insignificant traits, which lie upon the surface of life, the same psychic currents characteristically unite. Modern mind has become more and more calculating. The calculative exactness of practical life that the money economy has brought about corresponds to the ideal of natural science: to transform the world into an arithmetic problem, to fix every part of the world by mathematical formulas. Only money economy has filled the days of so many people with weighing, calculating, with numerical determinations, with a reduction of qualitative values to quantitative ones. Through the calculative nature of money a new precision, a certainty in the definition of identities and differences, unambiguousness in agreements and arrangements has been brought about in the relations of life-elements – just as externally this precision has been affected by the universal diffusion of pocket watches. However, the conditions of metropolitan life are at once cause and effect of this trait. The relationships and affairs of the typical metropolitan usually are so varied and complex that without the strictest punctuality in promises and services the whole structure would break down into an inextricable chaos. Above all, this necessity is brought about by the aggregation of so many people with such differentiated interests, who must integrate their relations and activities into a highly complex organism. If all clocks and watches in Berlin would suddenly go wrong in different ways, even if only by one hour, all economic life and communication of the city would be disrupted for a long time. In addition an apparently mere external factor: long distances, would make all waiting and broken appointments result in an ill-afforded waste of time. Thus, the technique of metropolitan life is unimaginable without the most punctual integration of all activities and mutual relations into a stable and impersonal time schedule. Here again the general conclusions of this entire task of reflection become obvious namely, that from each point on the surface of existence – however closely attached to the surface alone – one may drop a sounding into the depth of the psyche so that all the most banal externalities of life finally are connected with the ultimate decisions concerning the meaning and style of life. Punctuality, calculability, exactness are forced upon life by the complexity and extension of metropolitan existence and are not only most intimately connected with its money economy and intellectualist character. These traits must also color the contents of life and favor the exclusion of those irrational, instinctive, sovereign traits and impulses which aim at determining the mode of life from within, instead of receiving the general and precisely schematized form of life from without. Even though sovereign types of personality, characterized by irrational impulses, are by no means impossible in the city, they are nevertheless, opposed to typical city life. The passionate hatred of men like Ruskin and Nietzsche for the metropolis is understandable in these terms. Their natures discovered the value of life alone in the unschematized existence that cannot be defined with precision for all alike. From the same source of this hatred of the metropolis surged their hatred of money economy and of the intellectualism of modern existence.

The same factors that have thus coalesced into the exactness and minute precision of the form of life have coalesced into a structure of the highest impersonality; on the other hand, they have promoted a highly personal subjectivity. There is perhaps no psychic phenomenon that has been so unconditionally reserved to the metropolis as has the blasé attitude. The blasé attitude results first from the rapidly changing and closely compressed contrasting stimulations of the nerves. From this, the enhancement of metropolitan intellectuality, also, seems originally to stem. Therefore, stupid people who are not intellectually alive in the first place usually are not exactly blasé. A life in boundless pursuit of pleasure makes one blasé because it agitates the nerves to their strongest reactivity for such a long time that they finally cease to react at all. In the same way, through the rapidity and contradictoriness of their changes, more harmless impressions force such violent responses, tearing the nerves so brutally hither and thither that their last reserves of strength are spent; and if one remains in the same milieu they have no time to gather new strength. Incapacity thus emerges to react to new sensations with the appropriate energy. This constitutes that blasé attitude which, in fact, every metropolitan child shows when compared with children of quieter and less changeable milieus.

This physiological source of the metropolitan blasé attitude is joined by another source that flows from the money economy. The essence of the blasé attitude consists in the blunting of discrimination. This does not mean that the objects are not perceived, as is the case with the half-wit, but rather that the meaning and differing values of things, and thereby the things themselves, are experienced as insubstantial. They appear to the blasé person in an evenly flat and gray tone; no one object deserves preference over any other. This mood is the faithful subjective reflection of the completely internalized money economy. By being the equivalent to all the manifold things in one and the same way, money becomes the most frightful leveler. For money expresses all qualitative differences of things in terms of "how much?" Money, with all its colorlessness and indifference, becomes the common denominator of all values; irreparably it hollows out the core of things, their individuality, their specific value, and their incomparability. All things float with equal specific gravity in the constantly moving stream of money. All things lie on the same level and differ from one another only in the size of the area that they cover. In the individual case this coloration, or rather discoloration, of things through their money equivalence may be unnoticeably minute. However, through the relations of the rich to the objects to be had for money, perhaps even through the total character that the mentality of the contemporary public everywhere imparts to these objects, the exclusively pecuniary evaluation of objects has become quite considerable. The large cities, the main seats of the money exchange, bring the purchasability of things to the fore much more impressively than do smaller localities. That is why cities are also the genuine locale of the blasé attitude. In the blasé attitude the concentration of men and things stimulate the nervous system of the individual to its highest achievement so that it attains its peak. Through the mere quantitative intensification of the same conditioning factors this achievement is transformed into its opposite and appears in the peculiar adjustment of the blasé attitude. In this phenomenon the nerves find in the refusal to react to their stimulation the last possibility of accommodating to the contents and forms of metropolitan life. The self-preservation of certain personalities is brought at the price of devaluing the whole objective world, a devaluation that in the end unavoidably drags one's own personality down into a feeling of the same worthlessness.

Whereas the subject of this form of existence has to come to terms with it entirely for himself, his self-preservation in the face of the large city demands from him a no less negative behavior of a social nature. This mental attitude of metropolitans toward one another we may designate, from a formal point of view, as reserve. If so many inner reactions were responses to the continuous external contacts with innumerable people as are those in the small town, where one knows almost everybody one meets and where one has a positive relation to almost everyone, one would be completely atomized internally and come to an unimaginable psychic state. Partly this psychological fact, partly the right to distrust that men have in the face of the touch-and-go elements of metropolitan life, necessitates our reserve. As a result of this reserve we frequently do not even know by sight those who have been our neighbors for years. And it is this reserve that in the eyes of the small-town people makes us appear to be cold and heartless. Indeed, if I do not deceive myself, the inner aspect of this outer reserve is not only indifference but, more often than we are aware, it is a slight aversion, a mutual strangeness and repulsion, which will break into hatred and fight at the moment of a closer contact, however caused. The whole inner organization of such an extensive communicative life rests upon an extremely varied hierarchy of sympathies, indifferences, and aversions of the briefest as well as of the most permanent nature. The sphere of indifference in this hierarchy is not as large as might appear on the surface. Our psychic activity still responds to almost every impression of somebody else with a somewhat distinct feeling. The unconscious, fluid and changing character of this impression seems to result in a state of indifference. Actually this indifference would be just as unnatural as the diffusion of indiscriminate mutual suggestion would be unbearable. From both these typical dangers of the metropolis, indifference and indiscriminate suggestibility, antipathy protects us. A latent antipathy and the preparatory stage of practical antagonism affect the distances and aversions without which this mode of life could not at all be led. The extent and the mixture of this style of life, the rhythm of its emergence and disappearance, the forms in which it is satisfied – all these, with the unifying motives in the narrower sense, form the inseparable whole of the metropolitan style of life. What appears in the metropolitan style of life directly as dissociation is in reality only one of its elemental forms of socialization.

This reserve with its overtone of hidden aversion appears in turn as the form or the cloak of a more general mental phenomenon of the metropolis: it grants to the individual a kind and an amount of personal freedom which has no analogy whatsoever under other conditions. The metropolis goes back to one of the large developmental tendencies of social life as such, to one of the few tendencies for which an approximately universal formula can be discovered. The earliest phase of social formations found in historical as well as in contemporary social structures is this: a relatively small circle firmly closed against neighboring, strange, or in some way antagonistic circles. However, this circle is closely coherent and allows its individual members only a narrow field for the development of unique qualities and free, self-responsible movements. Political and kinship groups, parties and religious associations begin in this way. The self-preservation of very young associations requires the establishment of strict boundaries and a centripetal unity. Therefore they cannot allow the individual freedom and unique inner and outer development. From this stage social development proceeds at once in two different, yet corresponding, directions. To the extent to which the group grows – numerically, spatially, in significance and in content of life – to the same degree the group's direct, inner unity loosens, and the rigidity of the original demarcation against others is softened through mutual relations and connections. At the same time, the individual gains freedom of movement, far beyond the first jealous delimitation. The individual also gains a specific individuality to which the division of labor in the enlarged group gives both occasion and necessity. The state and Christianity, guilds and political parties, and innumerable other groups have developed according to this formula, however much, of course, the special conditions and forces of the respective groups have modified the general scheme. This scheme seems to me distinctly recognizable also in the evolution of individuality within urban life. The small-town life in Antiquity and in the Middle Ages set barriers against movement and relations of the individual toward the outside, and it set up barriers against individual independence and differentiation within the individual self. These barriers were such that under them modern man could not have breathed. Even today a metropolitan man who is placed in a small town feels a restriction similar, at least, in kind. The smaller the circle which forms our milieu is, and the more restricted those relations to others are which dissolve the boundaries of the individual, the more anxiously the circle guards the achievements, the conduct of life, and the outlook of the individual, and the more readily a quantitative and qualitative specialization would break up the framework of the whole little circle.

The ancient *polis* in this respect seems to have had the very character of a small town. The constant threat to its existence at the hands of enemies from near and afar effected strict coherence in political and military respects, a supervision of the citizen by the citizen, a jealousy of the whole against the individual whose particular life was suppressed to such a degree that he could compensate only by acting as a despot in his own household. The tremendous agitation and excitement, the unique colorfulness of Athenian life, can perhaps be understood in terms of the fact that a people of incomparably individualized personalities struggled against the constant inner and outer pressure of a deindividualizing small town. This produced a tense atmosphere in which the weaker individuals were suppressed and those of stronger natures were incited to prove themselves in the most passionate manner. This is precisely why it was that there blossomed in Athens what must be called, without defining it exactly, "the general human character" in the intellectual development of our species. For we maintain factual as well as historical validity for the following connection: the most extensive and the most general contents and forms of life are most intimately connected with the most individual ones. They have a preparatory stage in common, that is, they find their enemy in narrow formations and groupings the maintenance of which places both of them into a state of defense against expanse and generality lying without and the freely moving individuality within. Just as in the feudal age, the "free" man was the one who stood under the law of the land, that is, under the law of the largest social orbit, and the unfree man was the one who derived his right merely from the narrow circle of a feudal association and was excluded from the larger social orbit – so today metropolitan man is "free" in a spiritualized and refined sense, in contrast to the pettiness and prejudices which hem in the small-town man. For the reciprocal reserve and indifference and the intellectual life conditions of large circles are never felt more strongly by the individual in their impact upon his independence than in the thickest crowd of the big city. This is because the bodily proximity and narrowness of space makes the mental distance only the more visible. It is obviously only the obverse of this freedom if, under certain circumstances, one nowhere feels as lonely and lost as in the metropolitan crowd. For here as elsewhere it is by no means necessary that the freedom of man be reflected in his emotional life as comfort.

It is not only the immediate size of the area and the number of persons that, because of the universal historical correlation between the enlargement of the circle and the personal inner and outer freedom, has made the metropolis the locale of freedom. It is rather in transcending this visible expanse that any given city becomes the seat of cosmopolitanism. The horizon of the city expands in a manner comparable to the way in which wealth develops; a certain amount of property increases in a quasi-automatical way in ever more rapid progression. As soon as a certain limit has been passed, the economic, personal, and intellectual relations of the citizenry, the sphere of intellectual predominance of the city over its hinterland, grow as in geometrical progression. Every gain in dynamic extension becomes a step, not for an equal, but for a new and larger extension. From every thread spinning out of the city, ever-new threads grow as if by themselves, just as within the city the unearned increment of ground rent, through the mere increase in communication, brings the owner automatically increasing profits. At this point, the quantitative aspect of life is transformed directly into qualitative traits of character. The sphere of life of the small town is, in the main, self-contained and autarchic. For it is the decisive nature of the metropolis that its inner life overflows by waves into a far-flung national or international area. Weimar is not an example to the contrary, since its significance was hinged upon individual personalities and died with them; whereas the metropolis is indeed characterized by its essential independence even from the most eminent individual personalities. This is the counterpart to the independence, and it is the price the individual pays for the independence, which he enjoys in the metropolis. The most significant characteristic of the metropolis is this functional extension beyond its physical boundaries. And this efficacy reacts in turn and gives weight, importance, and responsibility to metropolitan life. Man does not end with the limits of his body or the area comprising his immediate activity. Rather is the range of the person constituted by the sum of effects emanating from him temporally and spatially. In the same way, a city consists of its total effects that extend beyond its immediate confines. Only this range is the city's actual extent in which its existence is expressed. This fact makes it obvious that individual freedom, the logical and historical complement of such extension, is not to be understood only in the negative sense of mere freedom of mobility and elimination of prejudices and petty philistinism. The essential point is that the particularity and incomparability, which ultimately every human being possesses, be somehow expressed in the working-out of a way of life. That we follow the laws of our own nature – and this after all is freedom – becomes obvious and convincing to ourselves and to others only if the expressions of this nature differ from the expressions of others. Only our unmistakability proves that our way of life has not been superimposed by others.

Cities are, first of all, seats of the highest economic division of labor. They produce thereby such extreme phenomena as in Paris the remunerative occupation of the *quatorzième*. They are persons who identify themselves by signs on their residences and who are ready at the dinner hour in correct attire, so that they can be quickly called upon if a dinner party should consist of thirteen persons. In the measure of its expansion, the city offers more and more the decisive conditions of the division of labor. It offers a circle that through its size can absorb a highly diverse variety of services. At the same time, the concentration of individuals and their struggle for customers compel the individual to specialize in a function from which he cannot be readily displaced by another. It is decisive that city life has transformed the struggle with nature for livelihood into an inter-human struggle for gain, which here is not granted by nature but by other men. For specialization does not flow only from the competition for gain but also from the underlying fact that the seller must always seek to call forth new and differentiated needs of the lured customer. In order to find a source of income that is not yet exhausted, and to find a function that cannot readily be displaced, it is necessary to specialize in one's services. This process promotes differentiation, refinement, and the enrichment of the public's needs, which obviously must lead to growing personal differences within this public.

All this forms the transition to the individualization of mental and psychic traits that the city occasions in proportion to its size. There is a whole series of obvious causes underlying this process. First, one must meet the difficulty of asserting his own personality within the dimensions of metropolitan life. Where the quantitative increase in importance and the expense of energy reach their limits, one seizes upon qualitative differentiation in order somehow to attract the attention of the social circle by playing upon its sensitivity for differences. Finally, man is tempted to adopt the most tendentious peculiarities, that is, the specifically metropolitan extravagances of mannerism, caprice, and preciousness. Now, the meaning of these extravagances does not at all lie in the contents of such behavior, but rather in its form of "being different," of standing out in a striking manner and thereby attracting attention. For many character types, ultimately the only means of saving for themselves some modicum of self-esteem and the sense of filling a position is indirect, through the awareness of others. In the same sense a seemingly insignificant factor is operating, the cumulative effects of which are, however, still noticeable. I refer to the brevity and scarcity of the inter-human contacts granted to the metropolitan man, as compared with social intercourse in the small town. The temptation to appear "to the point," to appear concentrated and strikingly characteristic, lies much closer to the individual in brief metropolitan contacts than in an atmosphere in which frequent and prolonged association assures the personality of an unambiguous image of himself in the eyes of the other.

The most profound reason, however, why the metropolis conduces to the urge for the most individual personal existence – no matter whether justified and successful – appears to me to be the following: the development of modern culture is characterized by the preponderance of what one may call the "objective spirit" over the "subjective spirit." This is to say, in language as well as in law, in the technique of production as well as in art, in science as well as in the objects of the domestic environment, there is embodied a sum of spirit. The individual in his intellectual development follows the growth of this spirit very imperfectly and at an ever-increasing distance. If, for instance, we view the immense culture that for the last hundred years has been embodied in things and in knowledge, in institutions and in comforts, and if we compare all this with the cultural progress of the individual during the same period – at least in high status groups – a frightful disproportion in growth between the two becomes evident. Indeed, at some points we notice retrogression in the culture of the individual with reference to spirituality, delicacy, and idealism. This discrepancy results essentially from the growing division of labor. For the division of labor demands from the individual an ever more one-sided accomplishment, and the greatest advance in a one-sided pursuit only too frequently means death to the personality of the individual. In any case, he can cope less and less with the overgrowth of objective culture. The individual is reduced to a negligible quantity, perhaps less in his consciousness than in his practice and in the totality of his obscure emotional states that are derived from this practice. The individual has become a mere cog in an enormous organization of things and powers which tear from his hands all progress, spirituality, and value in order to transform them from their subjective form into the form of a purely objective life. It needs merely to be pointed out that the metropolis is the genuine arena of this culture that outgrows all personal life. Here in buildings and educational institutions, in the wonders and comforts of space-conquering technology, in the formations of community life, and in the visible institutions of the state, is offered such an overwhelming fullness of crystallized and impersonalized spirit that the personality, so to speak, cannot maintain itself under its impact. On the one hand, life is made infinitely easy for the personality in that stimulations, interests, uses of time and consciousness are offered to it from all sides. They carry the person as if in a stream, and one needs hardly to swim for oneself. On the other hand, however, life is composed more and more of these impersonal contents and offerings that tend to displace the genuine personal colorations and incomparabilities. This results in the individual's summoning the utmost in uniqueness and particularization, in order to preserve his most personal core. He has to exaggerate this personal element in order to remain audible even to himself. The atrophy of individual culture through the hypertrophy of objective culture is one reason for the bitter hatred that the preachers of the most extreme individualism, above all Nietzsche, harbor against the metropolis. But it is, indeed, also a reason why these preachers are so passionately loved in the metropolis and why they appear to the metropolitan man as the prophets and saviors of his most unsatisfied yearnings.

If one asks for the historical position of the two forms of individualism that are nourished by the quantitative relation of the metropolis, namely, individual independence and the elaboration of individuality itself, then the metropolis assumes an entirely new rank order in the world history of the spirit. The eighteenth century found the individual in oppressive bonds that had become meaningless – bonds of a political, agrarian, guild, and religious character. They were restraints that, so to speak, forced upon man an unnatural form and outmoded, unjust inequalities. In this situation the cry for liberty and equality arose, the belief in the individual's full freedom of movement in all social and intellectual relationships. Freedom would at once permit the noble substance common to all to come to the fore, a substance which nature had deposited in every man and which society and history had only deformed. Besides this eighteenth-century ideal of liberalism, in the nineteenth century, through Goethe and Romanticism, on the one hand, and through the economic division of labor, on the other hand, another ideal arose: individuals liberated from historical bonds now wished to distinguish themselves from one another. The carrier of man's values is no longer the "general human being" in every individual, but rather man's qualitative uniqueness and irreplaceability. The external and internal history of our time takes its course within the struggle and in the changing entanglements of these two ways of defining the individual's role in the whole of society. It is the function of the metropolis to provide the arena for this struggle and its reconciliation. For the metropolis presents the peculiar conditions which are revealed to us as the opportunities and the stimuli for the development of both these ways of allocating roles to men. Therewith these conditions gain a unique place, pregnant with inestimable meanings for the development of psychic existence. The metropolis reveals itself as one of those great historical formations in which opposing streams that enclose life unfold, as well as join one another with equal right. However, in this process the currents of life, whether their individual phenomena touch us sympathetically or antipathetically, entirely transcend the sphere for which the judge's attitude is appropriate. Since such forces of life have grown into the roots and into the crown of the whole of the historical life in which we, in our fleeting existence, as a cell, belong only as a part, it is not our task either to accuse or to pardon, but only to understand.

"Urbanism as a Way of Life"

from *American Journal of Sociology* (1930)

Louis Wirth

Editors' Introduction

Published in 1938, Wirth's essay on urbanism, and the factors of size, density, and heterogeneity, is one of the foundational statements of the Chicago School of urban sociology. It is clearly influenced by Ferdinand Tönnies, Georg Simmel, and Robert E. Park. Like Tönnies, he views the theory of urbanism as an ideal type. Wirth's concept of the "schizoid" urban personality, beset by "segmental roles," is akin to Simmel's blasé and reserved metropolitan man. Simmel felt, however, that the cosmopolitanism of city life liberated urbanites from the prejudices and provincialities of rural life. Wirth was less impressed by the positive benefits of this emancipation from primary group controls. He drew our attention to the growth of Durkheimian *anomie*, which consequently engendered a host of modern social problems, including crime, deviance, and various kinds of mental illness that were seen to proliferate in the city. Wirth also informed our understanding of Robert Park's concept of the city as a "mosaic of social worlds" that increases social distance between people. He viewed this as an outcome of urban density and specialization. He was more sensitive to the practical implications of a theory of urbanism than Tönnies or Simmel, as he suggested that knowledge of the causes of urban social problems were important to apply to a range of social policy and urban planning practices.

Louis Wirth was born August 28, 1897, in Gemünden, a small village in the Rhineland district of Germany to a Jewish rural cattle farming family. He followed his maternal uncle to Omaha, Nebraska, in the United States, to take advantage of educational opportunities. He was a successful high school debater and eventually won a scholarship to the University of Chicago. He flirted for a while with leftist anti-war causes during World War I, and then worked with delinquent boys with the Jewish Charities of Chicago after college. He obtained a Ph.D. in Sociology at the University of Chicago in 1925. His doctoral thesis on the Jewish quarter of Chicago was published as *The Ghetto* (Chicago: University of Chicago Press, 1928). After various teaching posts and fellowships, he joined the Chicago faculty under the chairmanship of Robert E. Park in 1931. In *The Ghetto*, Wirth examined the consequences of centuries of discrimination on Jewish community life, ranging from Renaissance Italy to Chicago's Maxwell Street. The book served as a model for the University's researchers in ethnicity, many of whom later studied under Wirth when he joined the University's faculty.

As a professor at the University of Chicago, Wirth blended empirical research and theory in his work and contributed to the emergence of sociology as a profession. The advent of the Roosevelt administration gave many opportunities for sociologists to work with government in congressional testimony, consulting, and funded research. Wirth also played a significant role in organizing an introductory course in the social sciences and was popularly known as a persuasive lecturer. During the late 1930s he grew involved in community affairs in Chicago and was often invited to make public addresses on urban planning and race relations. He became a well-known radio speaker, acting as a moderator and discussant on a series of 62 University of Chicago "round tables" broadcast between 1937 and 1952.

As an academic committed to social action, Louis Wirth became involved in numerous groups, committees, and associations concerned with the effects of racial prejudice on community life. He was a founder and president of the Chicago-based American Council on Race Relations, which sponsored research into problems of fair employment, education, housing, and integration. In 1947, with funds from the Carnegie and Rockefeller Foundations, Wirth also established the Committee on Education, Training, and Research in Race Relations at the University of Chicago. Led by Wirth, demographer Philip Hauser, and anthropologist Sol Tax, the Committee played a key role in addressing the social and political factors underlying racial discrimination in the city of Chicago.

Wirth was president of the American Sociological Society and his Presidential Address, "Consensus and Mass Communication," was delivered at the organization's annual meeting in New York City in December 1947. He was also the first president (1949–52) of the International Sociological Association. Wirth died suddenly and unexpectedly one spring day in 1952 in Buffalo, New York at the young age of 55. He had been in Buffalo to speak at a conference on community relations; he collapsed and died following his presentation.

Wirth also published a book on the selected writings of Karl Mannheim, entitled *Ideology and Utopia*, which he co-edited with Edward Shils. A useful book on Louis Wirth's legacy is by Albert J. Reiss, Jr., *Louis Wirth: On Cities and Social Life* (Chicago: University of Chicago Press, 1964). This book includes an excellent biographical memorandum by Elizabeth Wirth Marvick.

THE CITY AND CONTEMPORARY CIVILIZATION

Just as the beginning of Western civilization is marked by the permanent settlement of formerly nomadic peoples in the Mediterranean basin, so the beginning of what is distinctively modern in our civilization is best signalized by the growth of great cities. Nowhere has mankind been farther removed from organic nature than under the conditions of life characteristic of these cities. The contemporary world no longer presents a picture of small isolated groups of human beings scattered over a vast territory as Sumner described primitive society. The distinctive feature of man's mode of living in the modern age is his concentration into gigantic aggregations around which cluster lesser centers and from which radiate the ideas and practices that we call civilization.

[. . .]

Since the city is the product of growth rather than of instantaneous creation, it is to be expected that the influences which it exerts upon the modes of life should not be able to wipe out completely the previously dominant modes of human association. To a greater or lesser degree, therefore, our social life bears the imprint of an earlier folk society, the characteristic modes of settlement of which were the farm, the manor, and the village. This historic influence is reinforced by the circumstances that the population of the city itself is in large measure recruited from the countryside, where a mode of life reminiscent of this earlier form of existence persists. Hence we should not expect to find abrupt and discontinuous variation between urban and rural types of personality. The city and the country may be regarded as two poles in reference to one or the other of which all human settlements tend to arrange themselves. In viewing urban-industrial and rural-folk society as ideal types of communities, we may obtain a perspective for the analysis of the basic models of human association as they appear in contemporary civilization.

SOCIOLOGICAL DEFINITION OF THE CITY

Despite the preponderant significance of the city in our civilization, our knowledge of the nature of urbanism and the process of urbanization is meager, notwithstanding many attempts to isolate the distinguishing characteristics of urban life. Geographers, historians, economists, and political scientists have incorporated the points of view of their respective disciplines into diverse definitions of the city. While in no sense intended to supersede these, the formulation of a sociological approach to the city may incidentally serve to call attention to the interrelations between them by emphasizing the peculiar characteristics of the city as a particular form of human association. A sociologically significant definition of the city seeks to select those elements of urbanism which mark it as a distinctive mode of human group life.

[. . .]

While urbanism, or that complex of traits which makes up the characteristic mode of life in cities, and urbanization, which denotes the development and extensions of these factors, are thus not exclusively found in settlements which are cities in the physical and demographic sense, they do, nevertheless, find their most pronounced expression in such areas, especially in metropolitan cities. In formulating a definition of the city it is necessary to exercise caution in order to avoid identifying urbanism as a way of life with any specific locally or historically conditioned cultural influences which, though they may significantly affect the specific character of the community, are not the essential determinants of its character as a city.

[. . .]

For sociological purposes a city may be defined as a relatively large, dense, and permanent settlement of socially heterogeneous individuals. On the basis of the postulates which this minimal definition suggests, a theory of urbanism may be formulated in the light of existing knowledge concerning social groups.

A THEORY OF URBANISM

In the rich literature on the city we look in vain for a theory systematizing the available knowledge concerning the city as a social entity. We do indeed have excellent formulations of theories on such special problems as the growth of the city viewed as a historical trend and as a recurrent process, and we have a wealth of literature presenting insights of social relevance and empirical studies offering detailed information on a variety of particular aspects of urban life. But despite the multiplication of research and textbooks on the city, we do not as yet have a comprehensive body of compendent hypotheses which may be derived from a set of postulates implicitly contained in a sociological definition of the city. Neither have we abstracted such hypotheses from our general sociological knowledge which may be substantiated through empirical research. The closest approximations to a systematic theory of urbanism are to be found in a penetrating essay, "Die Stadt," by Max Weber and in a memorable paper by Robert E. Park on "The City: Suggestions for the Investigation of Human Behavior in the Urban Environment." But even these excellent contributions are far from constituting an ordered and coherent framework of theory upon which research might profitably proceed.

[. . .]

To say that large numbers are necessary to constitute a city means, of course, large numbers in relation to a restricted area or high density of settlement. There are, nevertheless, good reasons for treating large numbers and density as separate factors, because each may be connected with significantly different social consequences. Similarly the need for adding heterogeneity to numbers of population as a necessary and distinct criterion of urbanism might be questioned, since we should expect the range of differences to increase with numbers. In defense, it may be said that the city shows a kind and degree of heterogeneity of population which cannot be wholly accounted for by the law of large numbers or adequately represented by means of a normal distribution curve. Because the population of the city does not reproduce itself, it must recruit its migrants from other cities, the countryside, and – in the United States until recently – from other countries. The city has thus historically been the melting-pot of races, peoples, and cultures, and a most favorable breeding-ground of new biological and cultural hybrids. It has not only tolerated but rewarded individual differences. It has brought together people from the ends of the earth *because* they are different and thus useful to one another, rather than because they are homogeneous and like-minded.

There are a number of sociological propositions concerning the relationship between (a) numbers of population, (b) density of settlement, (c) heterogeneity of inhabitants and group life can be formulated on the basis of observation and research.

Size of the population aggregate

Ever since Aristotle's *Politics*, it has been recognized that increasing the number of inhabitants in a settlement beyond a certain limit will affect the relationships between them and the character of the city. Large numbers involve, as has been pointed out, a greater range of individual variation. Furthermore, the greater the number of individuals participating in a process of interaction, the greater is the *potential* differentiation between them. The personal traits, the occupations, the cultural life, and the ideas of the members of an urban community may, therefore, be expected to range between more widely separated poles than those of rural inhabitants.

That such variations should give rise to the spatial segregation of individuals according to color, ethnic heritage, economic and social status, tastes and preferences, may readily be inferred. The bonds of kinship, of neighborliness, and the sentiments arising out of living together for generations under a common folk tradition are likely to be absent or, at best, relatively weak in an aggregate the members of which have such diverse origins and backgrounds. Under such circumstances competition and formal control mechanisms furnish the substitutes for the bonds of solidarity that are relied upon to hold a folk society together.

[. . .]

The multiplication of persons in a state of interaction under conditions which make their contact as full personalities impossible produces that segmentalization of human relationships which has sometimes been seized upon by students of the mental life of the cities as an explanation for the "schizoid" character of urban personality. This is not to say that the urban inhabitants have fewer acquaintances than rural inhabitants, for the reverse may actually be true; it means rather that in relation to the number of people whom they see and with whom they rub elbows in the course of daily life, they know a smaller proportion, and of these they have less intensive knowledge.

Characteristically, urbanites meet one another in highly segmental roles. They are, to be sure, dependent upon more people for the satisfactions of their life-needs than are rural people and thus are associated with a great number of organized groups, but they are less dependent upon particular persons, and their dependence upon others is confined to a highly fractionalized aspect of the other's round of activity. This is essentially what is meant by saying that the city is characterized by secondary rather than primary contacts. The contacts of the city may indeed be face to face, but they are nevertheless impersonal, superficial, transitory, and segmental. The reserve, the indifference, and the blasé outlook which urbanites manifest in their relationships may thus be regarded as devices for immunizing themselves against the personal claims and expectations of others.

The superficiality, the anonymity, and the transitory character of urban social relations make intelligible, also, the sophistication and the rationality generally ascribed to city-dwellers. Our acquaintances tend to stand in a relationship of utility to us in the sense that the role which each one plays in our life is overwhelmingly regarded as a means for the achievement of our own ends. Whereas the individual gains, on the one hand, a certain degree of emancipation or freedom from the personal and emotional controls of intimate groups, he loses, on the other hand, the spontaneous self-expression, the morale, and the sense of participation that comes with living in an integrated society. This constitutes essentially the state of *anomie*, or the social void, to which Durkheim alludes in attempting to account for the various forms of social disorganization in technological society.

The segmental character and utilitarian accent of interpersonal relations in the city find their institutional expression in the proliferation of specialized tasks which we see in their most developed form in the professions. The operations of the pecuniary nexus lead to predatory relationships which tend to obstruct the efficient functioning of the social order unless checked by professional codes and occupational etiquette. The premium put upon utility and efficiency suggests the adaptability of the corporate device for the organization of enterprises in which individuals can engage only in groups. The advantage that the corporation has over the individual entrepreneur and the partnership in the urban-industrial world derives not only from the possibility it affords of centralizing the resources of thousands of individuals or from the legal privilege of limited liability and perpetual succession, but from the fact that the corporation has no soul.

The specialization of individuals, particularly in their occupations, can proceed only, as Adam Smith pointed out, upon the basis of an enlarged market, which in turn accentuates the division of labor. This enlarged market is only in part supplied by the city's hinterland; in large measure it is found among the larger numbers that the city itself contains. The dominance of the city over the surrounding hinterland becomes explicable in terms of the division of labor which urban life occasions and promotes. The extreme degree of interdependence and the unstable equilibrium of urban life are closely associated with the division of labor and the specialization of occupations. This interdependence and this instability are increased by the tendency of each city to specialize in those functions in which it has the greatest advantage.

[. . .]

Density

As in the case of numbers, so in the case of concentration in limited space, certain consequences of relevance in sociological analysis of the city emerge. Of these only a few can be indicated.

As Darwin pointed out for flora and fauna and as Durkheim noted in the case of human societies, an increase in numbers when area is held constant (i.e., an increase in density) tends to produce differentiation and specialization, since only in this way can the area support increased numbers. Density thus reinforces the effect of numbers in diversifying men and their activities and in increasing the complexity of the social structure.

On the subjective side, as Simmel has suggested, the close physical contact of numerous individuals necessarily produces a shift in the media through which we orient ourselves to the urban milieu, especially to our fellowmen. Typically, our physical contacts are close but our social contacts are distant. The urban world puts a premium on visual recognition. We see the uniform which denotes the role of the functionaries, and are oblivious to the personal eccentricities hidden behind the uniform. We tend to acquire and develop a sensitivity to a world of artifacts, and become progressively farther removed from the world of nature.

We are exposed to glaring contrasts between splendor and squalor, between riches and poverty, intelligence and ignorance, order and chaos. The competition for space is great, so that each area generally tends to be put to the use which yields the greatest economic return. Place of work tends to become dissociated from place of residence, for the proximity of industrial and commercial establishments makes an area both economically and socially undesirable for residential purposes.

Density, land values, rentals, accessibility, healthfulness, prestige, aesthetic consideration, absence of nuisances such as noise, smoke, and dirt determine the desirability of various areas of the city as places of settlement for different sections of the population. Place and nature of work, income, racial and ethnic characteristics, social status, custom, habit, taste, preference, and prejudice are among the significant factors in accordance with which the urban population is selected and distributed into more or less distinct settlements. Diverse population elements inhabiting a compact settlement thus become segregated from one another in the degree in which their requirements and modes of life are incompatible and in the measure in which they are antagonistic. Similarly, persons of homogeneous status and needs unwittingly drift into, consciously select, or are forced by circumstances into the same area. The different parts of the city thus acquire specialized functions. The city consequently tends to resemble a mosaic of social worlds in which the transition from one to the other is abrupt. The juxtaposition of divergent personalities and modes of life tends to produce a relativistic perspective and a sense of toleration of differences which may be regarded as prerequisites for rationality and which lead toward the secularization of life.

The close living together and working together of individuals who have no sentimental and emotional ties foster a spirit of competition, aggrandizement, and mutual exploitation. Formal controls are instituted to counteract irresponsibility and potential disorder. Without rigid adherence to predictable routines a large compact society would scarcely be able to maintain itself. The clock and the traffic signal are symbolic of the basis of our social order in the urban world. Frequent close physical contact, coupled with great social distance, accentuates the reserve of unattached individuals toward one another and, unless compensated by other opportunities for response, gives rise to loneliness. The necessary frequent movement of great numbers of individuals in a congested habitat causes friction and irritation. Nervous tensions which derive from such personal frustrations are increased by the rapid tempo and the complicated technology under which life in dense areas must be lived.

Heterogeneity

The social interaction among such a variety of personality types in the urban milieu tends to break down the rigidity of caste lines and to complicate the class structure, and thus induces a more ramified and differentiated framework of social stratification than is found in more integrated societies. The heightened mobility of the individual, which brings him within the range of stimulation by a great number of diverse individuals and subjects him to fluctuating status in the differentiated social groups that compose the social structure of the city, brings him toward the acceptance of instability and insecurity in the world at large as a norm. This fact helps to account too for the sophistication and cosmopolitanism of the urbanite. No single group has the undivided allegiance of the individual. The groups with which he is affiliated do not lend themselves readily to a simple hierarchical arrangement. By virtue of his different interests arising out of different aspects of social life, the individual acquires membership in widely divergent groups, each of which functions only with reference to a certain segment of his personality. Nor do these groups easily permit of a concentric arrangement so that the narrower ones fall within the circumference of the more inclusive ones, as is more likely to be the case in the rural community or in primitive societies. Rather the groups with which the person typically is affiliated are tangential to each other or intersect in highly variable fashion.

Partly as a result of the physical footlooseness of the population and partly as a result of their social mobility, the turnover in group membership generally is rapid. Place of residence, place and character of employment, income, and interests fluctuate, and the task of holding organizations together and maintaining and promoting intimate and lasting acquaintanceship between the members is difficult. This applies strikingly to the local areas within the city into which persons become segregated more by virtue of differences in race, language, income, and social status than through choice or positive attraction to people like themselves. Overwhelmingly the city-dweller is not a home-owner, and since a transitory habitat does not generate binding traditions and sentiments, only rarely is he a true neighbor. There is little opportunity for the individual to obtain a conception of the city as a whole or to survey his place in the total scheme. Consequently he finds it difficult to determine what is to his own "best interests" and to decide between the issues and leaders presented to him by the agencies of mass suggestion. Individuals who are thus detached from the organized bodies which integrate society comprise the fluid masses that make collective behavior in the urban community so unpredictable and hence so problematical.

Although the city, through the recruitment of variant types to perform its diverse tasks and the accentuation of their uniqueness through competition and the premium upon eccentricity, novelty, efficient performance, and inventiveness, produces a highly differentiated population, it also exercises a leveling influence. Wherever large numbers of differently constituted individuals congregate, the process of depersonalization also enters. This leveling tendency inheres in part in the economic basis of the city. The development of large cities, at least in the modern age, was largely dependent upon the concentrative force of steam. The rise of the factory made possible mass production for an impersonal market. The fullest exploitation of the possibilities of the division of labor and mass production, however, is possible only with standardization of processes and products. A money economy goes hand in hand with such a system of production. Progressively as cities have developed upon a background of this system of production, the pecuniary nexus which implies the purchasability of services and things has displaced personal relations as the basis of association. Individuality under these circumstances must be replaced by categories. When large numbers have to make common use of facilities and institutions, those facilities and institutions must serve the needs of the average person rather than those of particular individuals. The services of the public utilities, of the recreational, educational, and cultural institutions, must be adjusted to mass requirements. Similarly, the cultural institutions, such as the schools, the movies, the radio, and the newspapers, by virtue of their mass clientele, must necessarily operate as leveling influences. The political process as it appears in urban life could not be understood unless one examined the mass appeals made through modern propaganda techniques. If the individual would participate at all in the social, political, and economic life of the city, he must subordinate some of his individuality to the demands of the larger community and in that measure immerse himself in mass movements.

THE RELATION BETWEEN A THEORY OF URBANISM AND SOCIOLOGICAL RESEARCH

By means of a body of theory such as that illustratively sketched above, the complicated and many-sided phenomena of urbanism may be analyzed in terms of a limited number of basic categories. The sociological approach to the city thus acquires an essential unity and coherence enabling the empirical investigator not merely to focus more distinctly upon the problems and processes that properly fall in his province but also to treat his subject matter in a more integrated and systematic fashion. A few typical findings of empirical research in the field of urbanism, with special reference to the United States, may be indicated to substantiate the theoretical propositions set forth in the preceding pages, and some of the crucial problems for further study may be outlined.

On the basis of the three variables, number, density of settlement, and degree of heterogeneity, of the urban population, it appears possible to explain the characteristics of urban life and to account for the differences between cities of various sizes and types.

Urbanism as a characteristic mode of life may be approached empirically from three interrelated perspectives: (1) as a physical structure comprising a population base, a technology, and an ecological order; (2) as a system of social organization involving a characteristic social structure, a series of social institutions, and a typical pattern of social relationships; and (3) as a set of attitudes and ideas, and a constellation of personalities engaging in typical forms of collective behavior and subject to characteristic mechanisms of social control.

Urbanism in ecological perspective

Since in the case of physical structure and ecological process we are able to operate with fairly objective indices, it becomes possible to arrive at quite precise and generally quantitative results. The dominance of the city over its hinterland becomes explicable through the functional characteristics of the city which derive in large measure from the effect of numbers and density. Many of the technical facilities and the skills and organizations to which urban life gives rise can grow and prosper only in cities where the demand is sufficiently great. The nature and scope of the services rendered by these organizations and institutions and the advantage which they enjoy over the less developed facilities of smaller towns enhance the dominance of the city, making ever wider regions dependent upon the central metropolis.

The composition of an urban population shows the operation of selective and differentiating factors. Cities contain a larger proportion of persons in the prime of life than rural areas, which contain more old and very young people. In this, as in so many other respects, the larger the city the more this specific characteristic of urbanism is apparent. With the exception of the largest cities, which have attracted the bulk of the foreign-born males, and a few other special types of cities, women predominate numerically over men. The heterogeneity of the urban population is further indicated along racial and ethnic lines. The foreign-born and their children constitute nearly two-thirds of all the inhabitants of cities of one million and over. Their proportion in the urban population declines as the size of the city decreases, until in the rural areas they comprise only about one-sixth of the total population. The larger cities similarly have attracted more Negroes and other racial groups than have the smaller communities. Considering that age, sex, race, and ethnic origin are associated with other factors such as occupation and interest, one sees that a major characteristic of the urban-dweller is his dissimilarity from his fellows. Never before have such large masses of people of diverse traits as we find in our cities been thrown together into such close physical contact as in the great cities of America. Cities generally, and American cities in particular, comprise a motley of peoples and cultures of highly differentiated modes of life between which there often is only the faintest communication, the greatest indifference, the broadest tolerance, occasionally bitter strife, but always the sharpest contrast.

The failure of the urban population to reproduce itself appears to be a biological consequence of a combination of factors in the complex of urban life, and the decline in the birth rate generally may be regarded as one of the most significant signs of the urbanization of the Western world. While the proportion of deaths in cities is slightly greater than in the country, the outstanding difference between the failure of present-day cities to maintain their population and that of cities of the past is that in former times it was due to the exceedingly high death rates in cities, whereas today, since cities have become more livable from a health standpoint, it is due to low birth rates. These biological characteristics of the urban population are significant sociologically, not merely because they reflect the urban mode of existence but also because they condition the growth and future dominance of cities and their basic social organization. Since cities are the consumers rather than the producers of men, the value of human life and the social estimation of the personality will not be unaffected by the balance between births and deaths. The pattern of land use, of land values, rentals, and ownership, the nature and functioning of the physical structures, of housing, of transportation and communication facilities, of public utilities – these and many other phases of the physical mechanism of the city are not isolated phenomena unrelated to the city as a social entity but are affected by and affect the urban mode of life.

Urbanism as a form of social organization

The distinctive features of the urban mode of life have often been described sociologically as consisting of the substitution of secondary for primary contacts, the weakening of bonds of kinship, and the declining social significance of the family, the disappearance of the neighborhood, and the undermining of the traditional basis of social solidarity. All these phenomena can be substantially verified through objective indices. Thus, for instance, the low and declining urban-reproduction rates suggest that the city is not conducive to the traditional type of family life, including the rearing of children and the maintenance of the home as the locus of a whole round of vital activities. The transfer of industrial, educational, and recreational activities to specialized institutions outside the home has deprived the family of some of its most characteristic historical functions. In cities mothers are more likely to be employed, lodgers are more frequently part of the household, marriage tends to be postponed, and the proportion of single and unattached people is greater. Families are smaller and more frequently without children than in the country. The family as a unit of social life is emancipated from the larger kinship group characteristic of the country, and the individual members pursue their own diverging interests in their vocational, educational, religious, recreational, and political life.

Such functions as the maintenance of health, the methods of alleviating the hardships associated with personal and social insecurity, the provisions for education, recreation, and cultural advancement have given rise to highly specialized institutions on a community-wide, statewide, or even national basis. The same factors which have brought about greater personal insecurity also underlie the wider contrasts between individuals to be found in the urban world. While the city has broken down the rigid caste lines of preindustrial society, it has sharpened and differentiated income and status groups. Generally, a larger proportion of the adult urban population is gainfully employed than is the case with the adult-rural population. The white-collar class, comprising those employed in trade, in clerical, and in professional work, are proportionately more numerous in large cities and in metropolitan centers and in smaller towns than in the country.

On the whole, the city discourages an economic life in which the individual in time of crisis has a basis of subsistence to fall back upon, and it discourages self-employment. While incomes of city people are on the average higher than those of country people, the cost of living seems to be higher in the larger cities. Home-ownership involves greater burdens and is rarer. Rents are higher and absorb a larger proportion of the income. Although the urban-dweller has the benefit of many communal services, he spends a large proportion of his income for such items as recreation and advancement and a smaller proportion for food. What the communal services do not furnish, the urbanite must purchase, and there is virtually no human need which has remained unexploited by commercialism. Catering to thrills and furnishing means of escape from drudgery, monotony, and routine thus become one of the major functions of urban recreation, which at its best furnishes means for creative self-expression and spontaneous group association, but which more typically in the urban world results in passive spectatorism, on the one hand, or sensational record-smashing feats, on the other.

Reduced to a stage of virtual impotence as an individual, the urbanite is bound to exert himself by joining with others of similar interest into groups organized to obtain his ends. This results in the enormous multiplication of voluntary organizations directed toward as great a variety of objectives as there are human needs and interests. While, on the one hand, the traditional ties of human association are weakened, urban existence involves a much greater degree of interdependence between man and man and a more complicated, fragile, and volatile form of mutual interrelations over many phases of which the individual as such can exert scarcely any control. Frequently there is only the most tenuous relationship between the economic position or other basic factors that determine the individual's existence in the urban world and the voluntary groups with which he is affiliated. In a primitive and in a rural society it is generally possible to predict on the basis of a few known factors who will belong to what and who will associate with whom in almost every relationship of life, but in the city we can only project the general pattern of group formation and affiliation, and this pattern will display many incongruities and contradictions.

Urban personality and collective behavior

It is largely through the activities of the voluntary groups, be their objectives economic, political, educational, religious, recreational, or cultural, that the urbanite expresses and develops his personality, acquires status, and is able to carry on the round of activities that constitutes his life. It may easily be inferred, however, that the organizational framework which these highly differentiated functions call into being does not of itself insure the consistency and integrity of the personalities whose interests it enlists. Personal disorganization, mental breakdown, suicide, delinquency, crime, corruption, and disorder might be expected under these circumstances to be more prevalent in the urban than in the rural community. This has been confirmed in so far as comparable indexes are available, but the mechanisms underlying these phenomena require further analysis.

Since for most group purposes it is impossible in the city to appeal individually to the large number of discrete and differentiated citizens, and since it is only through the organizations to which men belong that their interests and resources can be enlisted for a collective cause, it may be inferred that social control in the city should typically proceed through formally organized groups. It follows, too, that the masses of men in the city are subject to manipulation by symbols and stereotypes managed by individuals working from afar or operating invisibly behind the scenes through their control of the instruments of communication. Self-government either in the economic, or political, or the cultural realm is under these circumstances reduced to a mere figure of speech, or, at best, is subject to the unstable equilibrium of pressure groups. In view of the ineffectiveness of actual kinship ties, we create fictional kinship groups. In the face of the disappearance of the territorial unit as a basis of social solidarity, we create interest units. Meanwhile the city as a community resolves itself into a series of tenuous segmental relationships superimposed upon a territorial base with a definite center but without a definite periphery, and upon a division of labor which far transcends the immediate locality and is world-wide in scope. The larger the number of persons in a state of interaction with another, the lower is the level of communication and the greater is the tendency for communication to proceed on an elementary level, i.e., on the basis of those things which are assumed to be common or to be of interest to all.

It is obviously, therefore, to the emerging trends in the communication system and to the production and distribution technology that has come into existence with modern civilization that we must look for the symptoms which will indicate the probable development of urbanism as a mode of social life. The direction of the ongoing changes in urbanism will for good or ill transform not only the city but the world.

It is only in so far as the sociologist, with a workable theory of urbanism, has a clear conception of the city as a social entity that he can hope to develop a unified body of reliable knowledge – which what passes as "urban sociology" is certainly not at the present time. By taking his point of departure from a theory of urbanism such as that sketched in the foregoing pages, a theory to be elaborated, tested, and revised, in the light of further analysis and empirical research, the sociologist can hope to determine the criteria of relevance and validity of factual data. The miscellaneous assortment of disconnected information which has hitherto found its way into sociological treatises on the city may thus be sifted and incorporated into a coherent body of knowledge. Incidentally, only by means of some such theory will the sociologist escape the futile practice of voicing in the name of sociological science a variety of often unsupportable judgments about poverty, housing, city-planning, sanitation, municipal administration, policing, marketing, transportation, and other technical issues. Though the sociologist cannot solve any of these practical problems – at least not by himself – he may, if he discovers his proper function, have an important contribution to make to their comprehension and solution. The prospects for doing this are brightest through a general, theoretical, rather than through an *ad hoc* approach.

"Urbanism and Suburbanism as Ways of Life: A Reevaluation of Definitions"

from *People and Plans* (1968)

Herbert Gans

Editors' Introduction

In this selection, Herbert Gans issues a forceful critique of the Wirthian theory of urbanism, which he believes places an overemphasis on the social dynamics of the central or inner city, and cannot be generalized to the whole urban area. The decentralization of people and businesses to the suburbs also impels a reconsideration of Wirthian urbanism. Gans draws attention to the growth of *quasi-primary* relationships in the "outer city" or suburbs, where there is more privacy than in the inner city. The loss of kinship and primary community ties is superseded by the emergence of neighborliness. Neighborly ties are more intimate than professional and other secondary ties, but more guarded than primary ties of kinship and extended family. The concept of *quasi-primary* relationships invites intriguing speculation and discussion on what favors or obligations an individual would ask of a neighbor as compared with a close family member or colleague at work. His beliefs on suburban life are largely drawn from his 1967 book, *The Levittowners*, based on participant observation among the residents of a prototypical "package suburb" built by Levitt and Sons, Inc. on Long Island, New York in the first two years of its existence. Interestingly, Levittown became the stereotypic moniker of the dull and prefabricated suburb, and the town later changed its name to Willingboro. In his study he found there was greater sociability and less boredom in suburbs than originally thought. Gans contributes an alternative to the view that suburbs are places of conformity, alienation, or monotony. Gans believes the differences between the city and suburbs are overstated and even spurious. Suburbanism is just as legitimate a concept as urbanism.

Gans counters the Wirthian view of urban life as segmented, anonymous and anomic, offering a typology of five major types of residents in the central city: (a) cosmopolites, (b) the unmarried or childless, (c) ethnic villagers, (d) the deprived, and (e) the trapped. The last two groups suffer a transience and residential instability that bears the closest resemblance to the condition of Wirthian urbanism, but the other groups do not suffer from mental instability or community decline. Gans believes that demographic factors such as class, life-cycle stage, ethnicity, and culture, are more important variables affecting the neighborhood choices and social life of urbanites than ecological variables such as size, density, and heterogeneity.

The phenomena of ethnic villagers offer perhaps the clearest articulation of Gans' ideas concerning life in the inner city. The persistence of immigrant colonies in the city is testament to the continuing salience of primary ties and the community networks of *Gemeinschaft*. In his first major ethnographic study, *The Urban Villagers* (1962), he explored the social life of what he called the "peer group" society of second-generation Italian Americans in the West End of Boston. He found relations of gender and ethnicity to be partly manifested as deeply rooted class dynamics. The values inherent in this working or lower-class subculture clashed at times with middle-class expectations, but Gans believed they carried their own respectability. Despite some mental health problems and deterioration in housing quality that had drawn the concerns of social workers, he found considerable evidence of social and community vitality, countering the prevailing view of the West End as a socially disorganized slum in need of clearance and redevelopment. Many residents accepted location grants before a community defense network was able to effectively organize an opposition, and the area was razed and replaced with a district of hospitals, and middle-income and luxury housing. Gans became nationally known for his critical outrage against "slum clearance" policies, especially those involving residential communities. The failure of urban renewal was even starker with the realization that some urban villages gave way to high-density tower blocks that fomented new kinds of social problems and crime. High-density planning and development is the outgrowth of a Wirthian ecological formulation that treats people as a matter of statistics that ignores and violates their humanity and their culture. He articulates these and other issues in his book *People and Plans* (1968).

Gans became associated with national debates connected with William Julius Wilson and others concerning the emergence of the "underclass" and the "culture of poverty," and his writing was assembled in the book *The War against the Poor* (1995). In this book, he takes up the problem of labeling that besets the poor and other socially marginalized groups in America, connecting with another arena of research he has established on the sociology of culture and media, which he has taken up in *Popular Culture and High Culture* (1974) and *Deciding What's News* (1979). His latest book is *Making Sense of America* (1999).

Herbert Gans is one of the seminal sociologists of the contemporary era, with a durable legacy of research on urban and ethnic studies, the mass media, and urban public policy in America. He was born in 1927 in Cologne, Germany, and he immigrated to the U.S. in 1940 and became a U.S. citizen. He was trained as both a planner and sociologist; he received a bachelor's and master's degree in sociology at the University of Chicago, and a Ph.D. in planning and sociology from the University of Pennsylvania in 1957. He worked as a staff member or consultant for planning and social policy agencies for several years before beginning a career as an academic professor and public intellectual. After teaching at the University of Pennsylvania and the Massachusetts Institute of Technology, he joined the Columbia University Sociology Department in 1971. He is currently the Robert S. Lynd Professor of Sociology at Columbia University.

Herbert J. Gans served as the 78th President of the American Sociological Association. His Presidential Address, entitled "Sociology in America: The Discipline and the Public," was delivered at the Association's 1988 Annual Meeting in Atlanta, and was later published in the *American Sociological Review*. In 1992, he won the Robert and Helen Lynd Award for lifetime contribution in community and urban sociology from the American Sociological Association. In 1999, he received the Association's Award for Contributions to the Public Understanding of Sociology.

The contemporary sociological conception of cities and of urban life is based largely on the work of the Chicago School and its summary statement in Louis Wirth's essay "Urbanism as a Way of Life." In that paper, Wirth developed a "minimum sociological definition of the city" as "a relatively large, dense and permanent settlement of socially heterogeneous individuals." From these prerequisites, he then deduced the major outlines of the urban way of life. As he saw it, number, density, and heterogeneity created a social structure in which primary-group relationships were inevitably replaced by secondary contacts that were impersonal, segmental, superficial, transitory, and often predatory in nature. As a result, the city dweller became anonymous, isolated, secular, relativistic, rational, and sophisticated. In order to function in an urban society, he or she was forced to combine with others to organize corporations, voluntary associations, representative forms of government, and the impersonal mass media of communications. These replaced the primary groups and the integrated way of life found in rural and other preindustrial settlements.

Wirth's paper has become a classic in urban sociology, and most texts have followed his definition and description faithfully. In recent years, however, a considerable number of studies and essays have questioned his formulations. In addition, a number of changes have taken place in cities since the article was published in 1938, notably the exodus of white residents to low- and medium-priced houses in the suburbs and the decentralization of industry. The evidence from these studies and the changes in American cities suggest that Wirth's statement must be revised.

There is yet another and more important reason for such a revision. Despite its title and intent, Wirth's paper deals with urban-industrial society, rather than with the city. This is evident from his approach. Like other urban sociologists, Wirth based his analysis on a comparison of settlement types, but unlike his colleagues, who pursued urban-rural comparisons, Wirth contrasted the city to the folk society. Thus, he compared settlement types of preindustrial and industrial society. This allowed him to include in his theory of urbanism the entire range of modern institutions which are not found in the folk society, even though many such groups (for example, voluntary associations) are by no means exclusively urban. Moreover, Wirth's conception of the city dweller as depersonalized, atomized, and susceptible to mass movements suggests that his paper is based on, and contributes to, the theory of the mass society.

Many of Wirth's conclusions may be relevant to the understanding of ways of life in modern society. However, since the theory argues that all of society is now urban, his analysis does not distinguish ways of life in the city from those in other settlements within modern society. In Wirth's time, the comparison of urban and preurban settlement types was still fruitful, but today, the primary task for urban (or community) sociology seems to me to be the analysis of the similarities and differences between contemporary settlement types.

This paper is an attempt at such an analysis; it limits itself to distinguishing ways of life. A reanalysis of Wirth's conclusions from this perspective suggests that his characterization of the urban way of life applies only – and not too accurately – to the residents of the inner city. The remaining city dwellers, as well as most suburbanites, pursue a different way of life which I shall call "quasi-primary." This proposition raises some doubt about the mutual exclusiveness of the concepts of city and suburb and leads to a yet broader question: whether settlement concepts and other ecological concepts are useful for explaining ways of life.

THE INNER CITY

Wirth argued that number, density, and heterogeneity had two social consequences which explain the major features of urban life. On the one hand, the crowding of diverse types of people into a small area led to the segregation of homogeneous types of people into separate neighborhoods. On the other hand, the lack of physical distance between city dwellers resulted in social contact between them, which broke down existing social and cultural patterns and encouraged assimilation as well as acculturation – the melting-pot effect. Wirth implied that the melting-pot effect was far more powerful than the tendency toward segregation and concluded that, sooner or later, the pressures engendered by the dominant social, economic, and political institutions of the city would destroy the remaining pockets of primary-group relationships. Eventually, the social system of the city would resemble Tönnies' *Gesellschaft* – a way of life which Wirth considered undesirable.

Because Wirth had come to see the city as the prototype of mass society, and because he examined the city from the distant vantage point of the folk society – from the wrong end of the telescope, so to speak – his view of urban life is not surprising. In addition, Wirth found support for his theory in the empirical work of his Chicago colleagues. As Greer and Kube and Wilensky have pointed out, the Chicago sociologists conducted their most intensive studies in the inner city. At that time, it consisted mainly of slums recently invaded by new waves of European immigrants and rooming-house and skid-row districts, as well as the habitat of Bohemians and well-to-do "Gold Coast" apartment dwellers. Wirth himself studied the Maxwell Street Ghetto, a poor inner-city Jewish neighborhood then being dispersed by the acculturation and mobility of its inhabitants. Some of the characteristics of urbanism which Wirth stressed in his essay abounded in these areas.

Wirth's diagnosis of the city as *Gesellschaft* must be questioned on three counts. First, the conclusions derived from a study of the inner city cannot be generalized to the entire urban area. Second, there is as yet not enough evidence to prove – or, admittedly, to deny – that number, density, and heterogeneity result in the social consequences which Wirth proposed. Finally, even if the causal relationship could be verified, it can be shown that a significant proportion of the city's inhabitants were, and are, isolated from these consequences by social structures and cultural patterns which they either brought to the city or developed by living in it. Wirth conceived the urban population as consisting of heterogeneous individuals, torn from past social systems, unable to develop new ones, and therefore prey to social anarchy in the city. While it is true that a not insignificant proportion of the inner-city population was, and still is, made up of unattached individuals, Wirth's formulation ignores the fact that this population consists mainly of relatively homogeneous groups, with social and cultural moorings that shield it fairly effectively from the suggested consequences of number, density, and heterogeneity. This applies even more to the residents of the outer city, who constitute a majority of the total city population.

The social and cultural moorings of the inner-city population are best described by a brief analysis of the five major types of inner-city residents. These are: 1) the "cosmopolites"; 2) the unmarried or childless; 3) the "ethnic villagers"; 4) the "deprived"; and 5) the "trapped" and downward-mobile.

The "cosmopolites" include students, artists, writers, musicians, and entertainers, as well as other intellectuals and professionals. They live in the city in order to be near the special "cultural" facilities that can be located only near the center of the city. Many cosmopolites are unmarried or childless. Others rear children in the city, especially if they have the income to afford the aid of servants and governesses. The less affluent ones may move to the suburbs to raise their children, continuing to live as cosmopolites under considerable handicaps, especially in the lower-middle-class suburbs. Many of the very rich and powerful are also cosmopolites, although they are likely to have at least two residences, one of which is suburban or exurban.

The unmarried or childless must be divided into two subtypes, depending on the permanence or transience of their status. The temporarily unmarried or childless live in the inner city for only a limited time. Young adults may team up to rent an apartment away from their parents and close to job or entertainment opportunities. When they marry, they may move first to an apartment in a transient neighborhood, but if they can afford to do so, they leave for the outer city or the suburbs with the arrival of the first or second child. The permanently unmarried may stay in the inner city for the remainder of their lives, their housing depending on their income.

The "ethnic villagers" are ethnic groups which are found in such inner-city neighborhoods as New York's Lower East Side, living in some ways as they did when they were peasants in European or Puerto Rican villages. Although they reside in the city, they isolate themselves from significant contact with most city facilities, aside from workplaces. Their way of life differs sharply from Wirth's urbanism in its emphasis on kinship and the primary group, the lack of anonymity and secondary-group contacts, the weakness of formal organizations, and the suspicion of anything and anyone outside their neighborhood.

The first two types live in the inner city by choice: the third is there partly because of necessity, partly because of tradition. The final two types are in the inner city because they have no other choice. One is the "deprived" population: the emotionally disturbed or otherwise handicapped; broken families; and, most important, the poor-white and especially the nonwhite population. These urban dwellers must take the dilapidated housing and blighted neighborhoods to which the housing market relegates them, although among them are some for whom the slum is a hiding place or a temporary stopover to save money for a house in the outer city or the suburbs.

The "trapped" are the people who stay behind when a neighborhood is invaded by nonresidential land uses or lower-status immigrants, because they cannot afford to move or are otherwise bound to their present location." The "downward-mobiles" are a related type: they may have started life in a higher class position, but have been forced down in the socioeconomic hierarchy and in the quality of their accommodations. Many of them are old people, living out their existence on small pensions.

These five types may all live in dense and heterogeneous surroundings; yet they have such diverse ways of life that it is hard to see how density and heterogeneity could exert a common influence. Moreover, all but the last two types are isolated or detached from their neighborhood and thus from the social consequences that Wirth described.

When people who live together have social ties based on criteria other than mere common occupancy, they can set up social barriers, regardless of the physical closeness or the heterogeneity of their neighbors. The ethnic villagers are the best illustration. While a number of ethnic groups are usually found living together in the same neighborhood, they are able to isolate themselves from one another through a variety of social devices. Wirth himself recognized this when he wrote that "two groups can occupy a given area without losing their separate identity because each side is permitted to live its own inner life and each somehow fears or idealizes the other." Although it is true that the children in these areas were often oblivious of social barriers set up by their parents, at least until adolescence, it is doubtful whether their acculturation can be traced to the melting-pot effect as much as to the pervasive influence of the American culture that flowed into these areas from the outside.

The cosmopolites, the unnamed, and the childless are *detached* from neighborhood life. The cosmopolites possess a distinct subculture which causes them to be uninterested in all but the most superficial contacts with their neighbors, somewhat like the ethnic villagers. The unmarried and childless – who may also be cosmopolites – are detached from the neighborhood because of their life-cycle stage, which frees them from the routine family responsibilities that entail some relationship to the local area. In their choice of residence, the two types are therefore not always concerned about their neighbors or the availability and quality of local community facilities. Even the well-to-do can choose expensive apartments in or near poor neighborhoods, because if they have children, these are sent to special schools and summer camps which effectively isolate them from neighbors. In addition, the childless and unmarried are often transient. Therefore, they tend to live in areas marked by high population turnover, where their own mobility and that of their neighbors creates a universal detachment from the neighborhood.

The deprived and the trapped do seem to be affected by some of the consequences of number, density, and heterogeneity. The deprived population suffers considerably from overcrowding, but this is a consequence of low income, racial discrimination. and other handicaps and cannot be considered an inevitable result of the ecological makeup of the city. Because the deprived have no residential choice, they are also forced to live amid neighbors not of their own choosing, with ways of life different and even contradictory to their own. If familial defenses against the neighborhood climate are weak, as may happen among single-parent families and downward-mobile people, parents may lose their children to the culture of "the street." The trapped are the unhappy people who remain behind when their more advantaged neighbors move on; they must endure the heterogeneity which results from neighborhood change.

Wirth's description of the urban way of life fits best the transient areas of the inner city. Such areas are typically heterogeneous in population, partly because they are inhabited by transient types who do not require homogeneous neighbors or by deprived people who have no choice or may themselves be quite mobile. Under conditions of transience and heterogeneity, people interact only in terms of the segmental roles necessary for obtaining social services. Their social relationships may thus display anonymity, impersonality, and superficiality.

The social features of Wirth's concept of urbanism seem, therefore, to be a result of residential instability, rather than of number, density, or heterogeneity. In fact, heterogeneity is itself an effect of residential instability, resulting when the influx of transients causes landlords and realtors to stop acting as gatekeepers – that is, wardens of neighborhood homogeneity. Residential instability is found in all types of settlements, and presumably its social consequences are everywhere similar. These consequences cannot, therefore, be identified with the ways of life of the city.

THE OUTER CITY AND THE SUBURBS

The second effect which Wirth ascribed to number, density, and heterogeneity was the segregation of homogeneous people into distinct neighborhoods on the basis of "place and nature of work, income, racial and ethnic characteristics, social status, custom, habit, taste, preference and prejudice." This description fits the residential districts of the *outer city*. Although these districts contain the majority of the city's inhabitants, Wirth went into little detail about them. He made it clear, however, that the sociopsychological aspects of urbanism were prevalent there as well.

Because existing neighborhood studies deal primarily with the exotic sections of the inner city, very little is known about the more typical residential neighborhoods of the outer city. However, it is evident that the way of life in these areas bears little resemblance to Wirth's urbanism. Both the studies which question Wirth's formulation and my own observations suggest that the common element in the ways of life of these neighborhoods is best described as *quasi-primary*. I use this term to characterize relationships between neighbors. Whatever the intensity or frequency of these relationships, the interaction is more intimate than a secondary contact, but more guarded than a primary one.

There are actually few secondary relationships, because of the isolation of residential neighborhoods from economic institutions and workplaces. Even shopkeepers, store managers, and other local functionaries who live in the area are treated as acquaintances or friends, unless they are of a vastly different social status or are forced by their corporate employers to treat their customers as economic units. Voluntary associations attract only a minority of the population. Moreover, much of the organizational activity is of a sociable nature, and it is often difficult to accomplish the association's "business" because of the members' preference for sociability. Thus, it would appear that interactions in organizations, or between neighbors generally, do not fit the secondary-relationship model of urban life. As anyone who has lived in these neighborhoods knows, there is little anonymity, impersonality, or privacy. In fact, American cities have sometimes been described as collections of small towns. There is some truth to this description, especially if the city is compared to the actual small town, rather than to the romantic construct of antiurban critics.

Postwar suburbia represents the most contemporary version of the quasi-primary way of life. Owing to increases in real income and the encouragement of homeownership provided by the F.H.A., families in the lower middle class and upper working class can now live in modern single-family homes in low-density subdivisions, an opportunity previously available only to the upper and upper-middle classes.

The popular literature of the 1950s described the new suburbs as communities in which conformity, homogeneity, and other-direction are unusually rampant. The implication is that the move from city to suburb initiates a new way of life which causes considerable behavior and personality change in previous urbanites. My research in Levittown, New Jersey, suggests, however, that the move from the city to this predominantly lower-middle-class suburb does not result in any major behavioral changes for most people. Moreover, the changes which do occur reflect the move from the social isolation of a transient city or suburban apartment building to the quasi-primary life of a neighborhood of single-family homes. Also, many of the people whose life has changed report that the changes were intended. They existed as aspirations before the move or as reasons for it. In other words, the suburb itself creates few changes in ways of life.

A COMPARISON OF CITY AND SUBURB

If outer-urban and suburban areas are similar in that the way of life in both is quasi-primary, and if urban residents who move out to the suburbs do not undergo any significant changes in behavior, it is fair to argue that the differences in ways of life between the two types of settlements have been overestimated. Yet the fact remains that a variety of physical and demographic differences exist between the city and the suburb. However, upon closer examination, many of these differences turn out to be either spurious or of little significance for the way of life of the inhabitants.

The differences between the residential areas of cities and suburbs which have been cited most frequently are:

1 Suburbs are more likely to be dormitories.
2 They are further away from the work and play facilities of the central business districts.

3 They are newer and more modern than city residential areas and are designed for the automobile rather than for pedestrian and mass-transit forms of movement.
4 They are built up with single-family rather than multifamily structures and are therefore less dense.
5 Their populations are more homogeneous.
6 Their populations differ demographically: they are younger; more of them are married; they have higher incomes; and they hold proportionately more white-collar jobs.

Most urban neighborhoods are as much dormitories as the suburbs. Only in a few older inner-city areas are factories and offices still located in the middle of residential blocks, and even here many of the employees do not live in the neighborhood.

[. . .]

The fact that suburbs are smaller is primarily a function of political boundaries drawn long before the communities were suburban. This affects the kinds of political issues which develop and provides somewhat greater opportunity for citizen participation. Even so, in the suburbs as in the city, the minority who participate routinely are the professional politicians, the economically concerned businesspeople, lawyers, and salespeople, and the ideologically motivated middle- and upper-middle-class people with better than average education.

The social consequences of differences in density and house type also seem overrated. Single-family houses in quiet streets facilitate the supervision of children; this is one reason why middle-class parents who want to keep an eye on their children move to the suburbs. House type also has some effects on relationships between neighbors, insofar as there are more opportunities for visual contact between adjacent homeowners than between people on different floors of an apartment house. However, if occupants' characteristics are also held constant, the differences in actual social contact are less marked. Homogeneity of residents turns out to be more important than proximity as a determinant of sociability. If the population is heterogeneous, there is little social contact between neighbors, either on apartment-house floors or in single-family-house blocks; if people are homogeneous, there is likely to be considerable social contact in both house types. One need only contrast the apartment house located in a transient, heterogeneous neighborhood and exactly the same structure in a neighborhood occupied by a single ethnic group. The former is a lonely, anonymous building; the latter, a bustling microsociety. I have observed similar patterns in suburban areas: on blocks where people are homogeneous, they socialize; where they are heterogeneous, they do little more than exchange polite greetings.

Suburbs are usually described as being more homogeneous in house type than the city, but if they are compared to the outer city, the differences are small. Most inhabitants of the outer city, other than well-to-do homeowners, live on blocks of uniform structures as well; for example, the endless streets of row houses in Philadelphia and Baltimore or of two-story duplexes and six-flat apartment houses in Chicago. They differ from the new suburbs only in that they were erected through more primitve methods of mass production. Suburbs are, of course, more predominantly areas of owner-occupied single homes, though in the outer districts of most American cities homeownership is also extremely high.

Demographically, suburbs as a whole are clearly more homogeneous than cities as a whole, though probably not more so than outer cities. However, people do not live in cities or suburbs as a whole, but in specific neighborhoods. An analysis of ways of life would require a determination of the degree of population homogeneity within the boundaries of areas defined as neighborhoods by residents' social contacts. Such an analysis would no doubt indicate that many neighborhoods in the city as well as the suburbs are homogeneous. Neighborhood homogeneity is actually a result of factors having little or nothing to do with the house type, density, or location of the area relative to the city limits. Brand new neighborhoods are more homogeneous than older ones, because they have not yet experienced resident turnover, which frequently results in population heterogeneity. Neighborhoods of low- and medium-priced housing are usually less homogeneous than those with expensive dwellings because they attract families who have reached the peak of occupational and residential mobility, as well as young families who are just starting their climb and will eventually move to neighborhoods of higher status. The latter, being accessible only to high income people, are therefore more homogeneous with respect to other resident characteristics as well. Moreover, such areas have the economic and political power to slow down or prevent invasion.

The demographic differences between cities and suburbs cannot be questioned, especially since the suburbs have attracted a large number of middle-class child-rearing families. The differences are, however, much reduced if suburbs are compared only to the outer city. In addition, a detailed comparison of suburban and outer-city residential areas would show that neighborhoods with the same kinds of people can be found in the city as well as the suburbs. Once again, the age of the area and the cost of housing are more important determinants of demographic characteristics than the location of the area with respect to the city limits.

CHARACTERISTICS, SOCIAL ORGANIZATION, AND ECOLOGY

The preceding sections of the paper may be summarized in three propositions:

1 As concerns ways of life, the inner city must be distinguished from the outer city and the suburbs; and the latter two exhibit a way of life bearing little resemblance to Wirth's urbanism.
2 Even in the inner city, ways of life resemble Wirth's description only to a limited extent. Moreover, economic condition, cultural characteristics, life-cycle stage, and residential instability explain ways of life more satisfactorily than number, density, or heterogeneity.
3 Physical and other differences between city and suburb are often spurious or without much meaning for ways of life.

These propositions suggest that the concepts "urban" and "suburban" are neither mutually exclusive nor especially relevant for understanding ways of life. They – and number, density, and heterogeneity as well – are ecological concepts which describe human adaptation to the environment. However, they are not sufficient to explain social phenomena, because these phenomena cannot be understood solely as the consequences of ecological processes. Therefore, other explanations must be considered.

Ecological explanations of social life are most applicable if the subjects under study lack the ability to *make choices*, be they plants, animals, or human beings. Thus, if there is a housing shortage, people will live almost anywhere, and under extreme conditions of no choice, as in a disaster, married and single, old and young, middle and working class, stable and transient will be found side by side in whatever accommodations are available. At that time, their ways of life represent an almost direct adaptation to the environment. If the supply of housing and of neighborhoods is such that alternatives are available, however, people will make choices, and if the housing market is responsive, they can even make and satisfy explicit *demands.*

Choices and demands do not develop independently or at random: they are functions of the roles people play in the social system. These can best be understood in terms of the *characteristics* of the people involved; that is, characteristics can be used as indices to choices and demands made in the roles that constitute ways of life. Although many characteristics affect the choices and demands people make with respect to housing and neighborhoods, the most important ones seem to be *class* – in all its economic, social, and cultural ramifications – and *life-cycle stage.* If people have an opportunity to choose, these two characteristics will go far in explaining the kinds of housing and neighborhood they will occupy and the ways of life they will try to establish within them.

Many of the previous assertions about ways of life in cities and suburbs can be analyzed in terms of class and life-cycle characteristics. Thus, in the inner city, the unmarried and childless live as they do, detached from neighborhood, because of their life-cycle stage; the cosmopolites, because of a combination of life-cycle stage and a distinctive but class-based subculture. The way of life of the deprived and trapped can be explained by low socioeconomic level and related handicaps. The quasi-primary way of life is associated with the family stage of the life cycle and the norms of childrearing and parental role found in the upper working class, the lower-middle class, and the noncosmopolite portions of the upper-middle and upper classes.

[. . .]

A REEVALUATION OF DEFINITIONS

The argument presented here has implications for the sociological definition of the city. Such a definition relates ways of life to environmental features of the city qua settlement type. But if ways of life do not coincide with settlement types, and if these ways are functions of class and life-cycle stage rather than of the ecological attributes of the settlement, a sociological definition of the city cannot be formulated. Concepts such as "city" and "suburb" allow us to distinguish settlement types from each other physically and demographically, but the ecological processes and conditions which they synthesize have no direct or invariate consequences for ways of life. The sociologist cannot, therefore, speak of an urban or suburban way of life.

CONCLUSION

Many of the descriptive statements made here are as time bound as Wirth's. In the 1940s Wirth concluded that some form of urbanism would eventually predominate in all settlement types. He was, however, writing during a time of immigrant acculturation and at the end of a serious depression, an era of minimal choice. Today, it is apparent that high-density, heterogeneous surroundings are for most people a temporary place of residence; other than for the Park Avenue or Greenwich Village cosmopolites, they are a result of necessity, rather than choice. As soon as they can afford to do so, most Americans head for the single-family house and the quasi-primary way of life of the low-density neighborhood, in the outer city or the suburbs.

Changes in the national economy and in government housing policy can affect many of the variables that make up housing supply and demand. For example, urban sprawl may eventually outdistance the ability of present and proposed transportation systems to move workers into the city; further industrial decentralization can forestall it and alter the entire relationship between work and residence. The expansion of urban-renewal activities can perhaps lure a significant number of cosmopolites back from the suburbs, while a drastic change in renewal policy might begin to ameliorate the housing conditions of the deprived population. A serious depression could once again make America a nation of doubled-up tenants.

These events will affect housing supply and residential choice; they will frustrate, but not suppress, demands for the quasi-primary way of life. However, changes in the national economy, society, and culture can affect people's characteristics – family size, educational level, and various other concomitants of life-cycle stage and class. These in turn will stimulate changes in demands and choices. The rising number of college graduates, for example, is likely to increase the cosmopolite ranks. This might in turn create a new set of city dwellers, although it will probably do no more than encourage the development of cosmopolite facilities in some suburban areas.

The current revival of interest in urban sociology and in city planning suggests that data may soon be available to formulate a more adequate theory of the relationship between settlements and the ways of life within them. The speculations presented in this essay are intended to raise questions; they can be answered only by more systematic data collection and theorizing.

“Theories of Urbanism”

from *The Urban Experience*, second edition (1984)[1976]

Claude S. Fischer

Editors’ Introduction

Claude Fischer offers a seminal codification of academic thought on the theory of urbanism, and in the process makes a valuable reformulation of the theory through his attention to the emergence of subcultures. Fischer accepts some of the precepts of the prevailing determinism in urban sociology that acknowledges the primary and independent impact of urban effects, which Louis Wirth had identified as size, density, and heterogeneity. He differs from the compositional approach of Oscar Lewis and Herbert Gans, who rebuke urban effects and look to the cultural, demographic, and class characteristics of urbanites. Fischer believes that size and density of the population, or what he calls *critical mass* in cities, have independent effects in fostering subcultures. The emergence of subcultures fosters the further creation of more subcultures through the touch and recoil of more intensive interactions between more diverse populations and heterogeneous communities. That is to say that increasing size and density fosters greater heterogeneity. This recalls Robert Park’s concept that the modern city becomes a *mosaic of social worlds*. The larger the city, the greater there is the potential to produce subcultural communities.

Fischer reconstructs the theory of urbanism by downplaying the negative effects of Durkheimian *anomie* with reference to the crime, mental health, and social problems that are found in the metropolis. He gives another perspective on *differentiation* as a cultural process linked with specialization in the division of the labor. Fischer sees the city and its subcultures as a vital force for the amplifying of cultural experience and human creativity. Subcultures mark the emancipation of the individual from traditional controls and conventions, while providing a new set of subgroup identities and communities. In this way, they counterbalance some of the alienation and normlessness, the spiritual anxieties and social disorders found in our cities and marketplaces, which result from the breakdown of traditional customs and primary relationships.

Fischer makes an intriguing contribution to subcultural theory, which has also been explored from the standpoint of media and cultural studies, including Dick Hebdige (*Subculture: The Meaning of Style*. London: Methuen, 1979) and an edited reader by Ken Gelder and Sarah Thornton (*The Subcultures Reader*. London: Routledge, 1997). Subcultures, in these cultural studies, are seen as a creative force of communication or *bricolage*, which provide youth, sexual and racial/ethnic minorities with a means of defying and criticizing the established cultural hegemony. Fischer’s understanding of subculture is less associated with identity politics and more eclectic and wide-ranging in its definition, comprising groups as diverse as delinquents, criminals, artists, bohemians, new religious sects, hobbyists, dance aficionados, hippies, and construction workers.

Understanding the subcultural life of the city helps understand the impact of the social movements of the 1960s and 1970s, which mobilized youthful, racial/ethnic, and gender/sexual minorities in movements of political resistance and empowerment. Fischer enlightens our understanding of the emergence of artistic, bohemian and gay/lesbian neighborhoods in American cities, such as New York’s Greenwich Village and East Village, the Castro district of San Francisco, and the Hollywood and West Hollywood districts of Los Angeles. These subcultural neighborhoods are often a lure for the children of suburbia who are drawn to the central city in search of the authenticity, excitement of what is unfamiliar. This is a distinct contrast from the earlier generation of the postwar period, which escaped the city in search of privacy and open space. Understanding urban subcultures also connects with the growing interest in the creative and cultural life of cities (see Part 6 of this volume) with the onset of widespread gentrification and the emergence of an urban cultural economy.

This selection is extracted from Claude Fischer's book, *The Urban Experience* (1984)[1976]. The subcultural theory of urbanism is further articulated in the articles "Toward a Subcultural Theory of Urbanism," *American Journal of Sociology* 80, 6 (1975): 1319–1341 and "The Subcultural Theory of Urbanism: A Twentieth-Year Assessment," *American Journal of Sociology* 101, 3 (November 1995): 543–577.

Claude Fischer teaches in the Sociology Department at the University of California, Berkeley. He is currently the executive editor of *Contexts*, an official publication of the American Sociological Association. Fischer served as Chair of the Community and Urban Sociology Section in 1991–92. In 1996, he won the Robert and Helen Lynd Award for lifetime contribution to community and urban sociology from the American Sociological Association. Along with Michael Hout, Claude Fischer directs a project funded by the Russell Sage Foundation at the University of California, Berkeley, called "USA: A Century of Difference." Drawing upon a century of data up to the 2000 Census, this project will report on how Americans live, work, consume, and pray at the beginning of the twenty-first century. Among his books are *To Dwell Among Friends: Personal Networks in Town and City* (Chicago: University of Chicago Press, 1982), and *America Calling: A Social History of the Telephone to 1940* (Berkeley, CA: University of California Press, 1992).

In *To Dwell Among Friends*, Fischer considers the thesis of declining community by comparing the differences in peoples' personal relations in urban versus non-urban areas. He takes the view that people exercise considerable individual agency in building their personal ties and networks. Initial relations are given to us, such as parents and other kin, but as we grow into adults, we select which ties are maintained and which are dropped. He concludes that people living in large cities versus small towns have roughly the same amount of social ties; neither group is any more likely to be isolated. Small-town residents tend to be more involved with kin, city respondents with non-kin. Urbanites also tend to have less dense networks, but more intense interpersonal relations that involve multiple exchanges with given individuals. Urban dwellers tend to display similar tendencies than the young and the educated. Cities also tend to contain younger, more educated, and more diverse populations than small towns.

We begin our inquiry into the nature of the urban experience by considering the theorizing of social scientists about the consequences of urbanism. The purpose of developing such theories before looking at the "real world" is to provide the investigator with a set of concepts needed to organize his or her perceptions of what would otherwise be a bewildering complexity. Properly developed, these concepts focus attention on the most critical features of the "real world." To begin a major study without a good theory or theories is like being dropped into a dark jungle with neither map nor compass.

But before reviewing those theories, we must consider, once again and more exactly, the problem of defining "urban." For it turns out that some of the disagreement and confusion about the nature of the urban experience stems from differences in interpretation of the terms "urban" and "city." Surely, little progress can be made in understanding city life if we do not understand what the word "city" means. The four broad types of definitions are: demographic, institutional, cultural, and behavioral.

Demographic definitions involve essentially the size and density of population. In the present book, a community is more or less "urban" depending on the size of its population; a "city," therefore, is a place with a relatively large population. *Institutional* definitions reserve the term "city" for communities with certain specific institutions. For example, to be a city, a community must have its own autonomous political elite; or, it must have specific economic institutions, such as a commercial market. *Cultural* definitions require that a community possess particular cultural features, such as a group of literate people. And *behavioral* definitions require certain distinctive and typical behavioral styles among the people of a community – for example, an impersonal style of social interaction – before the community is labeled a "city."

The demographic definition has at least three advantages: One, the numerical criterion is common to virtually all definitions of "urban" or "city"; even those focusing on other variables employ size. Two, the purely demographic definition does not beg the question as to whether any other factor is necessarily associated with size; that remains an open issue. And three, the demographic definition implies that "urban" and "city" refer to matters of degree; they are not all-or-nothing variables.

What theories are there about the social–psychological consequences of urbanism in the sense of the demographic definition–population concentration? Here and throughout the rest of the book, we shall center on three major theories of urbanism, two of which confront each other directly, and a third which attempts their synthesis:

1 *Determinist theory* (also called *Wirthian* theory or the *theory of urban anomie*) argues that urbanism increases social and personality disorders over those found in rural places.
2 *Compositional* (or *nonecological*) *theory* denies such effects of urbanism; it attributes differences between urban and rural behavior to the composition of the different populations.
3 *Subcultural theory* adopts the basic orientation of the compositional school but holds that urbanism does have certain effects on the people of the city, with consequences much like the ones determinists see as evidence of social disorganisation.

Before discussing each theory in detail, we should consider the history of social thought from which they all emerged.

The most influential and historically significant theory of urbanism received its fullest exposition in a 1938 paper by Louis Wirth (thus the term, "Wirthian") entitled "Urbanism as a Way of Life." This essay, one of the most often quoted, reprinted, and cited in the whole sociological literature, needs to be examined carefully. It is heir to a long tradition of sociological theory.

The events that formed the focal concern of social philosophers during the nineteenth and early twentieth centuries have been termed the "Great Transformation" (Karl Polanyi (1944) *The Great Transformation*. New York: Farrar and Rinehart). Western society was undergoing vast and dramatic changes as a result of the Industrial Revolution and its accompanying processes of urbanization, nationalization, and bureaucratization. These early social scientists (Marx, Durkheim, Weber, Simmel, Tönnies, and others) sought to understand the forms of social life and the psychological character of the emerging civilization – our civilization.

The analysis they developed greatly emphasized the matter of *scale*. Innovations in transportation and communication, together with rapid increases in population, meant that many more individuals than ever before were able to interact and trade with each other. Instead of a person's daily life being touched at most by only the few hundred people of one village, in modern society an individual is in virtually direct contact with thousands, and in indirect contact with millions.

This "dynamic density," to use Durkheim's term, in turn produces *social differentiation*, or diversification, the most significant aspect of which is an increased division of labor. In the preindustrial society, most workers engaged in similar activities; in modern society, they have very different and specialized occupations. In a small, undifferentiated population, where people know each other, perform the same sort of work, and have the same interests – where they look, act, and think alike – it is relatively easy to maintain a consensus on proper values and appropriate behavior. But in a large, differentiated society, where people differ in their work and do not know each other personally, they have divergent interests, views, and styles. A pipefitter and a ballet dancer have little in common. And so, there can be little consensus or cohesion in such a society, and the social order is precarious. Further ramifications of social differentiation, it was thought, included the development of formal institutions, such as contracts and bureaucracy; the rise of rational, scientific modes of understanding the world; an increase in individual freedom, at the cost of interpersonal estrangement; and a rise in the rate of deviant behavior and social disorganization.

The essence of this classic sociological analysis is the connection of the structural characteristics of a society, particularly its scale, to the quality of its "moral order." That turns out, not coincidentally, to parallel the focal interest of urban sociology: the interest between structural features of communities – particularly scale – and their moral orders. In fact, the city has long played a significant role in classic sociological theories. It was seen as modern society in microcosm, so that the ways of life in urban places were viewed as harbingers of life in the emerging civilization. At the same time, the classic theories have had a significant role in influencing the study of cities, having been borrowed from liberally in the formation of the determinist approach.

The development of urban theory moved from Europe to the University of Chicago during the first third of this century. There, the Department of Sociology, under the leadership of Robert Ezra Park, a former journalist and student of the classical German sociologist Georg Simmel, produced a vast and seminal array of theoretical and empirical studies of urban life, based on research conducted chiefly in the city of Chicago. In an influential essay (Robert Park (1915) "The City: Suggestions for the Investigation of Human Behavior in the City," *American Journal of Sociology* 20, 5: 577–612), Park followed the lead of the classic theorists by arguing that urbanism produced new ways of life and new types of people, and that sociologists should venture out to explore these new forms in their own cities, much in the style of anthropologists studying primitive tribes. Another strong motivation for such research was the social turmoil then accompanying the rapid growth and industrialization of Western cities, a realm of civic activity in which Chicago about 1910 was no doubt a leader. The serious social problems accompanying these developments demanded study and explanation.

The varied studies of Chicago resulted in a remarkable series of descriptions of urban ways of life. The "natural histories" depicted many different groups and areas: taxi-hall dancers, hobos, Polish-Americans, juvenile gangs, the Jewish ghetto, pickpockets, police, and so on. A theme running through the findings of these various studies was that the groups, whether "normal" or "deviant," formed their own "social worlds." That is, they tended to be specialized social units in which the members associated mainly with each other, held their own rather distinctive set of beliefs and values, spoke in a distinctive argot, and displayed characteristic styles of behavior. Together these studies described a city that was, to quote Park's famous phrase, "a mosaic of social worlds which touch but do not interpenetrate." As we shall see, the explanation for the urban phenomena observed by Chicago's sociologists was drawn largely from the classic theories of the Great Transformation.

DETERMINIST THEORY

Some leads to a determinist theory of urbanism can be found in Park's 1915 paper, but the full exposition of this theory was achieved in Wirth's essay 23 years later. Wirth begins with a definition of the city as "a relatively large, dense, and permanent settlement of socially heterogeneous individuals" – an essentially demographic definition. He then seeks to demonstrate how these inherent, essential features of urbanism produce social disorganization and personality disorders – the dramatic aspects of the city scene that had captured the attention of the Chicago School. Wirth's analysis operates on essentially two levels, one a psychological argument, the other an argument of social structure.

The psychological analysis draws heavily upon a 1905 paper by Georg Simmel, a teacher of both Park and Wirth. In his essay "The Metropolis and Mental Life" Simmel centered on the ways that living in the city altered individuals' minds and personalities. The key, he thought, lay in the sensations which life in the city produces: "The psychological basis of the metropolitan type of individuality consists in the *intensification of nervous stimulation* which results from the swift and uninterrupted change of inner and outer stimuli." The city's most profound effects, Simmel maintained, are its profusion of sensory stimuli – sights, sounds, smells, actions of others, their demands and interferences. The onslaught is stressful; individuals must protect themselves, they must adapt. Their basic mode of adaptation is to react with their heads instead of their hearts. This means that urban dwellers tend to become intellectual, rationally calculating, and emotionally distant from one another. At the same time, these changes promote freedom for self-development and creativity. . . .

Wirth's treatment of this process follows Simmel's and begins with the assumption that the large, dense, and heterogeneous environment of the city assaults the hapless city dweller with profuse and varied stimuli. Horns blare, signs flash, solicitors tug at coattails, poll-takers telephone, newspaper headlines try to catch the eye, strange-looking and strange-behaving persons distract attention – all these features of the urban milieu claim a different response from the individual. Adaptations to maintain mental equilibrium are necessary and they appear. These adaptations liberate urbanites from the claims being pressed upon them. They also insulate them from the other people. City dwellers become aloof, brusque, impersonal in their dealings with others, emotionally buffered in their human relationships. Even these protective devices are not enough, so that "psychic overload" exacts at least a partial toll in irritation, anxiety, and nervous strain.

The interpersonal estrangement that follows from urbanites' adaptations produces further consequences. The bonds that connect people to one another are loosened – even sundered – and without them people are left both unsupported and unrestrained. At the worst, they must suffer through material and emotional crises without assistance, must deal with them alone; being alone, they are more likely to fail, to suffer physical deterioration or mental illness, or both. The typical picture is one of an elderly pensioner living in a seedy hotel without friends or kin, suffering loneliness, illness, and pain. But this same estrangement permits people in the city to spin the wildest fantasies – and to act upon those fantasies, whether they result in feats of genius or deeds of crime and depravity. The typical picture here is one of a small-town boy suddenly unshackled by conventional constraints, and possessing unlimited options including a life of creative art or a life of crime. Ultimately, interpersonal estrangement produces a decline of community cohesion and a corresponding loss of "sense of community." These are the psychological changes and further consequences, Wirth argued, that follow from increases in urbanism.

In his analysis of social structure, Wirth reaches essentially the same conclusion as he does in his psychological analysis, but he posits different processes. Through economic processes of competition, comparative advantage, and specialization, the size, density, and heterogeneity of a population produce the multi-faceted community differentiation mentioned earlier. This is manifested most significantly in the division of labor, but it exists in other forms as well: in the diversity of locales – business districts, residential neighborhoods, "bright-lights" areas, and so on; in people's places of activity, with work conducted in one place, family life in another, recreation in yet a third; in people's social circles, with one set of persons co-workers, another set neighbors, another friends, and still another kin; in institutions, with the alphabetized diversity of government agencies, specialized school systems, and media catering to every taste. An important aspect of this community differentiation is that it is reflected in people's activities. Their time and attention come to be divided among different and disconnected places, and people. For example, a business executive might move from breakfast with her family, to discussions with office co-workers, to lunch with business contacts, to a conference with clients, to golf with friends from the club, and finally to dinner with neighbors.

The differentiation of the social structure and of the lives of individuals living within that structure weakens social bonds in two ways. At the community level, people differ so much from each other in such things as their jobs, their neighborhoods, and their life-styles that moral consensus becomes difficult. With divergent interests, styles, and views of life, groups in the city cannot agree on values or beliefs, on ends or on means. As community-wide cohesion is weakened, so is the cohesion of the small, intimate, "primary" groups of society, such as family, friends, and neighbors – the ones on which social order and individual balance depend. These groups are weakened because, as a result of the differentiation of urban life, each encompasses less of an individual's time or needs. For instance, people work outside the family and increasingly play outside the family, so that the family becomes less significant in their lives. Similarly, they can leave the neighborhood for shopping or recreation, so that the neighbors become less important. Claiming less of people's attention, controlling less of their lives, the primary groups become debilitated. Thus, by dividing the community and by weakening its primary groups, differentiation produces a general loosening of social ties.

This situation in turn results in *anomie,* a social condition in which the norms – the rules and conventions of proper and permissible behavior – are feeble. People do not agree about the norms, do not endorse them, and tend to challenge or ignore them. Yet some degree of social order must be, indeed is, maintained even in the largest cities. Since personal means of providing order have been weakened, other means must be used. These other means – rational and impersonal procedures that arise to prevent or to moderate anomie – are called *formal integration.* For example, instead of controlling the behavior of unruly teenagers by talking to them or their parents personally, neighbors call in the police. Instead of settling a community problem through friendly and informal discussions, people organize lobbying groups and campaign in formal elections.

This sort of formal integration avoids chaos and can even maintain a well-functioning social order. However, according to the classic theories that Wirth applied in his analysis of cities, such an order can never fully replace a communal order based on consensus and the moral strength of small, primary groups. Consequently, more anomie must develop in urban than in nonurban places.

The behavioral consequences of anomie and of the shedding of social ties are similar to those eventually resulting from overstimulation. People are left unsupported to suffer their difficulties alone; and they are unrestrained by social bonds or rules from committing all sorts of acts, from the simply "odd" to the dangerously criminal.

These, then, are the arguments with which Wirth explained what seemed to the Chicago School to be peculiarly urban phenomena – stress, estrangement, individualism, and especially social disorganization. On the psychological level, urbanism produces threats to the nervous system that then lead people to separate themselves from each other. On the level of social structure, urbanism induces differentiation, which also has the consequences of isolating people. A society in which social relationships are weak provides freedom for individuals, but it also suffers from a debilitated moral order, a weakness that permits social disruption and promotes personality disorders.

COMPOSITIONAL THEORY

The determinist approach has been challenged on a number of fronts. The most significant challenge has been posed by compositional theory, perhaps best represented by the work of Herbert Gans and Oscar Lewis. . . . Compositionalists emerged from the same Chicago School tradition as the determinists, but they derived their inspiration largely from that part of the Chicago orientation that describes the city as a "mosaic of social worlds." These "worlds" are intimate social circles based on kinship, ethnicity, neighborhood, occupation, life-style, or similar personal attributes. They are exemplified by enclaves such as immigrant neighborhoods ("Little Italy") and upper-class colonies ("Nob Hill"). . . . The crux of the compositional argument is that these private milieus endure even in the most urban of environments.

In contrast to determinists, social scientists such as Gans and Lewis do not believe that urbanism weakens small, primary groups. They maintain that these groups persist undiminished in the city. Not that people are torn apart because they must live simultaneously in different social worlds, but instead that people are enveloped and protected by their social worlds. This point of view denies that ecological factors – particularly the size, density, and heterogeneity of the wider community – have any serious, direct consequences for personal social worlds. In this view, it matters little to the average kith-and-kin group whether there are 100 people in the town or 100,000; in either case the basic dynamics of that group's social relationships and its members' personalities are unaffected.

In compositionalist terms, the dynamics of social life depend largely on the nonecological factors of social class, ethnicity, and stage in the life-cycle. Individuals' behavior is determined by their economic position, cultural characteristics, and by their marital and family status. The same attributes also determine who their associates are and what social worlds they live in. It is these attributes – not the size of the community or its density – that shape social and psychological experience.

Compositionalists do not suggest that urbanism has *no* social-psychological consequences, but they do argue that both the *direct* psychological effects on the individual and the *direct* anomic effects on social worlds are insignificant. If community size does have any consequences, these theorists stipulate, they result from ways in which size affects positions of individuals in the economic structure, the ethnic mosaic, and the life-cycle. For example, large communities may provide better-paying jobs, and the people who obtain those jobs will be deeply affected. But they will be affected by their new economic circumstances, not directly by the urban experience itself. Or, a city may attract a disproportionate number of males, so that many of them cannot find wives. This will certainly affect their behavior, but not because the city has sundered their social ties. Thus, the compositional approach can acknowledge urban–rural social–psychological differences, and can account for them insofar as these differences reflect variations in class, ethnicity, or life-cycle. But the compositional approach does not expect such differences to result from the psychological experience of city life or from an alteration in the cohesion of social groups.

The contrast between the determinist and compositional approaches can be expressed this way: Both emphasize the importance of social worlds in forming the experiences and behaviors of individuals, but they disagree sharply on the relationship of urbanism to the viability of those personal milieus. Determinist theory maintains that urbanism has a direct impact on the coherence of such groups, with serious consequences for individuals. Compositional theory maintains that these social worlds are largely impervious to ecological factors, and that urbanism thus has no serious, *direct* effects on groups or individuals.

SUBCULTURAL THEORY

The third approach, *subcultural theory* (Claude Fischer (1975) "Toward a Subcultural Theory of Urbanism," *American Journal of Sociology* 80, 6: 1319–1341), contends that urbanism independently affects social life – not, however, by destroying social groups as determinism suggests, but instead by helping to create and strengthen them. The most significant social consequence of community size is the promotion of diverse *subcultures* (culturally distinctive groups, such as college students or Chinese-Americans). Like compositional theory, subcultural theory maintains that intimate social circles persist in the urban environment. But, like determinism, it maintains that ecological factors do produce significant effects in the social orders of communities, precisely by supporting the emergence and vitality of distinctive subcultures.

Like the Chicago School in certain of its works and like compositionalists, the subcultural position holds that people in cities live in meaningful social worlds. These worlds are inhabited by persons who share relatively distinctive traits (like ethnicity or occupation), and who tend to interact especially with one another, and who manifest a relatively distinct set of beliefs and behaviors. Social worlds and subcultures are roughly synonymous. Obvious examples of subcultures include ones like those described by the Chicago School: the country club set in Grosse Pointe, Michigan; the Chicano community in East Los Angeles; and hippies in urban communes. There are more complex subcultures as well. For example, on the south side of Chicago is an area heavily populated by workers in the nearby steel mills. These workers together form a community and occupational subculture, with particular habits, interests, and attitudes. But they are further divided into even more specific subcultures by ethnicity and neighborhood; thus there are, for example, the recently immigrated Serbo-Croatian steelworkers in one area and the earlier-generation ones elsewhere, each group somewhat different from the other. In both subcultural and compositional theory, these subcultures persist as meaningful environments for urban residents.

However, in contrast to the compositional analysis, which discounts any effects of urbanism, subcultural theory argues that these groups *are* affected directly by urbanism, particularly by the effects of "critical mass." Increasing scale on the rural-to-urban continuum creates new subcultures, modifies existing ones, and brings them into contact with each other. Thus urbanism has unique consequences, including the production of "deviance," but not because it destroys social worlds – as determinism argues – but more often because it creates them.

The subcultural theory holds, first, that there are two ways in which urbanism produces Park's "mosaic of little worlds which touch but do not interpenetrate": 1) Large communities attract migrants from wider areas than do small towns, migrants who bring with them a great variety of cultural backgrounds, and thus contribute to the formation of a diverse set of social worlds. And 2), large size produces the structural differentiation stressed by the determinists – occupational specialization, the rise of specialized institutions, and of special interest groups. To each of these structural units are usually attached subcultures. For example, police, doctors, and longshoremen tend to form their own milieus – as do students, or people with political interests or hobbies in common. In these ways, urbanism generates a variety of social worlds.

But urbanism does more: It intensifies subcultures. Again, there are two processes. One is based on *critical mass*, a population size large enough to permit what would otherwise be only a small group of individuals to become a vital, active subculture. Sufficient numbers allow them to support institutions – clubs, newspapers, and specialized stores, for example – that serve the group; allow them to have a visible and affirmed identity, to act together on their own behalf, and to interact extensively with each other. For example, let us suppose that one in every thousand persons is intensely interested in modern dance. In a small town of 5,000 that means there would be, on the average, five such persons, enough to do little else than engage in conversation about dance. But in a city of one million, there would be a thousand – enough to support studios, occasional ballet performances, local meeting places, and a special social milieu. Their activity would probably draw other people beyond the original thousand into the subculture (those quintets of dance-lovers migrating from the small towns). The same general process of critical mass operates for artists, academics, bohemians, corporate executives, criminals, computer programmers – as well as for ethnic and racial minorities.

The other process of intensification results from contacts between these subcultures. People in different social worlds often do "touch," in Park's language. But in doing so, they sometimes rub against one another only to recoil, with sparks flying upward. Whether the encounter is between blacks and Irish, hard-hats and hippies, or town and gown, people from one subculture often find people in another subculture threatening, offensive, or both. A common reaction is to embrace one's own social world all the more firmly, thus contributing to its further intensification. This is not to deny that there are often positive contacts between groups. There are; and there is a good deal of mutual influence – for example, the symbolism of young construction workers growing beards, or middle-class white students using black ghetto slang. It is, however, the contrast and recoil that intensify and help to define urban subcultures.

Among the subcultures spawned or intensified by urbanism are those which are considered to be either downright deviant by the larger society – such as delinquents, professional criminals, and homosexuals; or to be at least "odd" – such as artists, missionaries of new religious sects, and intellectuals; or to be breakers of tradition – such as life-style experimenters, radicals, and scientists. These flourishing subcultures, together with the conflict that arises among them and with mainstream subcultures, are both effects of urbanism, and they both produce what the Chicago School thought of as social "disorganization." According to subcultural theory, these phenomena occur not because social worlds break down, and people break down with them, but quite the reverse – because social worlds are formed and nurtured.

Subcultural theory is thus a synthesis of the determinist and compositional theories: like the compositional approach, it argues that urbanism does not produce mental collapse, anomie, or interpersonal estrangement; that urbanites at least as much as ruralites are integrated into viable social worlds. However, like the determinist approach, it also argues that cities *do* have effects on social groups and individuals – that the differences between rural and urban persons have other causes than the economic, ethnic, or life-style circumstances of those persons. Urbanism does have *direct* consequences.

PART TWO

The Form and Function of Cities

Plate 5 Manhattan, 1959 by Henri Cartier-Bresson. In the upper part of the city, around 103rd Street, some 61 blocks from Times Square, slums are being torn down with ruthless speed to make way for low cost housing projects such as these seen against the skyline (reproduced by permission of Magnum Photos).

INTRODUCTION TO PART TWO

American urban sociology was initially formulated around the analysis of the city of Chicago, the great urban crossroads of the early twentieth century. As the Midwestern agro-industrial capital, Chicago was a nodal center for transportation and trade, as well as a gateway for foreign immigrants into the American heartland. Chicago's status as a thriving international metropolis was established with its World's Columbian Exposition of 1893. The city emerged as a main terminal in the hub and spoke system of cities in the railroad era. An elevated train, the Loop, circled the downtown, heightening the significance of the urban center. The dominance of the center was significant to the theories of "human ecology" associated with the "Chicago School" of urban sociology, which Robert Park and his associates established at the University of Chicago beginning in 1913. Like London and New York, Chicago incited a literary and social work tradition that testified to the presence of economic Darwinism and social inequality. This included Upton Sinclair's *The Jungle* (1906), Theodore Dreiser's *Sister Carrie* (1932), and Jane Adams' Hull House for settlement work. The poet Carl Sandburg described Chicago as "hog butcher to the world."

The human ecologists applied ideas of Charles Darwin to the urban scene, justifying the presence of urban social inequality through comparison with the "struggle for existence" in the evolutionary life of plant and animal communities. Principles of "natural selection" explained the dominance of banks and corporations in the central business district (CBD), and dynamics of invasion-succession, now commonly identified as the gentrification and displacement process. Park, Burgess, and Zorbaugh each discussed how the land market "sifted and sorted" a population engaged in competition and conflict, eventually leading to a new social equilibrium. Their notion of "equilibrium" drew from physics and the laws of thermodynamics, as well as economic theory. Human ecology applied concepts of economics and market competition to the biological and social realm. Park had a more passive understanding of Darwinism than Herbert Spencer, who coined the phrase "survival of the fittest" to justify free market liberalism, colonialism, and racial eugenics. Park drew from the evolutionary naturalism of John Dewey, an advocate for educational reform. He was an apologist for the status quo, rarely critical of social inequality, and he sought to devalue the reform tradition in early twentieth century sociology in favor of scientific analysis incorporating active fieldwork.

Burgess presented Chicago as the spatial paradigm of the modern city, with a dominant CBD ringed by successive concentric zones of settlement. This is one of the more durable images in the annals of urban sociology, and has been the object of much discussion and criticism for being less applicable to the decentralized U.S. cities of the postwar era, which often have a more poly-nucleated or multi-nodal structure that arises out of a sprawling metropolitan network of freeway interchanges. Burgess also drew from biology, relating the metabolism of the human body to the moving equilibrium of human ecology between zones of organization and disorganization. He suggested that the "pulse" of a city is related to the incidence of increasing movement and mobility of people. Mobility was not only geographic, but also cultural, with reference to changing cultural attitudes and unconventional social behaviors. He linked mobility to social disorganization in breeding greater demoralization, promiscuity, and vice. Burgess considered social disorganization not as pathological, but as normal. Burgess identified the "zone-in-transition" as an area of high mobility that featured transitional housing, immigrant colonies, vice and criminal activity.

Zorbaugh codified the notion of the "natural area," an urban cultural district or neighborhood that emerged through the intersection of natural geographic boundaries and cultural characteristics of the population. Zorbaugh included man-made boundaries such as railroad tracks, which invites contemplation of the spatial and class implications of the phrase "life on the other side of the tracks." The competitive land market was seen as a sifting and sorting mechanism that segregated the population into natural areas such as Harlem, Chinatown and Little Italy, New York's Greenwich Village, or Chicago's Towertown and Gold Coast. Zorbaugh and other human ecologists saw segregation as a natural outcome of human ecological processes. Human ecology overlooked the effect of Jim Crow segregation laws, restrictive covenants and land restrictions on immigrants and minorities. Human ecology also did not account for the influence of government and real estate interests on land-use outcomes through policies such as zoning, transportation and other areas of urban infrastructural development.

Walter Firey's chapter on sentimental and symbolic land use in Central Boston was an important article in the sociocultural school of urban sociology. Initiated by Milla Alihan (*Social Ecology*. New York: Columbia University Press, 1939), the sociocultural school criticized the dualism of Parkian human ecology for drawing a false dichotomy between the "social" and "biotic" dimensions of urban society. Socioculturalists emphasized the relative importance of culture versus the market economy in determining urban spatial order. Walter Firey criticized the economic determinism of the human ecologists by drawing attention to certain places in central Boston, such as ancestral cemeteries, the Boston Commons (a park), historic neighborhoods such as Beacon Hill, and ethnic enclaves such as the Italian North End. These antiquated land-uses were anomalous in their abrupt juxtaposition amidst modern office towers, banks, and department stores in the central business district. They defied the logic of human ecology theory, which predicted displacement of less profitable land-uses through the competitive force of the real estate market. Firey instead noted the enduring influence of collective sentiment on urban development and change, a recuperative force with aesthetic, familial, and historical dimensions.

The persistence of these Boston neighborhoods during the early twentieth-century, interwar period was a harbinger of the landmarks and historic preservation movement. The sociocultural school was somewhat transitional in presaging the later political-economy critique that emerged after the onset of urban renewal programs in the late 1940s and social movements in the 1950s and 1960s. Herbert Gans can be somewhat associated with the sociocultural critique through his book, *The Urban Villagers* (New York: The Free Press, 1962), which brought attention to the durability of Italian immigrant peer group networks as evidence of the persistence of community and social organization in the West End area of Boston, an area that had been designated a slum and slated for redevelopment. Attention in urban sociology then shifted to urban political economy, but the sociocultural perspective has experienced a kind of revival as it has morphed into the "cultural turn" in urban studies since the 1980s.

Civil rights and the associated social movements of the 1950s and 1960s brought new attention in the social sciences to the problems of racial segregation, immigrant exclusion, poverty and social inequality in urban life. A concatenation of urban riots, protests and community action movements arose to address issues such as racial inequality, pollution and toxic waste, transit and expressway expansion, and rights to housing, education, and welfare. They defended the integrity of urban neighborhoods and their quality of life, advanced the civil rights and legal entitlements of urban citizens, and promoted the economic and social development of urban communities.

The spirit of revolutionary change in urban society also stimulated a paradigm shift in urban sociology, promoting a host of writing that was critical of human ecology and addressed issues of elite control, political and class interests, and social inequality in cities. John Walton gives the historical background to this paradigm shift in "Urban Sociology: The Contribution and Limits of Political Economy," *Annual Review of Sociology* 19 (1993): 301–320. Logan and Molotch offer the seminal interpretation of the political-economy approach by applying Marxian concepts of use and exchange value to the analysis of urban land markets. They note idiosyncratic, or special use values associated with real estate, including the unsubstitutable preciousness that stems from the place attachments people may associate with homes and small businesses. Another special use value is that urban property markets structure access to resources and determine relative "life chances" for people to succeed. They also recognize that real estate has special exchange values in being a limited resource that may be monopolized through the actions of a variety of place entrepreneurs. These place entrepreneurs self-aggrandize their property interests by forging pro-growth coalitions with local business elites and governmental units to effectively create an apparatus of interlocking interests that they call an urban "growth machine." They lobby with government to focus the development of urban infrastructure to their interests. Joe Feagin discusses the Houston growth elite known as the "Suite 8F" crowd in *Free Enterprise City: Houston in Political and Economic Perspective* (New Brunswick, NJ: Rutgers University Press, 1988).

Logan and Molotch identify William Ogden as an example of a place entrepreneur who began as a railroad magnate, moved into real estate development, then became mayor of Chicago. They discuss also the partnership of Leland Stanford, Charles Crocker, Collis Huntington, and Mark Hopkins in San Francisco who worked to mutually enhance their interlocking interests in railroads, property, and commerce by lobbying to block the routing of a second transcontinental railroad terminating in San Diego. Los Angeles eventually became the terminus for two competing transcontinental railways, which stimulated a land boom in the 1880s. Collis Huntington's son, Henry Huntington, became the baron of the interurban electrical railway system in Los Angeles. He later joined with a syndicate including M. H. Sherman (land), General Harrison Otis Gray (land), and Harry Chandler (publisher of the *Los Angeles Times*, and son-in-law to Otis), that rapidly acquired monopoly control over land in the arid San Fernando Valley on the strength of knowledge that William Mulholland, the head of the Los Angeles Department of Water and Power, was working on a plan to build an aqueduct (completed in 1913) that would divert water from the Owens river valley in central California to Los Angeles to permit the further expansion of the metropolis. Robert Towne based his screenplay for the film *Chinatown* on these real events, changing the dates and taking poetic license by expressing the "rape of the Owens valley" through the metaphor of the cardinal sin of paternal incest by William Mulholland.

Urban growth machines have become more multifaceted networks of interests in the contemporary era, with the entry of an array of new players, including quasi-public redevelopmental agencies, housing developers, and developers of convention centers and sports stadiums. In a "land grab" comparable with the slum clearance of the Italian American West End of Boston, the Mexican American community of Chavez Ravine in Los Angeles was declared blighted and cleared to build a new stadium for the Brooklyn Dodgers, the first East Coast baseball team to move to the West Coast. The muralist Judith Baca depicted this historical atrocity in "Division of the Barrios and Chavez Ravine," a segment of a larger mural called the Great Wall of Los Angeles located along the Tujunga Wash flood control channel of the Los Angeles River (see www.sparcmurals.org). Raul Villa analyzes the tragic evisceration of the Mexican American community in his book *Barrio-Logos: Space and Place in Urban Chicano Literature and Culture* (Austin: University of Texas Press, 2000). Local growth machines destroyed many inner city minority neighborhoods during their ascendance over the course of the twentieth century, in favor of freeway interchanges, convention centers and stadiums, and upmarket housing. In his book, *The Power Broker* (New York: Vintage, 1975), Robert Caro chronicles the work of master-planner Robert Moses in designing and constructing the Cross-Bronx Expressway, which demolished and severed a huge Jewish neighborhood, especially the "tragic mile through East Tremont" that was the heart of the community.

The urban growth machine operates under the cover of the local press and media, who are interlocking business interests and also benefit from local growth. They promote a booster spirit and a civic ideology of growth as a public good through poetry contests, parades, beauty contests, dedications, derbies, public celebrations, and historical spectacles. They cover up the fact there is a "ghost in the growth machine" that leads to the destruction of communities, growing social inequality, environmental destruction, and urban sprawl.

Finally the "L.A. School" of urban sociology extends the invective against human ecology that was launched by urban political economy. The British expatriate turned Angeleno, Michael Dear, has written widely in codifying the L.A. School perspective. Los Angeles replaces Chicago as the urban prototype in the era of the postmodern metropolis. The decentralized freeway system has replaced the hub-and-spoke system and created a sprawling polycentric, polyglot, and polycultural metropolitan region where the hinterlands can command as much authority as the center. The waning economic and cultural dominance of the city center correlates with the onset of a variety of other economic and social trends, such as deindustrialization and globalization, the growth of ethnic suburbs, and the rise of gated communities in the exurbs. The metropolitan sprawl of Los Angeles resembles a giant keno game board of ethnoburbs, theme parks, and gated communities.

"Human Ecology"

from *American Journal of Sociology* (1936)

Robert Ezra Park

Editors' Introduction

Robert Park was born in Harveyville, Pennsylvania while his father was serving in the Civil War. They later settled in Minnesota and his father became a prosperous grocer. Park entered the University of Michigan in 1882 and was particularly drawn to the philosophy courses of John Dewey. He acquired ideas of evolutionary naturalism from Dewey, coming to see society as set in the natural order, in a competitive arena, but also held together by cognitive and moral consensus.

Upon graduation he began a career as a newspaper reporter, moving from Minneapolis, to Detroit, to Denver, to New York, and finally to Chicago. He wrote on the corruption of urban political machines, the immigrant areas of the city, crime, and other urban affairs. Journalism, particularly in Manhattan, satisfied his thirst for adventure and multifarious experience, but a persisting interest in the grand questions of life led him to return to academia to study philosophy at Harvard University in 1898. He subsequently grew interested in social thought and thus was impelled to move to Germany and the University of Berlin, which was then seen by many to be the intellectual center of Europe. While in Berlin, he came under the influence of Georg Simmel, then a *Privatdozent* lecturing in sociology. He obtained a Ph.D. from the University of Heidelberg in 1904.

Park returned from Germany to Massachusetts in 1903 and became a teaching assistant at Harvard. Through a chance encounter with a missionary, however, he discovered the work of the Congo Reform Association, and soon accepted work as their secretary and chief publicity agent. Through his work lobbying Congress to take action on the state of brutality and exploitation in the Congo Free State, Park met Booker T. Washington, who in 1905 was at the height of his notoriety as an accommodationist spokesman for black causes among the political elites. He became Washington's stenographer/ghostwriter, counselor and press agent for the next seven years, working mainly at the Tuskegee Institute in Macon, Georgia, with regular visits to New England and occasional tours to Europe with Washington. This migratory lifestyle did not suit his family, however, and he decided to return to academic life, at the invitation of W. I. Thomas, then a professor of sociology at the University of Chicago.

In the fall of 1913, at the age of 49, Robert Park began the quarter-century of teaching and research leadership during which the University of Chicago sociology department became a celebrated center of the discipline in America. Through the tremendous surge of field research that he supervised, he was instrumental in drawing sociology away from a normative and reform-oriented focus of the Progressive era to a more scientific analysis that still accounted for the social importance of knowledge. His seminal essay titled, "The City: Suggestions for the Investigation of Human Behavior in the City," published in 1915 in the *American Journal of Sociology*, became a kind of manifesto for the use of city as a research laboratory. In it, he called for the study of urban life using the same ethnographic methods used by anthropologists to study the Native Americans. With Ernest W. Burgess, Park wrote and edited a textbook, *Introduction to the Science of Sociology* (1921), which became the most influential reader in the early history of American sociology. Park served as President of the American Sociological Society (later changed to Association) in 1925.

R. D. McKenzie provided the first exposition on human ecology in an essay titled "The Ecological Approach to the Study of the Human Community," published in the *American Journal of Sociology* in 1924. Robert Park codified his beliefs in the same journal in 1936, in a paper titled, "Human Ecology." In this essay, Park applied the principles of Charles Darwin's "web of life" and "struggle for existence" in plant and animal communities to the study of human communities. Through his explanation of concepts such as dominance, invasion-succession, and natural areas, Park provided a justification for urban inequality and free market competition that is often associated with the beliefs of Social Darwinists, from whom he drew some of his thinking.

Social Darwinism was a body of late nineteenth-century philosophy and social thought expounded by the British Herbert Spencer and American William Sumner that applied Charles Darwin's principles of "natural selection" to the analysis of human social evolution. While Darwin held a passive sense of the interplay between variation and heredity, the Social Darwinists were more akin to Jean-Baptiste Lamarck, who had a more active conception of the inheritance of acquired characteristics. Spencer coined the concept "survival of the fittest" to express the concept that the rich and powerful are rewarded for their greater intelligence, talents, ambition, and industriousness, while the poor are doomed to failure for their lack of these characteristics. Free market liberalism was promoted in the economy, while charitable and state redistributional programs were opposed. Social Darwinism was eventually used to justify colonialism, racial eugenics, and policies of cultural assimilation.

Park was inspired by Charles Darwin, but ultimately diverges from Social Darwinism through his recognition that human societies participate in a social and moral order that has no counterpart on the nonhuman level. There is a dualism in human ecology in that there is competition as well as cooperation and symbiosis, especially at higher levels of the interactional pyramid. Park furthermore accounted for process, or social change, and was concerned that ecological equilibrium could commonly be disrupted by external changes.

Robert Park was driven by the philosophy of pragmatism that he learned from John Dewey, who exhorted American educators to school their students to engage in active learning through direct service in communities. He was influenced by the turn-of-the-century social reform and Progressive movements, as evidenced by his early passion for journalistic muckraking and devotion to anti-colonialist and black causes, distinguishing him from the conservative and racist Social Darwinists. Though liberal-minded, he did not buck the status quo, as attested by his association with the accommodationism of Booker T. Washington. Park died in 1944 in Nashville.

For further writing on the legacy of Robert Park, see: Fred H. Matthews, *Quest for an American Sociology: Robert E. Park and the Chicago School* (Montreal: McGill-Queen's University Press, 1977) and Edward Shils, "Robert E. Park, 1864–1944," *The American Scholar* (Winter, 1991): 120–127. See also the section on Robert Park in Lewis A. Coser's *Masters of Sociological Thought: Ideas in Historical and Social Context* (2nd Edition) (Fort Worth, Texas: Harcourt Brace Jovanovich, 1977).

THE WEB OF LIFE

Naturalists of the last century were greatly intrigued by their observation of the interrelations and co-ordinations, within the realm of animate nature, of the numerous, divergent, and widely scattered species. Their successors, the botanists, and zoologists of the present day, have turned their attention to more specific inquiries, and the "realm of nature," like the concept of evolution, has come to be for them a notion remote and speculative.

The "web of life," in which all living organisms, plants and animals alike, are bound together in a vast system of interlinked and interdependent lives, is nevertheless, as J. Arthur Thompson puts it, "one of the fundamental biological concepts" and is "as characteristically Darwinian as the struggle for existence."

Darwin's famous instance of the cats and the clover is the classic illustration of this interdependence. He found, he explains, that bumblebees were almost indispensable to the fertilization of the heartsease, since other bees do not visit this flower. The same thing is true with some kinds of clover. Bumblebees alone visit red clover, as other bees cannot reach the nectar. The inference is that if the bumblebees became extinct or very rare in England, the heartsease and red clover would become very rare, or wholly disappear. However, the number of bumblebees in any district depends in a great measure on the number of field mice, which destroy their combs and nests. It is estimated that more than two-thirds of them are thus destroyed all over England. Near villages and small towns the nests of bumblebees are more numerous than elsewhere and this is attributed to the number of cats that destroy the mice. Thus next year's crop of purple clover in certain parts of England depends on the number of bumblebees in the district; the number of bumblebees depends upon the number of field mice, the number of field mice upon the number and the enterprise of the cats, and the number of cats – as someone has added – depends on the number of old maids and others in neighboring villages who keep cats.

These large food chains, as they are called, each link of which eats the other, have as their logical prototype the familiar nursery rhyme, "The House that Jack Built." You recall:

The cow with the crumpled horn,
That tossed the dog,
That worried the cat,
That killed the rat,
That ate the malt
That lay in the house that Jack built.

Darwin and the naturalists of his day were particularly interested in observing and recording these curious illustrations of the mutual adaptation and correlation of plants and animals because they seemed to throw light on the origin of the species. Both the species and their mutual interdependence, within a common habitat, seem to be a product of the same Darwinian struggle for existence.

It is interesting to note that it was the application to organic life of a sociological principle – the principle, namely, of "competitive co-operation" – that gave Darwin the first clue to the formulation of his theory of evolution.

[. . .]

The active principle in the ordering and regulating of life within the realm of animate nature is, as Darwin described it, "the struggle for existence." By this means the numbers of living organisms are regulated, their distribution controlled, and the balance of nature maintained. Finally, it is by means of this elementary form of competition that the existing species, the survivors in the struggle, find their niches in the physical environment and in the existing correlation or division of labor between the different species.

[. . .]

These manifestations of a living, changing, but persistent order among competing organisms – organisms embodying "conflicting yet correlated interests" – seem to be the basis for the conception of a social order transcending the individual species, and of a society based on a biotic rather than a cultural basis, a conception later developed by the plant and animal ecologists.

In recent years the plant geographers have been the first to revive something of the earlier field naturalists' interest in the interrelations of species. Haeckel, in 1878, was the first to give to these studies a name, "ecology," and by so doing gave them the character of a distinct and separate science.

The interrelation and interdependence of the species are naturally more obvious and more intimate within the common habitat than elsewhere. Furthermore, as correlations have multiplied and competition has decreased, in consequence of mutual adaptations of the competing species, the habitat and habitants have tended to assume the character of a more or less completely closed system.

Within the limits of this system the individual units of the population are involved in a process of competitive co-operation, which has given to their interrelations the character of a natural economy. To such a habitat and its inhabitants – whether plant, animal, or human – the ecologists have applied the term "community."

The essential characteristics of a community, so conceived, are those of: (1) a population, territorially organized, (2) more or less completely rooted in the soil it occupies, (3) its individual units living in a relationship of mutual interdependence that is symbiotic rather than societal, in the sense in which that term applies to human beings.

These symbiotic societies are not merely unorganized assemblages of plants and animals which happen to live together in the same habitat. On the contrary, they are interrelated in the most complex manner. Every community has something of the character of an organic unit. It has a more or less definite structure and it has a life history in which juvenile, adult and senile phases can be observed. If it is an organism, it is one of the organs which are other organisms. It is, to use Spencer's phrase, a superorganism.

What more than anything else gives the symbiotic community the character of an organism is the fact that it possesses a mechanism (competition) for (1) regulating the numbers, and (2) preserving the balance between the competing species of which it is composed. It is by maintaining this biotic balance that the community preserves its identity and integrity as an individual unit through the changes and the vicissitudes to which it is subject in the course of its progress from the earlier to the later phases of its existence.

THE BALANCE OF NATURE

The balance of nature, as plant and animal ecologists have conceived it, seems to be largely a question of numbers. When the pressure of population upon the natural resources of the habitat reaches a certain degree of intensity, something invariably happens. In the one case the population may swarm and relieve the pressure of population by migration. In another, where the disequilibrium between population and natural resources is the result of some change, sudden or gradual, in the conditions of life, the pre-existing correlation of the species may be totally destroyed.

Change may be brought about by a famine, an epidemic, or an invasion of the habitat by some alien species. Such an invasion may result in a rapid increase of the invading population and a sudden decline in the numbers if not the destruction of the original population. Change of some sort is continuous, although the rate and pace of change sometimes vary greatly.

[. . .]

Under ordinary circumstances, such minor fluctuations in the biotic balance as occur are mediated and absorbed without profoundly disturbing the existing equilibrium and routine of life. When, on the other hand, some sudden and catastrophic change occurs – it may be a war, a famine, or pestilence – it upsets the biotic balance, breaks "the cake of custom," and releases energies up to that time held in check. A series of rapid and even violent changes may ensue which profoundly alter the existing organization of communal life and give a new direction to the future course of events.

The advent of the boll weevil in the southern cotton fields is a minor instance but illustrates the principle. The boll weevil crossed the Rio Grande at Brownsville in the summer of 1892. By 1894 the pest had spread to a dozen counties in Texas, bringing destruction to the cotton and great losses to the planters. From that point it advanced, with every recurring season, until by 1928 it had covered practically all the cotton producing area in the United States. Its progress took the form of a territorial succession. The consequences to agriculture were catastrophic but not wholly for the worse, since they served to give an impulse to changes in the organization of the industry long overdue. It also hastened the northward migration of the Negro tenant farmer.

The case of the boll weevil is typical. In this mobile modern world, where space and time have been measurably abolished, not men only but all the minor organisms (including the microbes) seem to be, as never before, in motion. Commerce, in progressively destroying the isolation upon which the ancient order of nature rested, has intensified the struggle for existence over an ever widening area of the habitable world. Out of this struggle a new equilibrium and a new system of animate nature, the new biotic basis of the new world society, is emerging.

[. . .]

The conditions which affect and control the movements and numbers of populations are more complex in human societies than in plant and animal communities, but they exhibit extraordinary similarities.

The boll weevil, moving out of its ancient habitat in the central Mexican plateau and into the virgin territory of the southern cotton plantations, incidentally multiplying its population to the limit of the territories and resources, is not unlike the Boers of Cape Colony, South Africa, trekking out into the high veldt of the central South African plateau and filling it, within a period of one hundred years, with a population of their own descendants.

Competition operates in the human (as it does in the plant and animal) community to bring about and restore the communal equilibrium, when, either by the advent of some intrusive factor from without or in the normal course of its life-history, that equilibrium is disturbed.

Thus every crisis that initiates a period of rapid change, during which competition is intensified, moves over finally into a period of more or less stable equilibrium and a new division of labor. In this manner competition brings about a condition in which competition is superseded by co-operation.

It is when, and to the extent that, competition declines that the kind of order which we call society may be said to exist. In short, society, from the ecological point of view, and in so far as it is a territorial unit, is just the area within which biotic competition has declined and the struggle for existence has assumed higher and more sublimated forms.

COMPETITION, DOMINANCE AND SUCCESSION

There are other and less obvious ways in which competition exercises control over the relations of individuals and species within the communal habitat. The two ecological principles, dominance and succession, which operate to establish and maintain such communal order as here described are functions of, and dependent upon, competition.

In every life-community there is always one or more dominant species. In a plant community this dominance is ordinarily the result of struggle among the different species for light. In a climate which supports a forest the dominant species will invariably be trees. On the prairie and steppes they will be grasses.

[. . .]

But the principle of dominance operates in the human as well as in the plant and animal communities. The so-called natural or functional areas of a metropolitan community – for example, the slum, the rooming-house area, the central shopping section and the banking center – each and all owe their existence directly to the factor of dominance, and indirectly to competition.

The struggle of industries and commercial institutions for a strategic location determines in the long run the main outlines of the urban community. The distribution of population, as well as the location and limits of the residential areas which they occupy, are determined by another similar but subordinate system of forces.

The area of dominance in any community is usually the area of highest land values. Ordinarily there are in every large city two such positions of highest land value – one in the central shopping district, the other in the central banking area. From these points land values decline at first precipitantly and then more gradually toward the periphery of the urban community. It is these land values that determine the location of social institutions and business enterprises. Both the one and the other are bound up in a kind of territorial complex within which they are at once competing and interdependent units.

As the metropolitan community expands into the suburbs the pressure of professions, business enterprises, and social institutions of various sorts destined to serve the whole metropolitan region steadily increases the demand for space at the center. Thus not merely the growth of the suburban area, but any change in the method of transportation which makes the central business area of the city more accessible, tends to increase the pressure at the center. From thence this pressure is transmitted and diffused, as the profile of land values discloses, to every other part of the city.

Thus the principle of dominance, operating within the limits imposed by the terrain and other natural features of the location, tends to determine the general ecological pattern of the city and the functional relation of each of the different areas of the city to all others.

Dominance is, furthermore, in so far as it tends to stabilize either the biotic or the cultural community, indirectly responsible for the phenomenon of succession.

The term "succession" is used by ecologists to describe and designate that orderly sequence of changes through which a biotic community passes in the course of its development from a primary and relatively unstable to a relatively permanent or climax stage. The main point is that not merely do the individual plants and animals within the communal habitat grow but the community itself, i.e., the system of relations between the species, is likewise involved in an orderly process of change and development.

The fact that, in the course of this development, the community moves through a series of more or less clearly defined stages is the fact that gives this development the serial character which the term "succession" suggests.

The explanation of the serial character of the changes involved in succession is the fact that at every stage in the process a more or less stable equilibrium is achieved, and as a result of progressive changes in life-conditions, possibly due to growth and decay, the equilibrium achieved in the earlier stages is eventually undermined. In this case the energies previously held in balance will be released, competition will be intensified, and change will continue at a relatively rapid rate until a new equilibrium is achieved.

The climax phase of community development corresponds with the adult phase of an individual's life.

[. . .]

The cultural community develops in comparable ways to that of the biotic, but the process is more complicated. Inventions, as well as sudden or catastrophic changes, seem to play a more important part in bringing about serial changes in the cultural than in the biotic community. But the principle involved seems to be substantially the same. In any case, all or most of the fundamental processes seem to be functionally related and dependent upon competition.

Competition, which on the biotic level functions to control and regulate the interrelations of organisms, tends to assume on the social level the form of conflict. The intimate relation between competition and conflict is indicated by the fact that wars frequently, if not always, have, or seem to have, their source and origin in economic competition which, in that case, assumes the more sublimated form of a struggle for power and prestige. The social function of war, on the other hand, seems to be to extend the area over which it is possible to maintain peace.

BIOLOGICAL ECONOMICS

If population pressure, on the one hand, co-operates with changes in local and environmental conditions to disturb at once the biotic balance and social equilibrium, it tends at the same time to intensify competition. In so doing it functions, indirectly, to bring about a new, more minute and, at the same time, territorially extensive division of labor.

Under the influence of an intensified competition, and the increased activity which competition involves, every individual and every species, each for itself, tends to discover the particular niche in the physical and living environment where it can survive and flourish with the greatest possible expansiveness consistent with its necessary dependence upon its neighbors.

It is in this way that a territorial organization and a biological division of labor, within the communal habitat, is established and maintained. This explains, in part at least, the fact that the biotic community has been conceived at one time as a kind of superorganism and at another as a kind of economic organization for the exploitation of the natural resources of its habitat.

In their interesting survey, *The Science of Life*, H. G. Wells and his collaborators, Julian Huxley and G. P. Wells, have described ecology as "biological economics," and as such very largely concerned with "the balances and mutual pressures of species living in the same habitat."

"Ecology," as they put it, is "an extension of Economics to the whole of life." On the other hand the science of economics as traditionally conceived, though it is a whole century older, is merely a branch of a more general science of ecology which includes man with all other living creatures. Under the circumstances what has been traditionally described as economics and conceived as restricted to human affairs, might very properly be described as Barrows some years ago described geography, namely as human ecology. It is in this sense that Wells and his collaborators would use the term.

Since human ecology cannot be at the same time both geography and economics, one may adopt, as a working hypothesis, the notion that it is neither one nor the other but something independent of both. Even so the motives for identifying ecology with geography on the one hand, and economics on the other, are fairly obvious.

From the point of view of geography, the plant, animal, and human population, including their habitations and other evidence of man's occupation of the soil, are merely part of the landscape, of which the geographer is seeking a detailed description and picture.

On the other hand ecology (biologic economics), even when it involves some sort of unconscious co-operation and a natural, spontaneous, and non-rational division of labor, is something different from the economics of commerce; something quite apart from the bargaining of the market place. Commerce, as Simmel somewhere remarks, is one of the latest and most complicated of all the social relationships into which human beings have entered. Man is the only animal that trades and traffics.

Ecology, and human ecology, if it is not identical with economics on the distinctively human and cultural level is, nevertheless, something more than and different from the static order which the human geographer discovers when he surveys the cultural landscape.

The community of the geographer is not, for one thing, like that of the ecologist, a closed system, and the web of communication which man has spread over the earth is something different from the "web of life" which binds living creatures all over the world in a vital nexus.

SYMBIOSIS AND SOCIETY

Human ecology, if it is neither economics on one hand nor geography on the other, but just ecology, differs, nevertheless, in important respects from plant and animal ecology. The interrelations of human beings and interactions of man and his habitat are comparable but not identical with interrelations of other forms of life that live together and carry on a kind of "biological economy" within the limits of a common habitat.

For one thing man is not so immediately dependent upon his physical environment as other animals. As a result of the existing world-wide division of labor, man's relation to his physical environment has been mediated through the intervention of other men. The exchange of goods and services have co-operated to emancipate him from dependence upon his local habitat.

Furthermore man has, by means of inventions and technical devices of the most diverse sorts, enormously increased his capacity for reacting upon and remaking, not only his habitat but his world. Finally, man has erected upon the basis of the biotic community an institutional structure rooted in custom and tradition.

Structure, where it exists, tends to resist change, at least change coming from without; while it possibly facilitates the cumulation of change within. In plant and animal communities structure is biologically determined, and so far as any division of labor exists at all it has a physiological and instinctive basis. The social insects afford a conspicuous example of this fact, and one interest in studying their habits is that they show the extent to which social organization can be developed on a purely physiological and instinctive basis, as is the case among human beings in the natural as distinguished from the institutional family.

In a society of human beings, however, this communal structure is reinforced by custom and assumes an institutional character. In human as contrasted with animal societies, competition and the freedom of the individual is limited on every level above the biotic by custom and consensus.

The incidence of this more or less arbitrary control which custom and consensus imposes upon the natural social order complicates the social process but does not fundamentally alter it – or, if it does, the effects of biotic competition will still be manifest in the succeeding social order and the subsequent course of events.

The fact seems to be, then, that human society, as distinguished from plant and animal society, is organized on two levels, the biotic and the cultural. There is a symbiotic society based on competition and a cultural society based on communication and consensus. As a matter of fact the two societies are merely different aspects of one society, which, in the vicissitudes and changes to which they are subject remain, nevertheless, in some sort of mutual dependence each upon the other. The cultural superstructure rests on the basis of the symbiotic substructure, and the emergent energies that manifest themselves on the biotic level in movements and actions reveal themselves on the higher social level in more subtle and sublimated forms.

However, the interrelations of human beings are more diverse and complicated than this dichotomy, symbiotic and cultural, indicates. This fact is attested by the divergent systems of human interrelations which have been the subject of the special social sciences. Thus human society, certainly in its mature and more rational expression, exhibits not merely an ecological, but an economic, a political, and a moral order. The social sciences include not merely human geography and ecology, but economics, political science, and cultural anthropology.

It is interesting also that these divergent social orders seem to arrange themselves in a kind of hierarchy. In fact they may be said to form a pyramid of which the ecological order constitutes the base and the moral order the apex. Upon each succeeding one of these levels, the ecological, economic, political, and moral, the individual finds himself more completely incorporated into and subordinated to the social order of which he is a part than upon the preceding.

Society is everywhere a control organization. Its function is to organize, integrate, and direct the energies resident in the individuals of which it is composed. One might, perhaps, say that the function of society was everywhere to restrict competition and by so doing bring about a more effective co-operation of the organic units of which society is composed.

Competition, on the biotic level, as we observe it in the plant and animal communities, seems to be relatively unrestricted. Society, so far as it exists, is anarchic and free. On the cultural level, this freedom of the individual to compete is restricted by conventions, understandings, and law. The individual is more free upon the economic level than upon the political, more free on the political than the moral.

As society matures control is extended and intensified and free commerce of individuals restricted, if not by law then by what Gilbert Murray refers to as "the normal expectation of mankind." The mores are merely what men, in a situation that is defined, have come to expect.

Human ecology, in so far as it is concerned with a social order that is based on competition rather than consensus, is identical, in principle at least, with plant and animal ecology. The problems with which plant and animal ecology have been traditionally concerned are fundamentally population problems. Society, as ecologists have conceived it, is a population settled and limited to its habitat. The ties that unite its individual units are those of a free and natural economy, based on a natural division of labor. Such a society is territorially organized and the ties which hold it together are physical and vital rather than customary and moral.

Human ecology has, however, to reckon with the fact that in human society competition is limited by custom and culture. The cultural superstructure imposes itself as an instrument of direction and control upon the biotic substructure.

Reduced to its elements the human community, so conceived, may be said to consist of a population and a culture, including in the term culture (1) a body of customs and beliefs and (2) a corresponding body of artifacts and technological devices.

To these three elements or factors – (1) population, (2) artifact (technological culture), (3) custom and beliefs (non-material culture) – into which the social complex resolves itself, one should, perhaps, add a fourth, namely, the natural resources of the habitat.

It is the interaction of these four factors – (1) population, (2) artifacts (technological culture), (3) custom and beliefs (non-material culture), and (4) the natural resources that maintain at once the biotic balance and the social equilibrium, when and where they exist.

The changes in which ecology is interested are the movements of population and of artifacts (commodities) and changes in location and occupation – any sort of change, in fact, which affects an existing division of labor or the relation of the population to the soil.

Human ecology is, fundamentally, an attempt to investigate the processes by which the biotic balance and the social equilibrium (1) are maintained once they are achieved and (2) the processes by which, when the biotic balance and the social equilibrium are disturbed, the transition is made from one relatively stable order to another.

"The Growth of the City: An Introduction to a Research Project"

from Robert Park *et al.* (eds), *The City* (1925)

Ernest W. Burgess

Editors' Introduction

Ernest W. Burgess (1886–1966), together with Robert Park, established a distinctive program of urban research in the sociology department at the University of Chicago in the early twentieth century. Ernest Watson Burgess was born on May 16, 1886 in Tilbury, Ontario, Canada. He obtained his Ph.D. from the University of Chicago in 1913 and served on the faculty there from 1916 to 1951. One of the important concepts he disseminated was succession, a term borrowed from plant ecology. Burgess was the originator of concentric zone theory, which predicted that cities would take the form of five concentric rings growing outwards, with a zone of deterioration immediately surrounding the city center, succeeding to increasingly prosperous residential zones moving out to the city's edge. Burgess understood the invasion–succession process as a "moving equilibrium" of the social order, a "process of distribution takes place which sifts and sorts and relocates individuals and groups by residence and occupation."

The human ecological research program also involved the extensive use of mapping to reveal the spatial distribution of social problems and to permit comparison between areas. Burgess was particularly interested in maps and used them extensively, requiring all his students to acquire proficiency in basic mapmaking techniques. Burgess and his students scoured the city of Chicago for data that could be used for maps, gleaning information from city agencies and making more extensive use of census data than any other social scientists of the time. This was one of the most important legacies of the urban ecology studies undertaken at the University of Chicago in the 1920s as mapmaking became part of the methodological toolkit of the developing disciplines of sociology, criminology, and public policy. Burgess was not a systematic theoretician but an eclectic promoter of theory and methodology. He sought to develop reliable tools for the prediction of social phenomena such as delinquency, parole violation, divorce, city growth, and adjustment in old age.

Human ecology drew criticism for its formalism in the postwar era, and newer sunbelt cities became more decentralized with the decline of railroad transportation and the onset of automobile travel and highways. In particular, Park and Burgess' search for "natural" or "organic" process was criticized as a superficial undertaking that neglected both the social and cultural dimensions of urban life and the political-economic impact of industrialization on urban geography. Overall, the urban ecology studies of the 1920s were largely oblivious to issues of class, race, gender, and ethnicity. However, the concentric rings model has become one of the more famous formulations in urban sociology and is still applied creatively to studies of urban processes. In *The Ecology of Fear* (1992), Mike Davis adapts the concentric rings model to describe the exclusionary fortress of Los Angeles, a metropolis with zones of "homeless containment" and drug enforcement and neighborhood watches in the central city and near suburbs, with gated communities and prisons on the urban periphery.

Burgess drew from biology more than did Robert Park, through his references to concepts like "metabolism" and "pulse." Through the organic metaphor, the city can be compared to a body, with metabolic and circulatory processes. He thought of mobility as the "pulse of the community." He connected the growth of cities with increasing mobility and movement of people and cultures. Geographic and cultural mobility put people in contact with increasingly diverse and unconventional behaviors. Mobility could thus breed social disorganization in the form of crime, deviance, and promiscuous behavior. Burgess viewed mobility and social disorganization not as pathological but as normal. Unusual behaviors were particularly focused in the central city and "zone-in-transition" encompassing both the "bright light" and "red light" areas of the city. As displayed in Times Square in New York and Hollywood in Los Angeles, the respectable theater district of a city attracts a critical mass of nocturnal pedestrians that can also support more licentious activities or cheap amusements. Like a body, the city can be seen as having a "heart," erogenous zones, and something like a hormonal metabolism. The city is a site of excitement, adventure, and thrills. The bright light and red light areas of the city are crucial components of the metropolitan mosaic of social worlds. They give a city its cultural and subcultural identity and are the focal points for the emergence of artistic and bohemian communities. While Burgess mainly fretted about the negative effects of mobility on social disorganization, we may see mobility can also produce positive effects on cultural life and community.

Ernest W. Burgess served as the 24th President of the American Sociological Association. His Presidential Address "Social Planning and the Mores" was delivered at the organization's annual meeting in Chicago in December 1934. His editing roles were extensive, and he was editor of the *American Journal of Sociology* from 1936 to 1940. He founded the Family Study Center at the University of Chicago, and was involved in a number of professional associations.

The outstanding fact of modern society is the growth of great cities. Nowhere else have the enormous changes which the machine industry has made in our social life registered themselves with such obviousness as in the cities. In the United States the transition from a rural to an urban civilization, though beginning later than in Europe, has taken place, if not more rapidly and completely, at any rate more logically in its most characteristic forms.

All the manifestations of modern life which are peculiarly urban – the skyscraper, the subway, the department store, the daily newspaper, and social work – are characteristically American. The more subtle changes in our social life, which in their cruder manifestations are termed "social problems," problems that alarm and bewilder us, as divorce, delinquency, and social unrest, are to be found in their most acute forms in our largest American cities. The profound and "subversive" forces which have wrought these changes are measured in the physical growth and expansion of cities. That is the significance of the comparative statistics of Weber, Bücher, and other students.

These statistical studies, although dealing mainly with the effects of urban growth, brought out into clear relief certain distinctive characteristics of urban as compared with rural populations. The larger proportion of women to men in the cities than in the open country, the greater percentage of youth and middle-aged, the higher ratio of the foreign-born, the increased heterogeneity of occupation increase with the growth of the city and profoundly alter its social structure. These variations in the composition of population are indicative of all the changes going on in the social organization of the community. In fact, these changes are part of the growth of the city and suggest the nature of the processes of growth.

The only aspect of growth adequately described by Bücher and other students of Weber was the rather obvious process of the *aggregation* of urban population. Almost as overt a process, that of *expansion*, has been investigated from a different and very practical point of view by groups interested in city planning, zoning, and regional surveys. Even more significant than the increasing density of urban population is its correlative tendency to overflow, and so to extend over wider areas, and to incorporate these areas into a larger communal life. This paper, therefore, will treat first of the expansion of the city, and then of the less-known processes of urban metabolism and mobility which are closely related to expansion.

EXPANSION AS PHYSICAL GROWTH

The expansion of the city from the standpoint of the city plan, zoning, and regional surveys is thought of almost wholly in terms of its physical growth. Traction studies have dealt with the development of transportation in its relation to the distribution of population throughout the city. The surveys made by the Bell Telephone Company and other public utilities have attempted to forecast the direction and the rate of growth of the city in order to anticipate the future demands for the extension of their services. In the city plan the location of parks and boulevards, the widening of traffic streets, the provision for a civic center, are all in the interest of the future control of the physical development of the city.

This expansion in area of our largest cities is now being brought forcibly to our attention by the Plan for the Study of New York and Its Environs, and by the formation of the Chicago Regional Planning Association, which extends the metropolitan district of the city to a radius of 50 miles, embracing 4,000 square miles of territory. Both are attempting to measure expansion in order to deal with the changes that accompany city growth. In England, where more than one-half of the inhabitants live in cities having a population of 100,000 and over, the lively appreciation of the bearing of urban expansion on social organization is thus expressed by C. B. Fawcett:

> One of the most important and striking developments in the growth of the urban populations of the more advanced peoples of the world during the last few decades has been the appearance of a number of vast urban aggregates, or conurbations, far larger and more numerous than the great cities of any preceding age. These have usually been formed by the simultaneous expansion of a number of neighboring towns, which have grown out toward each other until they have reached a practical coalescence in one continuous urban area. Each such conurbation still has within it many nuclei of denser town growth, most of which represent the central areas of the various towns from which it has grown, and these nuclear patches are connected by the less densely urbanized areas which began as suburbs of these towns. The latter are still usually rather less continuously occupied by buildings, and often have many open spaces.

These great aggregates of town dwellers are a new feature in the distribution of man over the earth. At the present day there are from thirty to forty of them, each containing more than a million people, whereas only a hundred years ago there were, outside the great centers of population on the waterways of China, not more than two or three. Such aggregations of people are phenomena of great geographical and social importance; they give rise to new problems in the organization of the life and well-being of their inhabitants and in their varied activities. Few of them have yet developed a social consciousness at all proportionate to their magnitude, or fully realized themselves as definite groupings of people with many common interests, emotions and thoughts.

In Europe and America the tendency of the great city to expand has been recognized in the term "the metropolitan area of the city," which far overruns its political limits, and in the case of New York and Chicago, even state lines. The metropolitan area may be taken to include urban territory that is physically contiguous, but it is coming to be defined by that facility of transportation that enables a business man to live in a suburb of Chicago and to work in the loop, and his wife to shop at Marshall Field's and attend grand opera in the Auditorium.

EXPANSION AS A PROCESS

No study of expansion as a process has yet been made, although the materials for such a study and intimations of different aspects of the process are contained in city planning, zoning, and regional surveys. The typical processes of the expansion of the city can best be illustrated, perhaps, by a series of concentric circles, which may be numbered to designate both the successive zones of urban extension and the types of areas differentiated in the process of expansion.

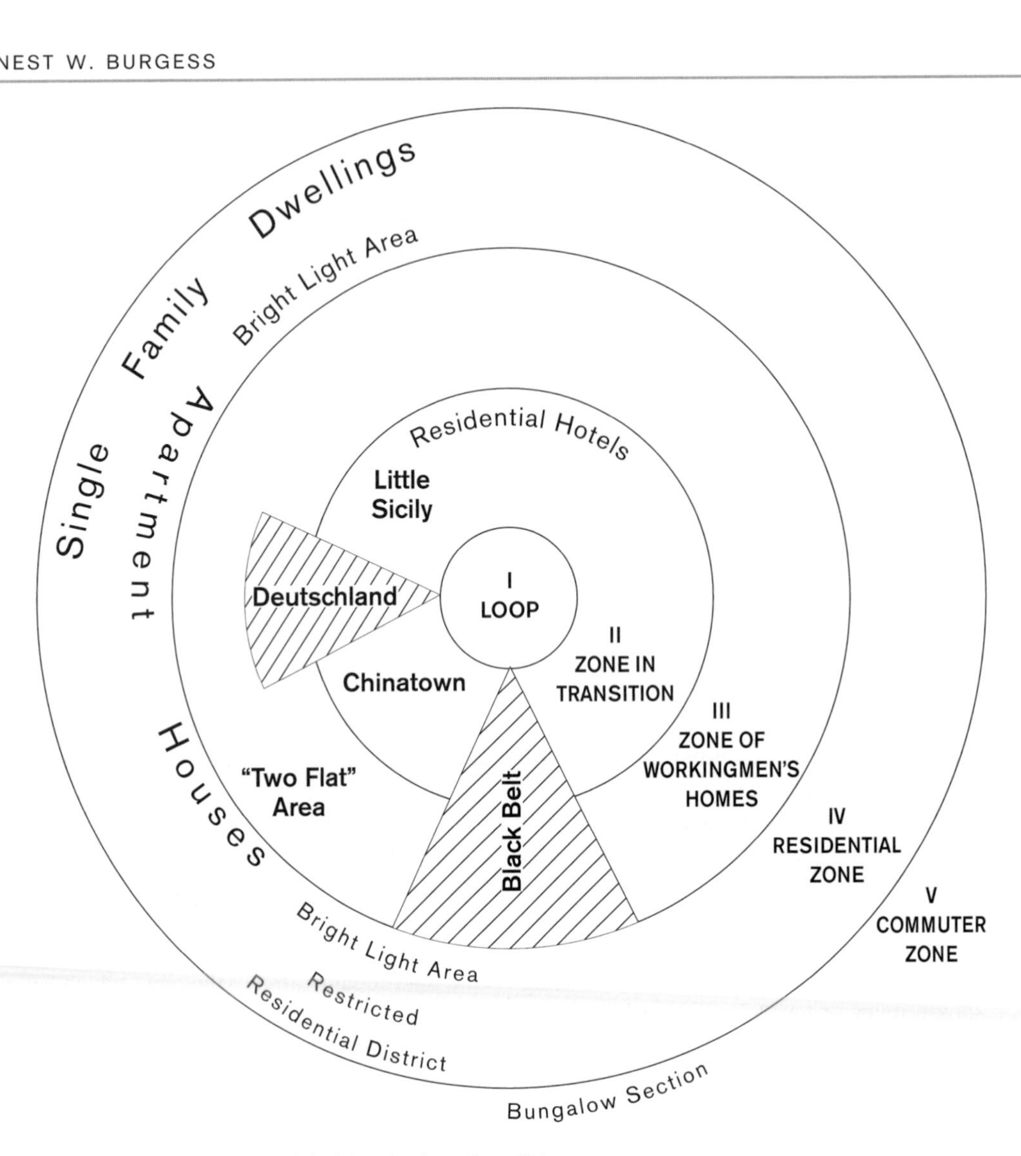

Figure 1 The concentric zone model of the city, based on Chicago

This chart represents an ideal construction of the tendencies of any town or city to expand radially from its central business district – on the map "The Loop" (I). Encircling the downtown area there is normally an area in transition, which is being invaded by business and light manufacture (II). A third area (III) is inhabited by the workers in industries who have escaped from the area of deterioration (II) but who desire to live within easy access of their work. Beyond this zone is the "residential area" (IV) of high-class apartment buildings or of exclusive "restricted" districts of single family dwellings. Still farther, out beyond the city limits, is the commuters' zone – suburban areas, or satellite cities – within a thirty- to sixty-minute ride of the central business district.

This chart brings out clearly the main fact of expansion, namely, the tendency of each inner zone to extend its area by the invasion of the next outer zone. This aspect of expansion may be called *succession*, a process which has been studied in detail in plant ecology. If this chart is applied to Chicago, all four of these zones were in its early history included in the circumference of the inner zone, the present business district. The present boundaries of the area of deterioration were not many years ago those of the zone now inhabited by independent wage-earners, and within the memories of thousands of Chicagoans contained the residences of the "best families." It hardly needs to be added that neither Chicago nor any other city fits perfectly into this ideal scheme. Complications are introduced by the lake front, the Chicago River, railroad lines, historical factors in the location of industry, the relative degree of the resistance of communities to invasion, etc.

Besides extension and succession, the general process of expansion in urban growth involves the antagonistic and yet complementary processes of concentration and decentralization. In all cities there is the natural tendency for local and outside transportation to converge in the central business district. In the downtown section of every large city we expect to find the department stores, the skyscraper office buildings, the railroad stations, the great hotels, the theaters, the art museum, and the city hall. Quite naturally, almost inevitably, the economic, cultural, and political life centers here. The relation of centralization to the other processes of city life may be roughly gauged by the fact that over half a million people daily enter and leave Chicago's "loop." More recently sub-business centers have grown up in outlying zones. These "satellite loops" do not, it seems, represent the "hoped for" revival of the neighborhood, but rather a telescoping of several local communities into a larger economic unity. The Chicago of yesterday, an agglomeration of country towns and immigrant colonies, is undergoing a process of reorganization into a centralized decentralized system of local communities coalescing into sub-business areas visibly or invisibly dominated by the central business district. The actual processes of what may be called centralized decentralization are now being studied in the development of the chain store, which is only one illustration of the change in the basis of the urban organization.

Expansion, as we have seen, deals with the physical growth of the city, and with the extension of the technical services that have made city life not only livable, but comfortable, even luxurious. Certain of these basic necessities of urban life are possible only through a tremendous development of communal existence. Three millions of people in Chicago are dependent upon one unified water system, one giant gas company, and one huge electric light plant. Yet, like most of the other aspects of our communal urban life, this economic co-operation is an example of co-operation without a shred of what the "spirit of co-operation" is commonly thought to signify. The great public utilities are a part of the mechanization of life in great cities, and have little or no other meaning for social organization.

Yet the processes of expansion, and especially the rate of expansion, may be studied not only in the physical growth and business development, but also in the consequent changes in the social organization and in personality types. How far is the growth of the city, in its physical and technical aspects, matched by a natural but adequate readjustment in the social organization. What, for a city, is a normal rate of expansion, a rate of expansion with which controlled changes in the social organization might successfully keep pace?

SOCIAL ORGANIZATION AND DISORGANIZATION AS PROCESSES OF METABOLISM

These questions may best be answered, perhaps, by thinking of urban growth as a resultant of organization and disorganization analogous to the anabolic and katabolic processes of metabolism in the body. In what way are individuals incorporated into the life of a city? By what process does a person become an organic part of his society? The natural process of acquiring culture is by birth. A person is born into a family already adjusted to a social environment – in this case the modern city. The natural rate of increase of population most favorable for assimilation may then be taken as the excess of the birth-rate over the death-rate, but is this the normal rate of city growth? Certainly, modern cities have increased and are increasing in population at a far higher rate. However, the natural rate of growth may he used to measure the disturbances of metabolism caused by any excessive increase, as those which followed the great influx of southern Negroes into northern cities since the war. In a similar way all cities show deviations in composition by age and sex from a standard population such as that of Sweden, unaffected in recent years by any great emigration or immigration. Here again, marked variations, as any great excess of males over females, or of females over males, or in the proportion of children, or of grown men or women, are symptomatic of abnormalities in social metabolism.

Normally the processes of disorganization and organization may be thought of as in reciprocal relationship to each other, and as co-operating in a moving equilibrium of social order toward an end vaguely or definitely regarded as progressive. So far as disorganization points to reorganization and makes for more efficient adjustment, disorganization must be conceived not as pathological, but as normal. Disorganization as preliminary to reorganization of attitudes and conduct is almost invariably the lot of the newcomer to the city, and the discarding of the habitual, and often of what has been to him the moral, is not infrequently accompanied by sharp mental conflict and sense of personal loss. Oftener, perhaps, the change gives sooner or later a feeling of emancipation and an urge toward new goals.

In the expansion of the city a process of distribution takes place which sifts and sorts and relocates individuals and groups by residence and occupation. The resulting differentiation of the cosmopolitan American city into areas is typically all from one pattern, with only interesting minor modifications. Within the central business district or on an adjoining street is the "main stem" of "hobohemia," the teeming Rialto of the homeless migratory man of the Middle West.[1] In the zone of deterioration encircling the central business section are always to be found the so-called "slums" and "bad lands," with their submerged regions of poverty, degradation, and disease, and their underworlds of crime and vice. Within a deteriorating area are rooming-house districts, the purgatory of "lost souls." Near by is the Latin Quarter, where creative and rebellious spirits resort. The slums are also crowded to overflowing with immigrant colonies – the Ghetto, Little Sicily, Greektown, Chinatown – fascinatingly combining old world heritages and American adaptations. Wedging out from here is the Black Belt, with its free and disorderly life. The area of deterioration, while essentially one of decay, of stationary or declining population, is also one of regeneration, as witness the mission, the settlement, the artists' colony, radical centers – all obsessed with the vision of a new and better world.

The next zone is also inhabited predominately by factory and shop workers, but skilled and thrifty. This is an area of second immigrant settlement, generally of the second generation. It is the region of escape from the slum, the *Deutschland* of the aspiring Ghetto family. For *Deutschland* (literally "Germany") is the name given, half in envy, half in derision, to that region beyond the Ghetto where successful neighbors appear to be imitating German Jewish standards of living. But the inhabitant of this area in turn looks to the "Promised Land" beyond, to its residential hotels, its apartment-house region, its "satellite loops," and its "bright light" areas.

This differentiation into natural economic and cultural groupings gives form and character to the city. For segregation offers the group, and thereby the individuals who compose the group, a place and a role in the total organization of city life. Segregation limits development in certain directions, but releases it in others. These areas tend to accentuate certain traits, to attract and develop their kind of individuals, and so to become further differentiated.

The division of labor in the city likewise illustrates disorganization, reorganization, and increasing differentiation. The immigrant from rural communities in Europe and America seldom brings with him economic skill of any great value in our industrial, commercial, or professional life. Yet interesting occupational selection has taken place by nationality, explainable more by racial temperament or circumstance than by old-world economic background, as Irish policemen, Greek ice-cream parlors, Chinese laundries, Negro porters, Belgian janitors, etc.

The facts that in Chicago one million (996,589) individuals gainfully employed reported 509 occupations, and that over 1,000 men and women in *Who's Who* gave 116 different vocations, give some notion of how in the city the minute differentiation of occupation "analyzes and sifts the population, separating and classifying the diverse elements." These figures also afford some intimation of the complexity and complication of the modern industrial mechanism and the intricate segregation and isolation of divergent economic groups. Interrelated with this economic division of labor is a corresponding division into social classes and into cultural and recreational groups. From this multiplicity of groups, with their different patterns of life, the person finds his congenial social world and – what is not feasible in the narrow confines of a village – may move and live in widely separated, and perchance conflicting, worlds. Personal disorganization may be but the failure to harmonize the canons of conduct of two divergent groups.

If the phenomena of expansion and metabolism indicate that a moderate degree of disorganization may and does facilitate social organization, they indicate as well that rapid urban expansion is accompanied by excessive increases in disease, crime, disorder, vice, insanity, and suicide, rough indexes of social disorganization. But what are the indexes of the causes, rather than of the effects, of the disordered social metabolism of the city? The excess of the actual over the natural increase of population has already been suggested as a criterion. The significance of this increase consists in the immigration into a metropolitan city like New York and Chicago of tens of thousands of persons annually. Their invasion of the city has the effect of a tidal wave inundating first the immigrant colonies, the ports of first entry, dislodging thousands of inhabitants who overflow into the next zone, and so on and on until the momentum of the wave has spent its force on the last urban zone. The whole effect is to speed up expansion, to speed up industry, to speed up the "junking" process in the area of deterioration (II). These internal movements of the population become the more significant for study. What movement is going on in the city, and how may this movement be measured? It is easier, of course, to classify movement within the city than to measure it. There is the movement from residence to residence, change of occupation, labor turnover, movement to and from work, movement for recreation and adventure. This leads to the question: What is the significant aspect of movement for the study of the changes in city life? The answer to this question leads directly to the important distinction between movement and mobility.

MOBILITY AS THE PULSE OF THE COMMUNITY

Movement, per se, is not an evidence of change or of growth. In fact, movement may be a fixed and unchanging order of motion, designed to control a constant situation, as in routine movement. Movement that is significant for growth implies a change of movement in response to a new stimulus or situation. Change of movement of this type is called *mobility*. Movement of the nature of routine finds its typical expression in work. Change of movement, or mobility, is characteristically expressed in adventure. The great city, with its "bright lights," its emporiums of novelties and bargains, its palaces of amusement, its underworld of vice and crime, its risks of life and property from accident, robbery, and homicide, has become the region of the most intense degree of adventure and danger, excitement and thrill.

Mobility, it is evident, involves change, new experience, stimulation. Stimulation induces a response of the person to those objects in his environment which afford expression for his wishes. For the person, as for the physical organism, stimulation is essential to growth. Response to stimulation is wholesome so long as it is a correlated *integral* reaction of the entire personality. When the reaction is *segmental*, that is, detached from, and uncontrolled by, the organization of personality, it tends to become disorganizing or pathological. That is why stimulation for the sake of stimulation, as in the restless pursuit of pleasure, partakes of the nature of vice.

The mobility of city life, with its increase in the number and intensity of stimulations, tends inevitably to confuse and to demoralize the person. For an essential element in the mores and in personal morality is consistency, consistency of the type that is natural in the social control of the primary group. Where mobility is the greatest, and where in consequence primary controls break down completely, as in the zone of deterioration in the modern city, there develop areas of demoralization, of promiscuity, and of vice.

In our studies of the city it is found that areas of mobility are also the regions in which are found juvenile delinquency, boys' gangs, crime, poverty, wife desertion, divorce, abandoned infants, vice.

These concrete situations show why mobility is perhaps the best index of the state of metabolism of the city. Mobility may be thought of in more than a fanciful sense, as the "pulse of the community." Like the pulse of the human body, it is a process which reflects and is indicative of all the changes that are taking place in the community, and which is susceptible of analysis into elements which may be stated numerically.

The elements entering into mobility may be classified under two main heads: (1) the state of mutability of the person, and (2) the number and kind of contacts or stimulations in his environment. The mutability of city populations varies with sex and age composition, the degree of detachment of the person from the family and from other groups. All these factors may be expressed numerically. The new stimulations to which a population responds can be measured in terms of change of movement or of increasing contacts. Statistics on the movement of urban population may only measure routine, but an increase at a higher ratio than the increase of population measures mobility. In 1860 the horse-car lines of New York City carried about 50,000,000 passengers; in 1890 the trolley cars (and a few surviving horse-cars) transported about 500,000,000; in 1921, the elevated, subway, surface, and electric and steam suburban lines carried a total of more than 2,500,000,000 passengers. In Chicago the total annual rides per capita on the surface and elevated lines were 164 in 1890; 215 in 1900; 320 in 1910; and 338 in 1921. In addition, the rides per capita on steam and electric suburban lines almost doubled between 1916 (23) and 1921 (41), and the increasing use of the automobile must not be overlooked. For example, the number of automobiles in Illinois increased from 131,140 in 1915 to 833,920 in 1923.

Mobility may be measured not only by these changes of movement, but also by increase of contacts. While the increase of population of Chicago in 1912–22 was less than 25 per cent (23.6 per cent), the increase of letters delivered to Chicagoans was double that (49.6 per cent) – (from 693,084,196 to 1,038,007,854). In 1912 New York had 8.8 telephones; in 1922, 16.9 per 100 inhabitants. Boston had, in 1912, 10.1 telephones; ten years later, 19.5 telephones per 100 inhabitants. In the same decade the figures for Chicago increased from 12.3 to 21.6 per 100 population. But increase of the use of the telephone is probably more significant than increase in the number of telephones. The number of telephone calls in Chicago increased from 606,131,928 in 1914 to 944,010,586 in 1922, an increase of 55.7 per cent, while the population increased only 13.4 per cent.

Land values, since they reflect movement, afford one of the most sensitive indexes of mobility. The highest land values in Chicago are at the point of greatest mobility in the city, at the corner of State and Madison Streets, in the Loop. A traffic count showed that at the rush period 31,000 people an hour, or 210,000 men and women in sixteen and one-half hours, passed the southwest corner. For over ten years land values in the Loop have been stationary, but in the same time they have doubled, quadrupled, and even sextupled in the strategic corners of the "satellite loops," an accurate index of the changes which have occurred. Our investigations so far seem to indicate that variations in land values, especially where correlated with differences in rents, offer perhaps the best single measure of mobility, and so of all the changes taking place in the expansion and growth of the city.

In general outline, I have attempted to present the point of view and methods of investigation which the department of sociology is employing in its studies in the growth of the city, namely, to describe urban expansion in terms of extension, succession, and concentration; to determine how expansion disturbs metabolism when disorganization is in excess of organization; and, finally, to define mobility and to propose it as a measure both of expansion and metabolism, susceptible to precise quantitative formulation, so that it may be regarded almost literally as the pulse of the community. In a way, this statement might serve as an introduction to any one of five or six research projects under way in the department: The project, however, in which I am directly engaged is an attempt to apply these methods of investigation to a cross-section of the city – to put this area, as it were, under the microscope, and so to study in more detail and with greater control and precision the processes which have been described here in the large. For this purpose the West Side Jewish community has been selected. This community includes the so-called "Ghetto," or area of first settlement, and Lawndale, the so-called "Deutschland," or area of second settlement. This area has certain obvious advantages for this study, from the standpoint of expansion, metabolism, and mobility. It exemplifies the tendency to expansion radially from the business center of the city. It is now relatively a homogeneous cultural group. Lawndale is itself an area in flux, with the tide of migrants still flowing in from the Ghetto and a constant egress to more desirable regions of the residential zone. In this area, too, it is also possible to study how the expected outcome of this high rate of mobility in social and personal disorganization is counteracted in large measure by the efficient communal organization of the Jewish community.

NOTE

1 For a study of this cultural area of city life see Nels Anderson (1923) *The Hobo.* Chicago: University of Chicago Press.

"The Natural Areas of the City"

from Ernest W. Burgess (ed.), *The Urban Community* (1926)

Harvey Zorbaugh

Editors' Introduction

Harvey Zorbaugh distinguished himself from his mentors Robert Park and Ernest Burgess through his interest in neighborhoods and the historical life of cities. He described the city as not an artifact of human creation, but a natural phenomenon with a natural history. Zorbaugh codified the concept of "natural areas" in urban sociology. The structure of the city is built about a framework of railroad lines, streets, highways, industrial infrastructure, and topographical features such as rivers, swamps, and hills. They overlay the concentric zone dartboard of the city and fragment into a patchwork of numerous smaller units or natural areas. Zorbaugh understood them to be the unplanned, natural product of the city's growth. The natural areas of the city tend to be distinct cultural areas as well, each with its complex of institutions, customs, beliefs, mores, traditions, attitudes, sentiments, and interests. They may express racial/ethnic differences, such as a "black belt" or Harlem, Little Italy, Little Havana, Chinatown, Koreatown or Little Tokyo. There may be affluent enclaves such as the Gold Coast, Beverly Hills, or Park Avenue. Bohemias, hobohemias and rooming-house areas include Greenwich Village, the Barbary Coast, Hell's Kitchen and the Bowery. In some senses the natural area and the neighborhood are equivalent; on other occasions neighborhoods represent subunits of natural areas.

According to Zorbaugh, the land market sifted and sorted the urban population into small enclaves that offered a means for cultural segregation of the population. In the competition for position, the population becomes segregated over the natural areas of the city. He understood segregation to be a cultural force that emerged through the natural history of the city. The natural areas emerged through a coincidence of physical boundaries and cultural forces. Zorbaugh said *a natural area is a geographical area characterized both by a physical individuality and by the cultural characteristics of the people who live in it.* Zorbaugh has a view of segregation as a purely natural and cultural force and not something that is economically or racially determined. Consider a poor neighborhood in the "ghetto" or "barrio" to a rich neighborhood on the "other side of the tracks." The tracks can serve to reinforce class, racial or cultural differences as well as distances. Yet Zorbaugh believed these differences, distances, and boundaries are naturally evolved through the forces of market competition. Zorbaugh has a very benign understanding of the causes of residential segregation.

Zorbaugh said that *natural areas collect the particular individuals of the city predestined to it.* This concept of predestination serves as an explanation and justification for the presence of affluent areas such as Chicago's Gold Coast, which naturally selected the most talented individuals from the economic and social elite of the city. This same sense of determinism was applied to the poor areas of Chicago's Near North Side, in areas such as Little Hell, Towertown and the Rialto. He understood urban slums as socially disorganized or unstable areas, which produced and further intensified unstable personalities. The anonymous rooming-house areas and tenements of the Near North Side attracted and created restless, lonely, and neurotic people, and these people further undermined the social order and group identity of the area. Some of the most striking ethnographic reportage in his book, *The Gold Coast and the Slum: A Sociological Study of Chicago's Near North Side* (University of Chicago Press, 1929), concerns the exploits of prostitutes, erotic dancers, hobos and panhandlers. He calls these people "shipwrecked humanity" living in "the world of furnished rooms and "human derelicts" in the "Rialto of the half-world." Like Burgess, he emphasizes the more negative consequences of human mobility in the zone-in-transition, and downplays the creative impact of artistic personalities and bohemian communities. He dismisses the artists, homosexuals, and rebels of bohemian Towertown as "egocentric poseurs" and "dilettantes" and incorrectly predicts that the subcultural area will pass out of existence because it has no further role to play in the life of the city. He misunderstands the importance of bohemian natural areas in cultivating the cosmopolitan and unconventional mindsets that lead to cultural and subcultural creativity in our society.

While Burgess emphasized the concept of ecological succession and the moving equilibrium of the concentric zone structure of the city, Zorbaugh's concept of natural areas suggests somewhat greater stability especially when considering affluent neighborhoods and ethnic enclaves. Zorbaugh understood natural areas, however, as only relatively stable, subject to the ongoing flux of human mobility and the changing land market. The Near North Side represented an epitome of instability. Even on the Gold Coast, he detected the movement of elites from the Esplanade to Sheridan Road. He believes, however, that natural areas *seem to change in a predictable manner, a succession like that observable in plant communities.*

Zorbaugh felt that urban planners and public officials had done a bad job coinciding their administrative units with the natural areas of the city. City planning and zoning can be more economically and efficiently run when we recognize the natural groupings of the population and the natural processes of urban growth. Statistical and fieldwork studies in urban natural areas are crucial to improving the planning and administration of cities. He also felt these studies would be valuable to community leaders seeking to improve and reform socially disorganized areas such as the Near North Side of Chicago.

Harvey Warren Zorbaugh was born in Cleveland, Ohio in 1896. He received a B.A. from Vanderbilt University and studied at the University of Chicago between 1923 and 1926 as a Laura Spelman Rockefeller Memorial fellow. He spent the rest of his career on the faculty of New York University, where he became a leading specialist in the social adjustment of gifted children. During fieldwork in Chicago's Near North Side, Zorbaugh became attuned to the challenges of youth gang leaders who were stigmatized as lawbreakers without channels to display their talents. He continued to work with clinics, committees, and other public service around the problems of children. He was an outspoken opponent of racial prejudice in public schools. He died in 1965.

THE CITY AS ARTIFACT AND AS NATURAL PHENOMENON

To the philosophically minded the city has often seemed to be the most colossal artifact of man's creation. The towering skyscrapers of a New York or a Chicago, palatial banking houses, the frenzied stock exchange, a Fifth or a Michigan Avenue with its ceaseless stream of automobiles and busses, its smart shops, and its brilliant hotels, underground tubes with roaring trains, or elevated railroads clattering overhead, great belts of smoking industries, miles of canyon-like streets flanked with magnificent park and boulevard systems, water works besides which the Roman aqueducts fall into insignificance – all in all the city seems the most exotic and artificial flower of a man-made civilization, a product not alone of man's brawn, but of man's brain and man's will.

Yet the city is curiously resistant to the fiats of man. Like the Robot, created by man, it goes its own way indifferent to the will of its creator. Reformers have stormed, the avaricious have speculated, and thoughtful men have speculated, and thoughtful men have planned. But again and again their programs have met with obstacles. Human nature offers some opposition; traditions and institutions offer more; and – of especial significance – the very physical configuration of the city is unyielding change. It becomes apparent that the city has a natural organization that must be taken into account.

In the latter part of the present century a tidal wave of reform swept over the city, culminating in the "Man with the Muckrake" and the "Yellow Press." Jacob Riis painted the descent into the slum. Parkhurst crusaded against vice in New York; and Stead in *If Christ Came to Chicago*, lashed the lords of Customs House Place. Ida M. Tarbell and Upton Sinclair took the muckrake into industry, while Lincoln Steffens laid bare the rotten spots in city government. There was a tremendous stir, public interest was aroused, reforms were proposed, but little happened. Practically all these movements for social reform met with unexpected obstacles: influential persons, "bosses," "union leaders," "local magnates," and powerful groups such as party organizations, "vested interests," "lobbies," unions, manufacturers' associations, and the like. Candid recognition of the role of these persons and groups led writers on social, political, and economic questions to give them the impersonal designation of "social forces."

The concept of social forces was a commonsense generalization. But implicit in Steffen's book, *The Shame of the Cities*, was a far more sophisticated insight. Steffens maintained that with his knowledge of New York he could go into any city and quickly gauge conditions; that conditions in New York were not due to failure of institutions peculiar to itself, but to a condition incident to the growth of all cities. This was the first recognition of the fact that the city is a natural phenomenon and has a natural history.

Meantime, realtors, public utilities, city-planning and zoning commissions, and others interested in predicting the future of the city were discovering much about the way in which the city grows. Richard Hurd, in a small volume, *The Principles of City Land Values*, attempting to generalize fluctuations of city land values, formulated certain typical processes of the city's growth. Most instructive are the more recent statistical studies of the American Bell Telephone Company and other utilities for the purposes of extension in anticipation of future service. The city is discovered to be an organization displaying certain typical processes of growth. Knowledge of these processes makes possible prediction of the direction, rate, and nature of its growth. That is, the city is found to be not an artifact but a natural phenomenon.

A HUMAN ECOLOGY

In an address in 1922, before the meeting at which the Russell Sage Foundation's proposal for a regional plan for metropolitan New York was first outlined, Elihu Root recognized this fact of the natural organization of the city when he said: "A city is a growth. It is not the result of political decrees or control. You may have all the lines you please between counties and states; a city is a growth responding to forces not at all political, quite disregarding political lines. It is a growth like that of a crystal responding to forces inherent in the atoms that make it up." In the three years that have elapsed since Elihu Root wrote these words, a mass of material about the city has been gathered and analyzed that enables us to describe these "atoms" to which he referred.

Studies of the expansion of the city have shown that all American cities exhibit certain typical processes in their growth. To begin with, they segregate into broad zones as they expand radially from the center – a "loop," or central business district, a zone of transition between business and resident; an invasion by business, and light manufacturing, involving physical deterioration and social disorganization; a zone of working men's homes, cut through by rooming-house districts along focal lines of transportation; a zone of apartments and "restricted" districts of single family dwellings; and, farther out, beyond city limits, a commuters' zone of suburban areas. Ideally, this gross segregation may be represented by a series of concentric circles, and such tends to be the actual fact where there are no complicating geographical factors.

Such is a generalized description of the gross anatomy of the city – the typical structure of a modern American commercial and industrial city. Of course, no city quite conforms to this ideal scheme. Physical barriers such as rivers, lakes, and rises of land may modify the growth and structure of tile individual city, as is strikingly demonstrated in the cases of New York, Pittsburgh, and Seattle. Railroads, with their belts of industry, cut through this generalized scheme, breaking the city up into sections; and lines of local transportation, along the more travelled of which grow up retail business streets, further modify the structure of the city.

The structure of the individual city, then, while always exhibiting the generalized zones described above, is built about this framework of transportation, business organization and industry, park and boulevard systems, and topographical features. All of these break the city up into numerous smaller areas, which we may call natural areas, in that they are the unplanned, natural product of the city's growth. Railroad and industrial belts, park and boulevard systems, rivers and rises of land acting as barriers to movements of population tend to fix the boundaries of these natural areas, while their centers are usually intersections of two or more business streets. By virtue of proximity to industry, business, transportation, or natural advantages each area acquires a physical individuality accurately reflected in land values and rents.

Now, in the intimate economic relationships in which all people are in the city everyone is, in a sense, in competition with everyone else. It is an impersonal competition – the individual does not know his competitors. It is a competition for other values in addition to those represented by money. One of the forms it takes is competition for position in the community. We do not know all the factors involved, but each individual influences the ultimate position of every other individua1.

In this competition for position the population is segregated over the natural areas of the city. Land values, characterizing the various natural areas, tend to sift and sort the population. At the same time segregation recognizes trends in values.[1] Cultural factors also play a part in this segregation, creating repulsions and attractions. From the mobile competing stream of the city's population each natural area of the city tends to collect the particular individuals predestined to it. These individuals, in turn, give to the area a peculiar character. And as a result of this segregation, the natural areas of the city tend to become distinct cultural areas as well – a "black belt" or a Harlem, a Little Italy, a Chinatown, a "stem" of the "hobo," a rooming-house world, a "Towertown," or a "Greenwich Village," a "Gold Coast," and the like – each with its characteristic complex of institutions, customs, beliefs, standards of life, traditions, attitudes, sentiments, and interests. The physical individuality of the natural areas of the city is re-emphasized by the cultural individuality of the populations segregated over them. Natural areas and natural cultural groups tend to coincide.

A natural area is a geographical area characterized both by a physical individuality and by the cultural characteristics of the people who live in it. Studies in various cities have shown, to quote Robert E. Park, that "every American city of a given size tends to reproduce all the typical areas of all the cities, and that the people in these areas exhibit, from city to city, the same cultural characteristics, the same types of institutions, the same social types, with the same opinions, interests, and outlook on life." That is, just as there is a plant ecology whereby, in the struggle for existence, like geographical regions become associated with like "communities" of plants, mutually adapted, and adapted to the area, so there is a human ecology whereby, in the competition of the city and according to definable processes, the population of the city is segregated over natural areas into natural groups. And these natural areas and natural groups are the "atoms" of city growth, the units we try to control in administering and planning for the city.

ADMINISTRATIVE AREA AND NATURAL AREA

The distinction between the natural area and the administrative area is apparent. The city is broken up into administrative units, such as the ward, the school district, the police precinct, and the health district, for the purposes of administrative convenience. The object is usually to apportion either the population or area of the city into equal units. The natural area, on the other hand, is a unit in the physical structure of the city, typified by a physical individuality and the characteristic attitudes, sentiments, and interests of the people segregated within it. Administrative areas and natural areas may coincide. In practice they rarely do. Administrative lines cut across the boundaries of natural areas, ignoring their existence.

The contrast between administrative and natural areas is not new. Historians long ago pointed out the international complications that have arisen because state lines were not drawn with reference to natural groupings of population and natural geographical units. A historian in a recent volume devotes a chapter to "Natural Areas and Boundaries." The geographer talks of production in terms of natural "regions." Gras, in his *Introduction to Economic History*, reminds us that a stable banking system must be based, not on units of administrative convenience, but upon the basis of natural "metropolitan" areas of financial service. We are just beginning, however, to take account of the natural areas of the city.

Students of municipal affairs are coming to appreciate the relationship of the cultural individuality of the natural areas of the city to the problem of city government. For one thing, the theory and practice of American municipal government, evolved to meet the needs of village communities, makes no allowance for the existence of distinct areas within the city, each with an individuality, and unequally adapted to function politically under our present system. On the Lower North Side of Chicago, for example, is a rooming-house area affording dormitories to 25,000 people. This population is exceedingly mobile. It turns over every four months. There are no permanent contacts in such an area. No one knows anyone else. There are no permanent interests in the area, and no public opinion. The population are not "citizens" of the locality. There are few votes, and many of these are sold. Local self-government is a myth. The area is administered by the social agencies and the police, though this fact is but imperfectly recognized by these agencies. The situation should be frankly faced. Such an area should be disfranchised and administered from the city hall. Natural areas are unequally adapted to function politically under our present system of municipal government.

Again, administrative units cut across natural areas. Ward lines divide a "Little Sicily," or ward lines encompass a number of natural areas and natural groups. As a result, the ward vote frequently represents a stalemate among conflicting natural areas; and large parts of the city are politically impotent. The real issues of the areas that make up the city rarely get into politics; municipal government becomes a concession, a state of affairs that is rapidly assuming the proportion of a national scandal. One remedy would seem to be the political recognition of the natural areas of the city, and at least a geographical pluralism in city government.

There have been numerous extra-political attempts to solve the problems of local self-government in the city. Among these is the community organization movement. Looking to the village as a "golden age" of social life, and believing that if the neighborliness of the village could be restored in the city the city's problems would take care of themselves, the community organizers have set out to make "villages" of areas within the city. But in selecting the areas for the experiments they have usually but substituted one administrative arm for another, totally oblivious of the existence and significance of natural areas and natural groups. The Lower North Community Council of Chicago set out to make a "community" of a section of the city including a colony of 15,000 Sicilians, a colony of 6,000 Persians, a belt of some 4,000 Negroes, a colony of 1,000 Greeks, a rooming-house population of 25,000, "Tower-town" – Chicago's Greenwich Village – and Chicago's much-vaunted "Gold Coast."

A further complicating factor is introduced by the fact that the natural areas of a city are only relatively stable, either in respect to values or in respect to the cultural segregation upon them. Particularly is this true in a new or growing city. In older cities residence is more permanent; a historical sentiment enters in to stabilize residence, inclining people to cling to the old community. And in a city that is not growing competition for position tends to cease and values and groupings of the population to reach an equilibrium. But in the growing city, expanding as it grows, natural areas are only relatively stable. They seem to change in a predictable manner, a succession like that observable in plant communities. The laws of this succession are imperfectly known, however. One of the purposes of the studies of the Community Research Fund of the University of Chicago has been to analyze this succession. Chicago's "Gold Coast," again, offers an interesting example of succession in process. As more and more of Chicago's industrial kings achieve incomes worthy of evasion of the government tax, they crowd in upon the "Gold Coast." Chicago's first families find themselves increasingly aliens in their own land. And we view the spectacle, not without its pathos, of the perambulators of the leaders of future assemblies disappearing from the Esplanade to reappear along Sheridan Road.

These ecological facts – natural areas within the city, competition for position, segregation over natural areas, succession – are facts that must be taken into account by those who would control the city's growth as much as by those who would administer the city's government. We are interested here not in cities planned from their origin though there seems to be limits to what can be done in such instances. Berlin, for example, like Amsterdam and many other European cities, has grown since the time when it was a small city according to a carefully directed plan. The scheme is not called zoning in Berlin, but there is a city architect and everything is planned in advance. The city is solidly built; there are no vacant spaces that may serve as speculative holdings. There is absolute standardization of buildings – squares, fountains, apothecaries' shops are located in advance. Houses have shops on the first floor, with the rooms of the tradesmen in the rear. The well-to-do have the apartments above, facing the street. The lower middle class have the back apartments. All classes are represented in a block. It is known how many people will be in each block, and what shops will be needed. Yet with all this careful planning Berlin has gotten out of bounds. The wealthy want to live on the parks and boulevards. They get located on certain streets. These streets acquire reputation and prestige, become distinctive regions not called for in the city plan. Values rise. Speculation goes on. The city gets out of control. Especially is this true since the war, with its recent turnover of fortunes and breaking down of class distinctions.

The experience of the Chicago Zoning Commission affords an interesting example of an attempt to control the growth of a new, rapidly growing, unplanned city. The Chicago zoning ordinance has been approximately two years in operation. Mr. H. J. Frost, formerly of the engineering staff which gathered the data on which the ordinance is based, and now of the board of appeals, has kindly given me data on the Chicago situation. His data would seem to indicate that it is futile to impose a plan upon a city which involves the attempt to control land values and the natural groupings of the population. Where use districts cut across natural areas of the city there is a constant pressure upon the board of appeals, which invariably necessitates revision. That is, use districts are merely another form of administrative area where they ignore natural areas. In attempting to control a city's growth we are not merely rearranging our "blocks," refashioning an artifact, but are working with a natural organization and natural groupings within that organization. The ordinance can neither control this organization of the city nor the inevitable succession of the city. It can, however, taking this organization and succession into account, stabilize the processes of city growth and prevent the waste involved in scattering and uncontrolled speculation. Whatever we may think such evidence indicates, certainly it is apparent that city planning and zoning, which attempt to control the growth of the city, can only be economical and successful where they recognize the natural organization of the city, the natural groupings of the city's population, the natural processes of the city's growth. An ideal city is not likely to be the mold of a real city.

NATURAL AREAS AND A SIGNIFICANT STATISTICS

One of our crying needs in planning for and administering the city is a significant statistics of city life. But statistics, to be significant, must be based not only upon accurately defined and comparable units but upon units that are actual factors in the process under examination. Our statistics of city life are based, at the present time, upon administrative areas, which have no real correspondence with the natural areas of the city. Consequently, our statistics are of little significance for the problems of city life. Mowrer, in his recent study of family disorganization in Chicago, found that statistics of family disorganization meant nothing until they were prepared for natural areas. Similarly, Shaw, studying the problem of juvenile delinquency, found that statistics, revealing when compiled for the natural areas of the city, meant nothing when compiled for wards. The natural areas of the city are real units. They can be accurately defined. Facts that have a position and can be plotted serve to characterize them. Within the areas we can study the subtler phases of city life – politics, opinion, cultural conflicts, and all social attitudes. As this data accumulates it becomes possible to compare, check, and find out knowledge. With natural areas defined, with the processes going on within them analyzed, statistics based upon natural areas should prove diagnostic of real situations and processes, indicative of real trends. It is not improbable that statistical ratios might be worked out which would afford a basis for prediction beyond the mere agglomeration of population, making it possible to apply numerical measurement to that collective human behavior in the urban environment which is the growth of the city.

NOTE

1 The nature of "value" in city land is a more complex problem than the average text on economics admits. Other cultural factors so condition the economic as to make the process of "value" – for it is a process – one difficult to analyze and state in abstract terms as it applies to city land.

"Sentiment and Symbolism as Ecological Variables"

from *American Sociological Review* (1945)

Walter Firey

Editors' Introduction

Walter Firey's 1945 article was an important essay in the sociocultural critique of human ecology, which emerged in the interwar period of the early twentieth century, in the wake of the first wave of urban renewal programs in U.S. cities. The socioculturalists may be associated with the emerging landmarks and historic preservation movement in America. They criticized Robert Park for drawing a false dualism between the social and biotic arenas of urban life. They skewered the human ecologists for subsuming cultural factors to market forces in determining urban land processes. Firey drew attention to the persistence of certain spaces and landmarks in Central Boston, including the Boston Commons, ancestral cemeteries, historic neighborhoods like Beacon Hill, and ethnic villages like the Italian North End. These places were like anomalies or antiquities in their immediate proximity to modern skyscrapers, state houses, hotels, and department stores in the dense central business district. They challenged the presumptions of human ecology theory, which predicted only the supremacy of the dominant economic forces able to pay the highest rents in the central city. The power of these places was in their sentimental and symbolic importance in the hearts and minds of citizenry, as a focus for collective sentiments and culture. They were spaces that persisted through the recuperative power of aesthetics, family traditions, and history. The socioculturalists triumphed values of culture over market profitability in the form and function of cities.

The Boston Commons, ancestral burying grounds, and other colonial landmarks and churches, were valuable not only for their intrinsic spatial attributes but from their representation in the minds of citizens as a symbol of collective sentiments. The Commons was a survival from the revolutionary era, when many New England towns retained a grassy central commons for pasturage and training of the militia. The Commons also evolved into a central location for political demonstrations, oratory and free speech. The denizens of the upper-class community of Beacon Hill were ferociously loyal to the historical and literary traditions of their district. They found charm rather than arcane inconvenience in its cobble-stoned streets, which were inconvenient, steep and slippery. Beacon Hill had lost some upper-class residents to Back Bay at the turn of the century, but resisted further decline through the interventions of groups like the Beacon Hill Association, which enacted zoning for only residential and historic quality to protect against speculative commercial development and residential transition. The lower-class enclave of the Italian American North End utilized ethnic rituals, family networks, and festivals to bring social coherence to their neighborhoods. Though gradually assimilating into American society, the ethnic enclave persisted as a symbol of immigrant solidarity. Older residents remain because they retained Italian values. While the area was not the conscious object of sentimental attachment or heritage preservation as Beacon Hill or the Commons, it was nonetheless a symbol of Italian American solidarity.

Walter Irving Firey, Jr. received his B.A. at the University of Washington, Seattle, and Ph.D. in Sociology at Harvard in 1945. In 1956–57, he was President of the Southwestern Sociological Association. He is currently Professor Emeritus at the University of Texas, Austin. He published *Land Use in Central Boston* in 1947. He later turned to a career in regional planning and conservation. In 1960, he published *Man, Mind, and Land: A Theory of Resource Use.* In this book and related writings, Firey originated a holistic conceptual framework for informing regional and community planning. He sought to integrate social science knowledge with objectives of resource management and environmental sustainability.

The sociocultural critique of human ecology was initiated by Milla Alihan in *Social Ecology* (1939). Associated work includes Richard Wohl and Anselm Strauss, "Symbolic Representation and the Urban Milieu," *American Journal of Sociology* 63 (1958): 523–532. Also memorable is Gerald Suttles, "The Cumulative Texture of Local Urban Culture," *American Journal of Sociology* 90 (1984): 283–304. In this article, Suttles draws attention to the symbolic and locational power of popular cultural representations, such as literary texts and folklore, and material artifacts such as statues and street names in our urban life. David Maines and Jeffrey Bridger "Narratives, Community and Land Use Decisions," *The Social Science Journal* 29, 4 (1992): 363–380, further explore the use of oral history and heritage preservation among the Amish in slowing urban sprawl in Pennsylvania. Lyn Lofland, "History, the City, and the Interactionist: Anselm Strauss, City Imagery, and Urban Sociology," *Symbolic Interaction* 14, 2 (1991): 205–223, discusses similar trends in northern California. Socioculturalists have morphed into interactionist urbanists. See Jan Lin, "Ethnic Places, Postmodernism, and Urban Change," *The Sociological Quarterly* 36, 4 (Fall 1995): 629–647, for a discussion of the use of racial and ethnic representations and landmarks for defensive and proactive purposes to stimulate neighborhood revitalization. He interprets the dynamics and implications of ethnic place preservation from the standpoint of postmodernism as well as symbolic interactionism.

The sociocultural perspective was a harbinger of the current vogue among city planners to incorporate landmarks and cultural monuments, neighborhood preservation, and heritage tourism strategies into their urban planning and redevelopment schemes. This is a new kind of urban renewal that is sensitive to the benefits of the cultural economy of the city. The earlier interest in landmarking and historic preservation taken by older cities such as Boston and Philadelphia has now taken root in newer Sunbelt cities such as Houston and Los Angeles. Baseball stadiums are also an illustrative case. Older stadiums used to be torn down to make way for modern ones. The ardent sentimental attachments of some fans to historic arenas such as Boston's Fenway Park and New York's Yankee Stadium have stalled redevelopment. Some stadiums, such as Baltimore's Camden Yards, have incorporated historic elements into the park's architecture. In Houston, often seen as a metropolitan symbol of southwestern free enterprise and modernity, the Astrodome was not destroyed to make way for the new Reliant Stadium. Many baseball stadiums have been modernized, but many still cling to the notion of baseball as a "field of dreams." Baseball represents the persistence of the rural pastoral in the midst of the modern city.

Systemization of ecological theory has thus far proceeded on two main premises regarding the character of space and the nature of locational activities. The first premise postulates that the sole relation of space to locational activities is an impeditive and cost-imposing one. The second premise assumes that locational activities are primarily economizing, "fiscal" agents. On the basis of these two premises the only possible relationship that locational activities may bear to space is an economic one. In such a relationship each activity will seek to so locate as to minimize the obstruction put upon its functions by spatial distance. Since the supply of the desired locations is limited it follows that not all activities can be favored with choice sites. Consequently a competitive process ensues in which the scarce desirable locations are preempted by those locational activities which can so exploit advantageous location as to produce the greatest surplus of income over expenditure. Less desirable locations devolve to correspondingly less economizing land uses. The result is a pattern of land use that is presumed to be most efficient for both the individual locational activity and for the community.

Given the contractualistic milieu within which the modern city has arisen and acquires its functions, such an "economic ecology" has had a certain explanatory adequacy in describing urban spatial structure and dynamics. However, as any theory matures and approaches a logical closure of its generalizations it inevitably encounters facts which remain unassimilable to the theoretical scheme. In this paper it will be our purpose to describe certain ecological processes which apparently cannot be embraced in a strictly economic analysis. Our hypothesis is that the data to be presented, while in no way startling or unfamiliar to the research ecologist, do suggest an alteration of the basic premises of ecology. This alteration would consist, first, of ascribing to space not only an impeditive quality but also an additional property, *viz.*, that of being at times a symbol for certain cultural values that have become associated with a certain spatial area. Second, it would involve a recognition that locational activities are not only economizing agents but may also bear sentiments which can significantly influence the locational process.

A test case for this twofold hypothesis is afforded by certain features of land use in central Boston. In common with many of the older American cities Boston has inherited from the past certain spatial patterns and landmarks which have had a remarkable persistence and even recuperative power despite challenges from other more economic land uses. The persistence of these spatial patterns can only be understood in terms of the group values that they have come to symbolize. We shall describe three types of such patterns: first, an in-town upper class residential neighborhood known as Beacon Hill; second, certain "sacred sites," notably the Boston Common and the colonial buryinggrounds; and third, a lower class, Italian neighborhood known as the North End. In each of these land uses we shall find certain locational processes which seem to defy a strictly economic analysis.

The first of the areas, Beacon Hill, is located some five minutes' walking distance from the retail center of Boston. This neighborhood has for fully a century and a half maintained its character as a preferred upper class residential district, despite its contiguity to a low rent tenement area, the West End. During its long history Beacon Hill has become the symbol for a number of sentimental associations which constitute a genuine attractive force to certain old families of Boston. Some idea of the nature of these sentiments may be had from statements in the innumerable pamphlets and articles written by residents of the Hill. References to "this sacred eminence," "stately old-time appearance," and "age-old quaintness and charm," give an insight into the attitudes attaching to the area. One resident reveals rather clearly the spatial referability of these sentiments when she writes of the Hill:

> It has a tradition all its own, that begins in the hospitality of a book-lover, and has never lost that flavor. Yes, our streets are inconvenient, steep, and slippery. The corners are abrupt, the contours perverse. . . . It may well be that the gibes of our envious neighbors have a foundation and that these dear crooked lanes of ours were indeed traced in ancestral mud by absent-minded kine.

Behind such expressions of sentiment are a number of historical associations connected with the area. Literary traditions are among the strongest of these; indeed, the whole literary legend of Boston has its focus at Beacon Hill. Many of America's most distinguished literati have occupied homes on the Hill. Present day occupants of these houses derive a genuine satisfaction from the individual histories of their dwellings. One lady whose home had had a distinguished pedigree remarked:

> I like living here for I like to think that a great deal of historic interest has happened here in this room.

Not a few families are able to trace a continuity of residence on the Hill for several generations, some as far back as 1800 when the Hill was first developed as an upper class neighborhood. It is a point of pride to a Beacon Hill resident if he can say that he was born on the Hill or was at least raised there; a second best boast is to point out that his forebears once lived on the Hill.

Thus a wide range of sentiments – aesthetic, historical, and familial – have acquired a spatial articulation in Beacon Hill. The hearing of these sentiments upon locational processes is a tangible one and assumes three forms: retentive, attractive, and resistive. Let us consider each of these in order. To measure the retentive influence that spatially-referred sentiments may exert upon locational activities we have tabulated by place of residence all the families listed in the Boston Social Register for the years 1894, 1905, 1914, 1929, and 1943. This should afford a reasonably accurate picture of the distribution of upper class families by neighborhoods within Boston and in suburban towns. . . . The most apparent feature . . . is, of course, the consistent increase of upper class families in the suburban towns and the marked decrease (since 1905) in two of the in-town upper class areas, Back Bay and Jamaica Plain. Although both of these neighborhoods remain fashionable residential districts their prestige is waning rapidly. Back Bay in particular, though still surpassing in numbers any other single neighborhood, has undergone a steady invasion of apartment buildings, rooming houses, and business establishments which are destroying its prestige value. The trend of Beacon Hill has been different. Today it has a larger number of upper class families than it had in 1894. Where it ranked second among fashionable neighborhoods in 1894 it ranks third today, being but slightly outranked in numbers by the suburban city of Brookline and by the Back Bay. Beacon Hill is the only in-town district that has consistently retained its preferred character and has held to itself a considerable proportion of Boston's old families.

There is, however, another aspect to the spatial dynamics of Beacon Hill, one that pertains to the "attractive" locational role of spatially referred sentiments. From 1894 to 1905 the district underwent a slight drop, subsequently experiencing a steady rise for 24 years, and most recently undergoing another slight decline. These variations are significant, and they bring out rather clearly the dynamic ecological role of spatial symbolism. The initial drop is attributable to the development of the then new Back Bay. Hundreds of acres there had been reclaimed from marshland and had been built up with palatial dwellings. Fashion now pointed to this as the select area of the city and in response to its dictates a number of families abandoned Beacon Hill to take up more pretentious Back Bay quarters. Property values on the Hill began to depreciate, old dwellings became rooming houses, and businesses began to invade some of the streets. But many of the old families remained on the Hill and a few of them made efforts to halt the gradual deterioration of the district. Under the aegis of a realtor, an architect, and a few close friends there was launched a program of purchasing old houses, modernizing the interiors and leaving the colonial exteriors intact, and then selling the dwellings to individual families for occupancy. Frequently adjoining neighbors would collaborate in planning their improvements so as to achieve an architectural consonance. The results of this program may be seen in the drift of upper class families back to the Hill. From 1905 to 1929 the number of Social Register families in the district increased by 120. Assessed valuations showed a corresponding increase: from 1919 to 1924 there was a rise of 24 per cent; from 1924 to 1929 the rise was 25 per cent. The nature of the Hill's appeal, and the kind of persons attracted, may be gathered from the following popular write-up:

> To salvage the quaint charm of Colonial Architecture on Beacon Hill, Boston, is the object of a well-defined movement among writers and professional folk that promises the most delightful opportunities for the home seeker of moderate means and conservative tastes. Because men of discernment were able to visualize the possibilities presented by these architectural landmarks, and have undertaken the gracious task of restoring them to their former glory, this historic quarter of old Boston, once the centre of literary culture, is coming into its own.

The independent variable in this "attractive" locational process seems to have been the symbolic quality of the Hill, by which it constituted a referent for certain strong sentiments of upper class Bostonians.

While this revival was progressing there remained a constant menace to the character of Beacon Hill, in the form of business encroachments and apartment-hotel developments. Recurrent threats from this source finally prompted residents of the Hill to organize themselves into the Beacon Hill Association. Formed in 1922, the declared object of this organization was "to keep undesirable business and living conditions from affecting the hill district." At the time the city was engaged in preparing a comprehensive zoning program and the occasion was propitious to secure for Beacon Hill suitable protective measures. A systematic set of recommendations was drawn up by the Association regarding a uniform 65-foot height limit for the entire Hill, the exclusion of business from all but two streets, and the restriction of apartment house bulk. It succeeded in gaining only a partial recognition of this program in the 1924 zoning ordinance. But the Association continued its fight against inimical land uses year after year. In 1927 it successfully fought a petition brought before the Board of Zoning Adjustment to alter the height limits in one area so as to permit the construction of a four million dollar apartment-hotel 155 feet high. Residents of the Hill went to the hearing en masse. In spite of the prospect of an additional twenty million dollars worth of exclusive apartment hotels that were promised if the zoning restrictions were withheld the petition was rejected, having been opposed by 214 of the 220 persons present at the hearing. In 1930 the Association gained an actual reduction in height limits on most of Beacon Street and certain adjoining streets, though its leader was denounced by opponents as "a rank sentimentalist who desired to keep Boston a village." One year later the Association defeated a petition to rezone Beacon Street for business purposes. In other campaigns the Association successfully pressed for the rezoning of a business street back to purely residential purposes, for the lowering of height limits on the remainder of Beacon Street, and for several lesser matters of local interest. Since 1929, owing partly to excess assessed valuations of Boston real estate and partly to the effects of the depression upon families living on securities, Beacon Hill has lost some of its older families, though its decline is nowhere near so precipitous as that of the Back Bay.

Thus for a span of one and a half centuries there have existed on Beacon Hill certain locational processes that largely escape economic analysis. It is the symbolic quality of the Hill, not its impeditive or cost-imposing character, that most tangibly correlates with the retentive, attractive, and resistive trends that we have observed. And it is the dynamic force of spatially referred sentiments, rather than considerations of rent, which explains why certain families have chosen to live on Beacon Hill in preference to other in-town districts having equally accessible location and even superior housing conditions. There is thus a noneconomic aspect to land use on Beacon Hill, one which is in some respects actually diseconomic in its consequences. Certainly the large apartment-hotels and specialty shops that have sought in vain to locate on the Hill would have represented a fuller capitalization on potential property values than do residences. In all likelihood the attending increase in real estate prices would not only have benefited individual property holders but would have so enhanced the value of adjoining properties as to compensate for whatever depreciation other portions of the Hill might have experienced.

If we turn to another type of land use pattern in Boston, that comprised by the Boston Common and the old burying grounds, we encounter another instance of spatial symbolism which has exerted a marked influence upon the ecological organization of the rest of the city. The Boston Common is a survival from colonial days when every New England town allotted a portion of its land to common use as a cow pasture and militia field. Over the course of three centuries Boston has grown entirely around the Common so that today we find a 48-acre tract of land wedged directly into the heart of the business district. On three of its five sides are women's apparel shops, department stores, theaters and other high-rent locational activities. On the fourth side is Beacon Street, extending alongside Beacon Hill. Only the activities of Hill residents have prevented business from invading this side. The fifth side is occupied by the Public Garden. A land value map portrays a strip of highest values pressing upon two sides of the Common, on Tremont and Boylston Streets, taking the form of a long, narrow band.

Before considering the ecological consequences of this configuration let us see what attitudes have come to be associated with the Common. There is an extensive local literature about the Common and in it we find interesting sentiments expressed. One citizen speaks of:

> . . . the great principle exemplified in the preservation of the Common. Thank Heaven, the tide of money making must break and go around that.

Elsewhere we read:

> Here, in short, are all our accumulated memories, intimate, public, private.

> Boston Common was, is, and ever will be a source of tradition and inspiration from which the New Englanders may renew their faith, recover their moral force, and strengthen their ability to grow and achieve.

The Common has thus become a "sacred" object, articulating and symbolizing genuine historical sentiments of a certain portion of the community. Like all such objects its sacredness derives, not from any intrinsic spatial attributes, but rather from its representation in peoples' minds as a symbol for collective sentiments.

Such has been the force of these sentiments that the Common has become buttressed up by a number of legal guarantees. The city charter forbids Boston in perpetuity to dispose of the Common or any portion of it. The city is further prohibited by state legislation from building upon the Common, except within rigid limits, or from laying out roads or tracks across it. By accepting the bequest of one George F. Parkman, in 1908, accounting to over five million dollars, the city is further bound to maintain the Common, and certain other parks, "for the benefit and enjoyment of its citizens."

What all this has meant for the spatial development of Boston's retail center is clear from the present character of that district. Few cities of comparable size have so small a retail district in point of area. Unlike the spacious department stores of most cities, those in Boston are frequently compressed within narrow confines and have had to extend in devious patterns through rear and adjoining buildings. Traffic in downtown Boston has literally reached the saturation point, owing partly to the narrow one-way streets but mainly to the lack of adequate arterials leading into and out of the Hub. The American Road Builders Association has estimated that there is a loss of $81,000 per day in Boston as a result of traffic delay. Trucking in Boston is extremely expensive. These losses ramify out to merchants, manufacturers, commuters, and many other interests. Many proposals have been made to extend a through arterial across the Common, thus relieving the extreme congestion on Tremont and Beacon Streets, the two arterials bordering the park. Earlier suggestions, prior to the construction of the subway, called for street car tracks across the Common. But "the controlling sentiment of the citizens of Boston, and of large numbers throughout the State is distinctly opposed to allowing any such use of the Common." Boston has long suffered from land shortage and unusually high real estate values as a result both of the narrow confines of the peninsula comprising the city center and as a result of the exclusion from income-yielding uses of so large a tract as the Common. A further difficulty has arisen from the rapid southwesterly extension of the business district in the past two decades. With the Common lying directly in the path of this extension the business district has had to stretch around it in an elongated fashion, with obvious inconvenience to shoppers and consequent loss to businesses.

The Common is not the only obstacle to the city's business expansion. No less than three colonial burying-grounds, two of them adjoined by ancient church buildings, occupy downtown Boston. The contrast that is presented by 9-story office buildings reared up beside quiet cemeteries affords visible evidence of the conflict between "sacred" and "profane" that operates in Boston's ecological pattern. The dis-economic consequences of commercially valuable land being thus devoted to non-utilitarian purposes goes even further than the removal from business uses of a given amount of space. For it is a standard principle of real estate that business property derives added value if adjoining properties are occupied by other businesses. Just as a single vacancy will depreciate the value of a whole block of business frontage, so a break in the continuity of stores by a cemetery damages the commercial value of surrounding properties. But, even more than the Common, the colonial burying-grounds of Boston have become invested with a moral significance which renders them almost inviolable. Not only is there the usual sanctity which attaches to all cemeteries but in those of Boston there is an added sacredness growing out of the age of the grounds and the fact that the forebears of many of New England's most distinguished families as well as a number of colonial and Revolutionary leaders lie buried in these cemeteries. There is thus a manifold symbolism to these old burying-grounds, pertaining to family lineage, early nationhood, civic origins, and the like, all of which have strong sentimental associations. What has been said of the old burying-grounds applies with equal force to a number of other venerable landmarks in central Boston. Such buildings as the Old South Meeting-house, the Park Street Church, King's Chapel, and the Old State House – all foci of historical associations – occupy commercially valuable land and interrupt the continuity of business frontage on their streets. Nearly all of these landmarks have been challenged at various times by real estate and commercial interests which sought to have them replaced by more profitable uses. In every case community sentiments have resisted such threats.

In all these examples we find a symbol–sentiment relationship which has exerted a significant influence upon land use. Nor should it be thought that such phenomena are mere ecological "sports." Many other older American cities present similar locational characteristics. Delancey Street in Philadelphia represents a striking parallel to Beacon Hill, and certain in-town districts of Chicago, New York, and Detroit, recently revived as fashionable apartment areas, bear resemblances to the Beacon Hill revival. The role of traditionalism in rigidifying the ecological patterns of New Orleans has been demonstrated in a recent study. Further studies of this sort should clarify even further the true scope of sentiment and symbolism in urban spatial structure and dynamics.

As a third line of evidence for our hypothesis we have chosen a rather different type of area from those so far considered. It is a well known fact that immigrant ghettoes, along with other slum districts, have become areas of declining population in most American cities. A point not so well established is that this decline tends to be selective in its incidence upon residents and that this selectivity may manifest varying degrees of identification with immigrant values. For residence within a ghetto is more than a matter of spatial placement; it generally signifies acceptance of immigrant values and participation in immigrant institutions. Some light on this process is afforded . . . from the North End of Boston. This neighborhood, almost wholly Italian in population, has long been known as "Boston's classic land of poverty." Eighteen percent of the dwellings are eighty or more years old, and sixty percent are forty or more years old. Indicative of the dilapidated character of many buildings is the recent sale of a 20-room apartment building for only $500. It is not surprising then to learn that the area has declined in population from 21,111 in 1930 to 17,598 in 1940. To look for spatially referable sentiments here would seem futile. And yet, examination of certain emigration differentials in the North End reveals a congruence between Italian social structure and locational processes. [. . .] In brief, the North End is losing its young people to a much greater extent than its older people.

These differentials are in no way startling; what is interesting, however, is their congruence with basic Italian values, which find their fullest institutionalized expression in the North End. Emigration from the district may be viewed as both a cause and a symbol of alienation from these values. At the core of the Italian value system are those sentiments which pertain to the family and the paesani. Both of these put a high premium upon maintenance of residence in the North End.

Paesani, or people from the same village of origin, show considerable tendency to live near one another, sometimes occupying much of a single street or court. Such proximity, or at least common residence in the North End, greatly facilitates participation in the *paesani* functions which are so important to the first generation Italian. Moreover, it is in the North End that the *festas*, anniversaries, and other old world occasions are held, and such is their frequency that residence in the district is almost indispensable to regular participation. The social relationships comprised by these groupings, as well as the benefit orders, secret societies, and religious organizations, are thus strongly localistic in character. One second generation Italian, when asked if his immigrant parents ever contemplated leaving their North End tenement replied:

> No, because all their friends are there, their relatives. They know everyone around there.

It is for this reason that the first generation Italian is so much less inclined to leave the North End than the American born Italian.

Equally significant is the localistic character of the Italian family. So great is its solidarity that it is not uncommon to find a tenement entirely occupied by a single extended family: grandparents, matured children with their mates, and grandchildren. There are instances where such a family has overflowed one tenement and has expanded into an adjoining one, breaking out the partitions for doorways.

These are ecological expressions, in part, of the expected concern, which an Italian mother has for the welfare of her newly married daughter. The ideal pattern is for the daughter to continue living in her mother's house, with she and her husband being assigned certain rooms which they are supposed to furnish themselves. Over the course of time the young couple is expected to accumulate savings and buy their own home, preferably not far away. Preferential renting, by which an Italian who owns a tenement will let apartments to his relatives at a lower rental, is another manifestation of the localizing effects of Italian kinship values.

Departure from the North End generally signifies some degree of repudiation of the community's values. One Italian writes of an emigrant from the North End:

> I still remember with regret the vain smile of superiority that appeared on his face when I told him that I lived at the North End of Boston. "Io non vado Ira quella plebaglia." (I do not go among those plebeians).

As a rule the older Italian is unwilling to make this break, if indeed he could. It is the younger adults, American-born and educated, who are capable of making the transition to another value system with radically different values and goals.

Residence in the North End seems therefore to be a spatial corollary to integration with Italian values. Likewise emigration from the district signifies assimilation into American values, and is so construed by the people themselves. Thus, while the area is not the conscious object of sentimental attachment, as are Beacon Hill and the Common, it has nonetheless become a symbol for Italian ethnic solidarity. By virtue of this symbolic quality the area has a certain retentive power over those residents who most fully share the values which prevail there.

It is reasonable to suggest, then, that the slum is much more than "an area of minimum choice." Beneath the surface phenomenon of declining population there may be differential rates of decline which require positive formulation in a systematic ecological theory. Such processes are apparently refractory to analysis in terms of competition for least impeditive location. A different order of concepts, corresponding to the valuative, meaningful aspect of spatial adaptation, must supplement the prevailing economic concepts of ecology.

"The City as a Growth Machine"

from *Urban Fortunes: The Political Economy of Place* (1987)

John Logan and Harvey Molotch

Editors' Introduction

The concept of the "growth machine" was initially formulated by Harvey Molotch as an outgrowth of his denunciation of the environmentally destructive effects of a massive 1969 oil tanker spill off the beautiful coastline of Santa Barbara, California. The spill was considered by many to be a watershed for the national environmental movement, and an even more irrevocable turning point for California. Molotch contributed to the national debate with a 1969 article in *Ramparts* magazine titled "Oil in the Velvet Playground." In this and other articles he expressed the outrage of local businesses and residents, who perceived that they gained little wealth or tax benefits from the oil companies working in offshore federal waters. The water and air pollution from the drilling operations furthermore hampered the tourism industry, another major component of the regional economic base. In the ensuing years, Molotch would continue to reflect on the damaging environmental and social consequences of American capitalism and urbanization, and move towards a more generalized urban political economy that understands cities as growth machines that serve elite interests, promote social inequality and harm the environment. He eventually collaborated with John Logan and they published *Urban Fortunes: The Political Economy of Place* in 1987.

Logan and Molotch reject the human ecology view that places and land markets are the natural outcome of Darwinian and market processes in favor of a Marxist-influenced political economy perspective. They comprehend homes not just as places that are "lived," but also as commodities within real estate markets that can be bought and exchanged, generating use and exchange values for producers and consumers. They see prices and markets, importantly, as *social phenomena*, governed not by natural laws of competition, supply, and demand, but by inequalities of wealth, ownership, and power. Places organize and distribute life chances in a class stratification system. As commodities, places also acquire special use and exchange values. In terms of use value, homes, neighborhoods, local businesses, localities, and other places obtain a special preciousness for people characterized by intense sentiment, commitment, or attachment. In terms of exchange values, places are idiosyncratic and not substitutable in the fashion of other commodities such as cars, clothes, and food, because real estate is a more limited and finite resource. Because property markets are structured by access to infrastructure such as jobs, housing, transportation, schools, hospitals, and other resources, places determine life chances and are key components of the American class stratification system.

Place entrepreneurs such as landlords, businessmen, developers, transportation and utility companies, banks, and corporations gain profit from their control of land and from the proceeds of economic growth. There are serendipitous, active, and structural entrepreneurs, with varying levels of access to capital as well as place attachment. Place entrepreneurs form pro-growth coalitions with governmental units and other economic interests to focus infrastructure and urban development in areas that intensify the profitability of their own interests. They promote a good business climate and an ideology of growth as a public good through their influence on politicians and the media. They foster a booster spirit through partnerships with schools and civic organizations with essay contests, public celebrations and spectacles such as dedications, soapbox derbies, parade floats, and beauty contests. Growth machines have evolved from small groups of local power brokers to include a more multifaceted matrix of auxiliary interests and institutions that include universities, museums, convention centers, sports franchises, entertainment conglomerates, and tourism interests such as theme parks. The growth machine works to suppress or deflect public consciousness from the negative social and environmental consequences of urban development. Citizens' movements have sprung up to contest the negative externalities of growth through protests, lobbying, public hearings, and environmental impact reviews.

The political scientist John Mollenkopf offers a similar concept of "growth coalitions" in his 1983 book, *The Contested City*. Joe Feagin offers an excellent historical and empirical case study of urban political economy in his 1988 book, *Free Enterprise City*. He charts the succession of the Houston growth elite from the days of the "Suite 8F crowd" to the global corporate interests of the late twentieth century. Andrew Kirby and A. Karen Lynch examine the negative social consequences of rapid growth and urban sprawl in Houston in "A Ghost in the Growth Machine: The Aftermath of Rapid Population Growth in Houston," *Urban Studies* 24 (1987): 587–596. Mark Gottdiener and Joe Feagin "The Paradigm Shift in Urban Sociology," *Urban Affairs Quarterly* 24, 2 (1987): 163–187, offer another useful articulation of the new urban sociology. John Walton published a useful historical perspective on urban political economy in "Urban Sociology: The Contribution and Limits of Political Economy," *Annual Review of Sociology* 19 (1993): 301–320.

Harvey L. Molotch obtained his Ph.D. in Sociology from the University of Chicago in 1968. He was on the faculty of the University of California at Santa Barbara from 1967 to 2003, and also held a stint as Centennial Professor at the London School of Economics in 1998–99. He is now Professor of Metropolitan Studies and Sociology at New York University. In 2003, Molotch won the Robert and Helen Lynd Award for lifetime career contribution from the Community and Urban Section of the American Sociological Association. His latest book is *Where Stuff Comes From: How Toasters, Toilets, Cars, Computers and Many Other Things Come to Be as They Are* (New York and London: Routledge, 2003).

John Logan received his Ph.D. from the University of California, Berkeley, in 1974. He was on the faculty of the State University of New York, Stony Brook from 1972 to 1980; then he moved to the State University of New York, Albany, where he is now Distinguished Professor in the Department of Sociology and Department of Public Administration and Policy. He is also Director of the Lewis Mumford Center for Comparative Urban and Regional Research. He served as the Chair of the Community and Urban Sociology Section of the American Sociological Association in 1993–94. Logan has published hundreds of articles in the areas of urban sociology, race and ethnicity, political sociology, immigration, family, aging/gerontology, and social movements.

Urban Fortunes won the 1988 Robert Park Award for best book community and urban sociology from the American Sociological Association (ASA). It also won the 1990 Distinguished Scholarly Publication Award of the ASA. *Urban Fortunes: The Political Economy of Place* is one of the best articulations of urban political economy, sometimes described as the new urban sociology.

THE SOCIAL CONSTRUCTION OF CITIES

The earth below, the roof above, and the walls around make up a special sort of commodity: a place to be bought and sold, rented and leased, as well as used for making a life. At least in the United States, this is the standing of place in legal statutes and in ordinary people's imaginations. Places can (and should) be the basis not only for carrying on a life but also for exchange in a market. We consider this commodification of place fundamental to urban life and necessary in any urban analysis of market societies.

Yet in contrast to the way neoclassical economists (and their followers in sociology) have undertaken the task of understanding the property commodity, we focus on how markets work as social phenomena. Markets are not mere meetings between producers and consumers, whose relations are ordered by the impersonal "laws" of supply and demand. For us, the fundamental attributes of all commodities, but particularly of land and buildings, are the social contexts through which they are used and exchanged. Any given piece of real estate has both a use value and an exchange value.

[...]

PLACES AS COMMODITIES

For us, as for many of our intellectual predecessors, the market in land and buildings orders urban phenomena and determines what city life can be. This means we must show how real estate markets actually work and how their operations fail to meet the neoclassical economists' assumptions. In short, we will find the substance of urban phenomena in the actual operations of markets. Our goal is to identify the specific processes, the sociological processes, through which the pursuit of use and exchange values fixes property prices, responds to prices, and in so doing determines land uses and the distribution of fortunes. Since economic sociology is still without a clear analytical foundation, we must begin our work in this chapter by laying a conceptual basis for the empirical descriptions that will be presented later.

Special use values

People use place in ways contrary to the neoclassical assumptions of how commodities are purchased and consumed. We do not dispose of place after it has been bought and used. Places have a certain preciousness for their users that is not part of the conventional concept of a commodity. A crucial initial difference is that place is indispensable; all human activity must occur somewhere. Individuals cannot do without place by substituting another product. They can, of course, do with less place and less desirable place, but they cannot do without place altogether.

Even when compared to other indispensable commodities – food, for example – place is still idiosyncratic. The use of a particular place creates and sustains access to additional use values. One's home in a particular place, for example, provides access to school, friends, work place, and shops. Changing homes disrupts connections to these other places and their related values as well. Place is thus not a discrete element, like a toy or even food; the precise conditions of its use determine how other elements, including other commodities, will be used. Any individual residential location connects people to a range of complementary persons, organizations, and physical resources.

The stakes involved in the relationship to place can be high, reflecting all manner of material, spiritual, and psychological connections to land and buildings. Numerous scholars have shown that given places achieve significance beyond the more casual relations people have to other commodities. The connection to place can vary in intensity for different class, age, gender, and ethnic groups, individual relationships to place are often characterized by intense feelings and commitments appropriate to long-term and multifaceted social and material attachments.

This special intensity creates an asymmetrical market relation between buyers and sellers. People pay what the landlord demands, not because the housing unit is worth it, but because the property is held to have idiosyncratic locational benefits. Access to resources like friends, jobs, and schools is so important that residents (as continuous consumers–buyers) are willing to resort to all sorts of "extramarket" mechanisms to fight for their right to keep locational relations intact. They organize, protest, use violence, and seek political regulation. They strive not just for tenure in a given home but for stability in the surrounding neighborhood as well.

Location establishes a special collective interest among individuals. People who have "bought" into the same neighborhood share a quality of public services (garbage pickup, police behavior); residents have a common stake in the area's future. Residents also share the same fate when natural disasters such as floods and hurricanes threaten and when institutions alter the local landscape by creating highways, parks, or toxic dumps. Individuals are not only mutually dependent on what goes on inside a neighborhood (including "compositional effects"); they are affected by what goes on outside it as well. The standing of a neighborhood vis-à-vis other neighborhoods creates conditions that its residents experience in common. Each place has a particular political or economic standing vis-à-vis other places that affect the quality of life and opportunities available to those who live within its boundaries. A neighborhood with a critical voting bloc (for example, Chicago's Irish wards in the 1930s) may generate high levels of public services or large numbers of patronage jobs for its working-class residents, thereby aiding their well being. A rich neighborhood can protect its residents' life styles from external threats (sewer plants, public housing) in a way that transcends personal resources, even those typically associated with the affluent. The community in itself can be a local force.

Neighborhoods organize life chances in the same sense as do the more familiar dimensions of class and caste. . . . Like class and status groupings, and even more than many other associations, places create communities of fate. Thus we must consider the stratification of places along with the stratification of individuals in order to understand the distribution of life chances. People's sense of these dynamics, perceived as the relative "standing" of their neighborhood, gives them some of their spiritual or sentimental stake in place – thus further distinguishing home from other, less life-significant, commodities.

Contrary to much academic debate on the subject, we hold that the material use of place cannot be separated from psychological use; the daily round that makes physical survival possible takes on emotional meanings through that very capacity to fulfill life's crucial goals. The material and psychic rewards thus combine to create feelings of "community." Much of residents' striving as members of community organizations or just as responsible neighbors represents an effort to preserve and enhance their networks of sustenance. Appreciation of neighborhood resources, so varied and diffusely experienced, gives rise to "sentiment." Sentiment is the inadequately articulated sense that a particular place uniquely fulfills a complex set of needs. When we speak of residents' use values, we imply fulfillment of all these needs, material and non-material.

Homeownership gives some residents exchange value interests along with use value goals. Their houses are the basis of a lifetime wealth strategy. For those who pay rent to landlords, use values are the only values at issue. Owners and tenants can thus sometimes have divergent interests. When rising property values portend neighborhood transformation, tenants and owners may adopt different community roles; but ordinarily, the exchange interests of owners are not sufficiently significant to divide them from other residents.

[. . .]

Special exchange values

Exchange values from place appear as "rent." We use the term broadly to include outright purchase expenditures as well as payments that homebuyers or tenants make to landlords, realtors, mortgage lenders, real estate lawyers, title companies, and so forth. As with use values, people pursue exchange values in ways that differ from the manner in which they create other commodities. Suppliers cannot "produce" places in the usual sense of the term. All places consist, at least in part, of land, which "is only another name for nature, which is not produced by man" (Karl Polanyi, *The Great Transformation*. Boston: Beacon Press, 1944, p. 72) and obviously not produced for sale in a market. The quantity is fixed. It is not, says David Harvey (*The Limits to Capital*. Chicago: Chicago University Press, 1982, p. 357), "the product of labor." This makes the commodity description of land, in Marx's word, "fictitious." Michael Storper and Richard Walker ("The Theory of Labor and the Theory of Location," *International Journal of Urban and Regional Research* 7, 1 (1983): 43) describe land, like labor, as a "pseudocommodity."

Place as monopoly

Perhaps the fundamental "curiosity" is that land markets are inherently monopolistic, providing owners, as a class, with complete control over the total commodity supply. There can be no additional entrepreneurs or any new product. The individual owner also has a monopoly over a subsection of the marketplace. Every parcel of land is unique in the idiosyncratic access it provides to other parcels and uses, and this quality underscores the specialness of property as a commodity. Unlike widgets or Ford Pintos, more of the same product cannot be added as market demand grows. Instead the owner of a particular parcel controls all access to it and its given set of spatial relations. In setting prices and other conditions of use, the owner operates with this constraint on competition in mind.

Property prices do go down as well as up, but less because of what entrepreneurs do with their own holdings than because of the changing relations among properties. This dynamic accounts for much of the energy of the urban system as place entrepreneurs strive to increase their rent by revamping the spatial organization of the city. Rent levels are based on the location of a property vis-à-vis other places, on its particularity. In Marxian conceptual terms, entrepreneurs establish the rent according to the "differential" locational advantage of one site over another. Gaining "differential rent" necessarily depends on the fate of other parcels and those who own and use them. In economists' language, each property use "spills over" to other parcels and, as part of these "externality effects," crucially determines what every other property will be. The "web of externalities" affects an entrepreneur's particular holding. When a favorable relationship can be made permanent (for example, by freezing out competitors through restrictive zoning), spatial monopolies that yield even higher rents – "monopoly rents" in the Marxian lexicon – are created. But all property tends to have a monopolistic character. . . .

Nevertheless, property owners can and do inventively alter the content of their holdings. Sometimes they build higher and more densely, increasing the supply of dwellings, stores, or offices on their land. According to neoclassical thinking, this manner of increase should balance supply and demand, thus making property respond to market pressures as other commodities supposedly do. But new construction has less bearing on market dynamics than such reasoning would imply. New units on the same land can never duplicate previous products; condominiums stacked in a high-rise building are not the same as split-levels surrounded by lawn. Office space on the top roof of a skyscraper is more desirable than the same square footage just one floor lower. Conversely, the advantages of street-level retail space cannot be duplicated on a floor above. Each product, old or new, is different and unique, and each therefore reinforces the monopoly character property and the resulting price system.

Another curious aspect of the real estate market is its essentially second-hand nature. Buildings and land parcels are sold and resold, rented and rerented. In a typical area, no more than 3 percent of the product for sale or rent consists of new construction. Not only land, but even the structures on any piece of land can have infinite (for all practical purposes) lives; neither utility nor market price need decrease through continuous use. . . . Moreover, since the amount of "new" property on the market at any given moment is ordinarily only a small part of the total that is for sale, entrepreneurs' decisions to add to this supply by building additional structures will have a much more limited impact on price than would the same decisions with other types of commodities. Indeed, recent studies indicate that U.S. cities with more rapid rates of housing construction have higher, not lower, housing costs, even when demand factors are statistically controlled. Similarly, relatively high vacancy rates are not associated with lower rent levels, which suggests that new construction "leads" local markets to a new, higher pricing structure rather than equilibrating a previous one. Given the fixed supply of land and the monopolies over relational advantages, more money entering an area's real estate market not only results in more structures being built but also increases the price of land and, quite plausibly, the rents on previously existing "comparable" buildings. Thus higher investment levels can push the entire price structure upward.

[. . .]

GROWTH MACHINES

Those seeking exchange value often share interests with others who control property in the same block, city, or region. Like residents, entrepreneurs in similar situations also make up communities of fate, and they often get together to help fate along a remunerative path.

Whether the geographical unit of their interest is as small as a neighborhood shopping district or as large as a national region, place entrepreneurs attempt, through collective action and often in alliance with other business people, to create conditions that will intensify future land use in an area. There is an unrelenting search, even in already successful places, for more and more. An apparatus of interlocking progrowth associations and governmental units makes up . . . the "growth machine." Growth machine activists are largely free from concern for what goes on within production processes (for example, occupational safety), for the actual use value of the products made locally (for example, cigarettes), or for spillover consequences in the lives of residents (for example, pollution). They tend to oppose any intervention that might regulate development on behalf of use values. They may quarrel among themselves over exactly how rents will be distributed among parcels, over how, that is, they will share the spoils of aggregate growth. But virtually all place entrepreneurs and their growth machine associates, regardless of geographical or social location, easily agree on the issue of growth itself.

They unite behind a doctrine of value-free development – the notion that free markets alone should determine land use. In the entrepreneur's view, land-use regulation endangers both society at large and the specific localities favored as production sites.

[. . .]

Growth machines in U.S. history

The role of the growth machine as a driving force in U.S. urban development has long been a factor in U.S. history, and is nowhere more clearly documented than in the histories of eighteenth- and nineteenth-century American cities. Indeed, although historians have chronicled many types of mass opposition to capitalist organization (for example, labor unions and the Wobblie movement), there is precious little evidence of resistance to the dynamics of value-free city building characteristic of the American past. . . . The creators of towns and the builders of cities strained to use all the resources at their disposal, including crude political clout, to make great fortunes out of place. . . . Sometimes, the "communities" were merely subdivided parcels with town names on them, on whose behalf governmental actions could nonetheless be taken. The competition among them was primarily among growth elites.

These communities competed to attract federal land offices, colleges and academies, or installations such as arsenals and prisons as a means of stimulating development. . . . The other important arena of competition was also dependent on government decision making and funding: the development of a transportation infrastructure that would give a locality better access to raw materials and markets. First came the myriad efforts to attract state and federal funds to link towns to waterways through canals. Then came efforts to subsidize and direct the paths of railroads. Town leaders used their governmental authority to determine routes and subsidies, motivated by their private interest in rents.

The people who engaged in this city building have often been celebrated for their inspired vision and "absolute faith". . . . But more important than their personalities, these urban founders were in the business of manipulating place for its exchange values. Their occupations most often were real estate or banking. Even those who initially practiced law, medicine, or pharmacy were rentiers in the making.

[. . .]

The city-building activities of these growth entrepreneurs in frontier towns became the springboard for the much celebrated taming of the American wilderness. The upstart western cities functioned as market, finance, and administrative outposts that made rural pioneering possible. This conquering of the West, accomplished through the machinations of "the urban frontier," was critically bound up with a coordinated effort to gain rents. . . .

Perhaps the most spectacular case of urban ingenuity was the Chicago of William Ogden. When Ogden came to Chicago in 1835, its population was under four thousand. He succeeded in becoming its mayor, its great railway developer, and the owner of much of its best real estate. As the organizer and first president of the Union Pacific (among other railroads) and in combination with his other business and civic roles, he was able to make Chicago (as a "public duty") the crossroads of America, and hence the dominant metropolis of the Midwest. Chicago became a crossroads not only because it was "central" (other places were also in the "middle") but because a small group of people (led by Ogden) had the power to literally have the roads cross in the spot they chose. . . .

This tendency to use land and government activity to make money was not invented in nineteenth-century America, nor did it end then. The development of the American Midwest was only one particularly noticed (and celebrated) moment in the total process. One of the more fascinating instances, farther to the West and later in history, was the rapid development of Los Angeles, an anomaly to many because it had none of the "natural" features that are thought to support urban growth: no centrality, no harbor, no transportation crossroads, not even a water supply. Indeed, the rise of Los Angeles as the preeminent city of the West, eclipsing its rivals San Diego and San Francisco, can only be explained as a remarkable victory of human cunning over the so-called limits of nature. Much of the development of western cities hinged on access to a railroad; the termination of the first continental railroad at San Francisco, therefore, secured that city's early lead over other western towns. The railroad was thus crucial to the fortunes of the barons with extensive real estate and commercial interests in San Francisco – Stanford, Crocker, Huntington, and Hopkins. These men feared the coming of a second cross-country railroad (the southern route), for its urban terminus might threaten the San Francisco investments. San Diego, with its natural port, could become a rival to San Francisco, but Los Angeles, which had no comparable advantage, would remain forever in its shadow. Hence the San Francisco elites used their economic and political power to keep San Diego from becoming the terminus of the southern route. . . . Of course, Los Angeles won in the end, but here again the wiles of boosters were crucial: the Los Angeles interests managed to secure millions in federal funds to construct a port, today the world's largest artificial harbor – as well as federal backing to gain water.

The same dynamic accounts for the other great harbor in the Southwest. Houston beat out Galveston as the major port of Texas (ranked third in the country in 1979) only when Congressman Tom Ball of Houston successfully won, at the beginning of this century, a million-dollar federal appropriation to construct a canal linking landlocked Houston to the Gulf of Mexico. That was the crucial event that, capitalizing on Galveston's susceptibility to hurricanes, put Houston permanently in the lead.

In more recent times, the mammoth federal interstate highway system . . . has similarly made and unmade urban fortunes. To use one clear case, Colorado's leaders made Denver a highway crossroads by convincing President Eisenhower in 1956 to add three hundred miles to the system to link Denver to Salt Lake City by an expensive mountain route. A presidential stroke of the pen removed the prospects of Cheyenne, Wyoming, of replacing Denver as a major western transportation center. In a case reminiscent of the nineteenth-century canal era, the Tennessee-Tolnbigbee Waterway opened in 1985, dramatically altering the shipping distances to the Gulf of Mexico for many inland cities. The largest project ever built by the U.S. Corps of Engineers, the $2 billion project was questioned as a boondoggle in Baltimore, which will lose port business because of it, but praised in Decatur, Alabama, and Knoxville, Tennessee, which expect to profit from it. The opening of the canal cut by four-fifths the distance from Chattanooga, Tennessee, to the Gulf, but did almost nothing for places like Minneapolis and Pittsburgh, which were previously about the same nautical distance from the Gulf as Chattanooga.

Despite the general hometown hoopla of boosters who have won infrastructural victories, not everyone gains when the structural speculators of a city defeat their competition. Given the stakes, the rentier elites would obviously become engulfed by the "booster spirit". . . . Researchers have made little effort to question the linkage between public betterment and growth, even when they could see that specific social groups were being hurt. Zunz reports that in industrializing Detroit, city authorities extended utility service into uninhabitated areas to help development rather than into existing residential zones, whose working-class residents went without service even as they bore the costs (through taxes) of the new installations.

[. . .]

The modern-day good business climate

The jockeying for canals, railroads, and arsenals of the previous century has given way in this one to more complex and subtle efforts to manipulate space and redistribute rents. The fusing of public duty and private gain has become much less acceptable (both in public opinion and in the criminal courts); the replacing of frontiers by complex cities has given important roles to mass media, urban professionals, and skilled political entrepreneurs. The growth machine is less personalized, with fewer local heroes, and has become instead a multifaceted matrix of important social institutions pressing along complementary lines.

With a transportation and communication grid already in place, modern cities typically seek growth in basic economic functions, particularly job intensive ones. Economic growth sets in motion the migration of labor and a demand for ancillary production services, housing, retailing, and wholesaling ("multiplier effects"). Contemporary places differ in the type of economic base they strive to build (for example, manufacturing, research and development, information processing, or tourism). But any one of the rainbows leads to the same pot of gold: more intense land use and thus higher rent collections, with associated professional fees and locally based profits.

Cities are in a position to affect the factors of production that are widely believed to channel the capital investments that drive local growth. They can, for example, lower access costs of raw materials and markets through the creation of shipping ports and airfields (either by using local subsidies or by facilitating state and federal support). Localities can decrease corporate overhead costs through sympathetic policies on pollution abatement, employee health standards, and taxes. Labor costs can be indirectly lowered by pushing welfare recipients into low-paying jobs and through the use of police to constrain union organizing. Moral laws can be changed; for example, drinking alcohol can be legalized (as in Ann Arbor, Mich., and Evanston, Ill.) or gambling can be promoted (as in Atlantic City, N.J.) to build tourism and convention business. Increased utility costs caused by new development can be borne, as they usually are, by the public at large rather than by those responsible for the "excess" demand they generate. Federally financed programs can be harnessed to provide cheap water supplies; state agencies can be manipulated to subsidize insurance rates; local political units can forgive business property taxes. Government installations of various sorts (universities, military bases) can be used to leverage additional development by guaranteeing the presence of skilled labor, retailing customers, or proximate markets for subcontractors. For some analytical purposes, it doesn't even matter that a number of these factors have little bearing on corporate locational decisions (some certainly do; others are debated); just the possibility that they might matter invigorates local growth activism and dominates policy agendas.

Following the lead of St. Petersburg, Florida, the first city to hire a press agent (in 1918) to boost growth, virtually all major urban areas now use experts to attract outside investment. One city, Dixon, Illinois, has gone so far as to systematically contact former residents who might be in a position to help (as many as twenty thousand people) and offer them a finder's fee up to $10,000 for directing corporate investment toward their old home town. More pervasively, each city tries to create a "good business climate." The ingredients are well known in city-building circles and have even been codified and turned into "official" lists for each regional area. The much-used Fantus rankings of business climates are based on factors like taxation, labor legislation, unemployment compensation, scale of government, and public indebtedness (Fantus ranks Texas as number one and New York as number forty-eight). In 1975, the Industrial Development Research Council, made up of corporate executives responsible for site selection decisions, conducted a survey of its members. In that survey, states were rated more simply as "cooperative," "indifferent," or "antigrowth"; the results closely paralleled the Fantus rankings of the same year.

Any issue of a major business magazine is replete with advertisements from localities of all types (including whole countries) striving to portray themselves in a manner attractive to business. Consider these claims culled from one issue of *Business Week* (February 12, 1979):

> New York City is open for business. No other city in America offers more financial incentives to expand or relocate. . . .

The state of Louisiana advertises

> Nature made it perfect. We made it profitable.

On another page we find the claim that "Northern Ireland works" and has a work force with "positive attitudes toward company loyalty, productivity and labor relations." Georgia asserts, "Government should strive to improve business conditions, not hinder them." Atlanta headlines that as "A City Without Limits" it "has ways of getting people like you out of town" and then details its transportation advantages to business. Some places describe attributes that would enhance the life style of executives and professional employees (not a dimension of Fantus rankings); thus a number of cities push an image of artistic refinement. No advertisements in this issue (or in any other, we suspect) show city workers living in nice homes or influencing their working conditions.

While a good opera or ballet company may subtly enhance the growth potential of some cities, other cultural ingredients arc crucial for a good business climate. There should be no violent class or ethnic conflict. Racial violence in South Africa is finally leading to the disinvestment that reformers could not bring about through moral suasion. In the good business climate, the work force should be sufficiently quiescent and healthy to be productive; this was the rationale originally behind many programs in work place relations and public health. Labor must, in other words, he "reproduced," but only under conditions that least interfere with local growth trajectories.

Perhaps most important of all, local publics should favor growth and support the ideology of value-free development. This public attitude reassures investors that the concrete enticements of a locality will be upheld by future politicians. The challenge is to connect civic pride to the growth goal, tying the presumed economic and social benefits of growth in general to growth in the local area. Probably only partly aware of this, elites generate and sustain the place patriotism of the masses. . . . In the nineteenth-century cities, the great rivalries over canal and railway installations were the political spectacles of the day, with attention devoted to their public, not private, benefits. With the drama of the new railway technology, ordinary people were swept into the competition among places, rooting for their own town to become the new "crossroads" or at least a way station.

The celebration of local growth continues to be a theme in the culture of localities. Schoolchildren are taught to view local history as a series of breakthroughs in the expansion of the economic base of their city and region, celebrating its numerical leadership in one sort of production or another; more generally, increases in population tend to be equated with local progress. Civic organizations sponsor essay contests on the topic of local greatness. They encourage public celebrations and spectacles in which the locality name can be proudly advanced for the benefit of both locals and outsiders. They subsidize soapbox derbies, parade floats, and beauty contests to "spread around" the locality's name in the media and at distant competitive sites.

One case can illustrate the link between growth goals and cultural institutions. In the Los Angeles area, St. Patrick's Day parades are held at four different locales, because the city's Irish leaders can't agree on the venue for a joint celebration. The source of the difficulty (and much acrimony) is that these parades march down the main business streets in each locale, thereby making them a symbol of the life of the city. Business groups associated with each of the strips want to claim the parade as exclusively their own, leading to charges by still a fifth parade organization that the other groups are only out to make money. The countercharge, vehemently denied, was that the leader of the challenging business street was not even Irish. Thus even an ethnic celebration can receive its special form from the machinations of growth interests and the competitions among them.

The growth machine avidly supports whatever cultural institutions can play a role in building locality. Always ready to oppose cultural and political developments contrary to their interests (for example, black nationalism and communal cults), rentiers and their associates encourage activities that will connect feelings of community . . . to the goal of local growth. The overall ideological thrust is to deemphasize the connection between growth and exchange values and to reinforce the link between growth goals and better lives for the majority. We do not mean to suggest that the only source of civic pride is the desire to collect rents; certainly the cultural pride of tribal groups predates growth machines. Nevertheless, the growth machine coalition mobilizes these cultural motivations, legitimizes them, and channels them into activites that are consistent with growth goals.

"Los Angeles and the Chicago School: Invitation to a Debate"

from *City and Community* (2002)

Michael Dear

Editors' Introduction

Michael Dear discusses the emergence of the "L.A. School" of urban studies, a concatenation of geographers, sociologists, planners, and architects that considers Los Angeles as the prototype of a new kind of urbanism. The L.A. School is linked with urban political economy but has a closer engagement with postmodern social theory and cultural studies. Along the way, it has forcefully indicted Chicago School human ecology theory for being a narrative of modernist hegemony that served to justify the dominant political, economic and cultural interests in favor of a paradigm of Los Angeles as the quintessential postmodern metropolis, a polycentric, polyglot, and polycultural pastiche that suggests a different paradigm of urban development in the new millennium.

Dear attacks the classic concentric zone formulation of urban development promulgated by Ernest Burgess of the Chicago School. Chicago was the prototype of the world metropolis during the early twentieth century, during which railroad infrastructure concentrated the business district in a central location and transportation lines radiated outwards in a hub-and-spoke fashion into the suburbs. Racial/ethnic and subcultural minorities were subsumed, marginalized, or assimilated both socially and spatially into urban life, while the elite interests ruled the city center, controlling politics, land markets, and the cultural values of hegemonic interests. Since World War II, with the emergence of cities like Los Angeles built around a freeway infrastructure, the business district has become poly-nucleated or multi-polar, and the center has relinquished authority to the hinterlands. The forces of cultural assimilation have also become more fragmented with the arrival of new immigrants and overseas capital investment from Asia, Latin America, and other regions, and the ascendance of Los Angeles as a major command and manufacturing center in the new global economy. Immigrant colonies do not disappear so much anymore with the process of invasion–succession, while ethnic suburbs are emerging on the urban periphery. These "ethnoburbs" are emerging to become new poles of economic and cultural activity, acting as transaction nodes to other world trading regions.

Los Angeles is also a center of postmodern architecture, a showplace for architects such as Frank Gehry (the Disney Music Hall), John Portman (the Bonaventure Hotel), and Charles Moore. There is an "L.A. School" of architecture, design, arts, and even filmmaking. The L.A. School invites a cultural interpretation of Los Angeles as harbinger of twenty-first century urbanism, just as Walter Benjamin described Paris as the capital of the nineteenth century. Ed Soja, in *Thirdspace: Journeys to Los Angeles and Other Real-and-Imagined Places* (Oxford: Blackwell, 1996) and *Postmetropolis: Critical Studies of Cities and Regions* (Oxford: Blackwell, 2000) discusses Los Angeles from the standpoint of identity politics, globalization, and theme parks. Mike Davis pontificates on the experience of Los Angeles as a fortress or citadel for transnational capitalism, in which the underclass and the homeless are marginalized in *City of Quartz* (New York: Verso, 1990) and on the environmental disaster of Los Angeles urbanism in *Ecology of Fear* (New York: Metropolitan Books, 1998).

The German expatriate turned Angeleno, Roger Keil, also contributed to the L.A. School with his book, *Los Angeles: Globalization, Urbanization and Social Struggles* (New York: John Wiley and Sons, 1998). Allen Scott joined with Ed Soja to edit the important reader, *The City: Los Angeles and Urban Theory at the End of the Twentieth Century* (Berkeley, CA: University of California Press, 1996). As Michael Dear discusses, the decentralized mosaic of greater Los Angeles resembles a game board of "keno capitalism." Many white and middle-class residents have abandoned the insecurity and social unrest found in the city center for gated communities in the exurbs.

Michael Dear, along with Allen Scott, published an earlier edited text, *Urbanization and Urban Planning in Capitalist Society* (New York: Methuen, 1981), a cross-national reader on urban political economy that was effectively an early precursor to the new urban sociology in the U.S. He also published an edited reader, with H. Eric Schockman and Greg Hise, called *Rethinking Los Angeles* (Thousand Oaks, CA: Sage Publications, 1996) and, more recently, the book *The Postmodern Urban Condition* (Oxford: Blackwell, 2000). He published an edited reader on the L.A. School, with J. Dallas Dishman, *From Chicago to L.A.: Making Sense of Urban Theory* (Thousand Oaks, CA: Sage Publications, 2002). The selection presented in this reader was published initially in the inaugural issue of *City and Community*. There are five response essays engaging Michael Dear in lively debate.

Michael Dear is Professor of Geography at the University of Southern California, and Director of the Southern California Studies Center. He received his higher education in Regional Science, Town Planning, and Geography in both the United Kingdom and United States. Dear conducts research on Los Angeles, postmodern urbanism, and political and social geography. He is often cited as an authority in geography, and is the author or editor of ten books and over 100 journal articles and reports. He was a Fellow at the Center for Advanced Study in the Behavioral Sciences at Stanford in 1995–96, and held a Guggenheim Fellowship in 1989. He received Honors from the Association of American Geographers in 1995 and, in the same year, received the University of Southern California's Associates Award for highest honors for creativity in research, teaching and service.

More than 75 years ago, the University of Chicago Press published a book of essays entitled *The City: Suggestions for Investigation of Human Behavior in the Urban Environment*. The book is still in print. Six of its 10 essays are by Robert E. Park, then Chair of the University's Sociology Department. There are also two essays by Ernest W. Burgess, and one each from Roderick D. McKenzie and Louis Wirth. In essence, the book announced the arrival of the "Chicago School" of urban sociology, defining an agenda for urban studies that persists to this day. Shrugging off challenges from competing visions, the School has maintained a remarkable longevity that is a tribute to its model's beguiling simplicity, to the tenacity of its adherents who subsequently constructed a formidable literature, and to the fact that the model "worked" in its application to so many different cities over such a long period of time.

The present essay begins the task of defining an alternative agenda for urban studies, based on the precepts of what I shall refer to as the "Los Angeles School." Quite evidently, adherents of the Los Angeles School take many cues from the Los Angeles metropolitan region, or (more generally) from Southern California – a five-county region encompassing Los Angeles, Orange, Riverside, San Bernardino, and Ventura Counties. This exceptionally complex, fast-growing megalopolis is already home to more than 16 million people. It is likely soon to overtake New York as the nation's premier urban region. Yet, for most of its history it has been regarded as an exception to the rules governing American urban development, an aberrant outlier on the continent's western edge.

All this is changing. During the past two decades, Southern California has attracted increasing attention from scholars, the media, and other social commentators. The region has become not the exception to but rather a prototype of our urban future. For many current observers, L.A. is simply confirming what contemporaries knew throughout its history: that the city posited a set of different rules for understanding urban growth. An alternative urban metric is now overdue, since as Joel Garreau (1991, p. 3) observed in his study of edge cities: "Every American city that is growing, is growing in the fashion of Los Angeles."

Just as the Chicago School emerged at a time when that city was reaching new national prominence, Los Angeles is now making its impression on the minds of urbanists across the world. Few argue that the city is unique, or necessarily a harbinger of the future, even though both viewpoints are at some level demonstrable – true. However, at a very minimum, they all assert that Southern California is an unusual amalgam – a polycentric, polyglot, polycultural pastiche that is deeply involved in rewriting American urbanism. Moreover, their theoretical inquiries do not end with Southern California, but are also focused on more general questions concerning broader urban socio-spatial processes. The variety, volume, and pace of contemporary urban change requires the development of alternative analytical frameworks; one can no longer make an unchallenged appeal to a single model for the myriad global and local trends that surround us.

[...]

The particular conditions that have led now to the emergence of a Los Angeles School may be almost coincidental: (1) that an especially powerful intersection of empirical and theoretical research projects have come together in this particular place at this particular time; (2) that these trends are occurring in what has historically been the most understudied major city in the United States; (3) that these projects have attracted the attention of an assemblage of increasingly self-conscious scholars and practitioners; and (4) that the world is facing the prospect of a Pacific century, in which Southern California is likely to become a global capital. The vitality and potential of the Los Angeles School derive from the intersection of these events, and the promise they hold for a renaissance of urban theory.

[...]

THE LOS ANGELES SCHOOL EMERGES

[...]

It was during the 1980s that a group of loosely associated scholars, professionals, and advocates based in Southern California became convinced that what was happening in the region was somehow symptomatic of a broader socio-geographic transformation taking place within the United States as a whole. Their common, but then unarticulated, project was based on certain shared theoretical assumptions, as well as on the view that L.A. was emblematic of a more general urban dynamic. One of the earliest expressions of the emergent "Los Angeles School" came with the appearance in 1986 of a special issue of the journal *Society and Space*, devoted entirely to understanding Los Angeles. In their prefatory remarks to that issue, Allen Scott and Edward Soja (1986, p. 249) referred to L.A. as the "capital of the twentieth century," deliberately invoking Walter Benjamin's designation of Paris as capital of the 19th. They predicted that the volume of scholarly work on Los Angeles would quickly overtake that on Chicago, the dominant model of the American industrial metropolis.

Ed Soja's celebrated tour of Los Angeles (which first appeared in the 1986 *Society and Space* issue, and was later incorporated into his 1989 *Postmodern Geographies*) most effectively achieved the conversion of L.A. from the exception to the rule – the prototype of late 20th-century postmodern geographies:

> What better place can there be to illustrate and synthesize the dynamics of capitalist spatialization? In so many ways, Los Angeles is the place where "it all comes together"... one might call the sprawling urban region ... a prototopos, a paradigmatic place; or a mesocosm, an ordered world in which the micro and the macro, the idiographic and the nomothetic, the concrete and the abstract, can be seen simultaneously in an articulated and interactive combination.
>
> (Soja, 1989, p. 191)

Soja went on to assert that L.A. "insistently presents itself as one of the most informative palimpsests and paradigms of twentieth-century urban development and popular consciousness," comparable to Borges's *Aleph*: "the only place on earth where all places are seen from every angle, each standing clear, without any confusion or blending" (Soja, 1989, p. 248).

As ever, Charles Jencks (1993, p. 132) quickly picked up on the trend toward an L.A.-based urbanism, taking care to distinguish its practitioners from the L.A. school of architecture:

> The L.A. School of geographers and planners had quite a separate and independent formulation in the 1980s, which stemmed from the analysis of the city as a new post-modern urban type. Its themes vary from L.A. as the post-Fordist, post-modern city of many fragments in search of a unity, to the nightmare city of social inequities.

This same group of geographers and planners (accompanied by a few dissidents from other disciplines) gathered at Lake Arrowhead in the San Bernardino Mountains on October 11–12, 1987, to discuss the wisdom of engaging in a Los Angeles School. The participants included, if memory serves, Dana Cuff, Mike Davis, Michael Dear, Margaret FitzSimmons, Rebecca Morales, Allen Scott, Ed Soja, Michael Storper, and Jennifer Wolch. Mike Davis (1989, p. 9) later provided the first description of the putative school:

> I am incautious enough to describe the "Los Angeles School." In a categorical sense, the twenty or so researchers I include within this signatory are a new wave of Marxist geographers – or, as one of my friends put it, "political economists with their space suits on" – although a few of us are also errant urban sociologists, or, in my case, a fallen labor historian. The "School," of course, is based in Los Angeles, at UCLA and USC, but it includes members in Riverside, San Bernardino, Santa Barbara, and even Frankfurt, West Germany.

[. . .]

Mike Davis was, to the best of my knowledge, the first to mention a specific L.A. school of urbanism, and he repeated the claim in his popular contemporary history of Los Angeles, *City of Quartz* (1990).

[. . .]

FROM CHICAGO TO L.A.

The basic primer of the Chicago school was *The City*. Originally published in 1925, the book retains a tremendous vitality far beyond its interest as a historical document. I regard the book as emblematic of a modernist analytical paradigm that remained popular for most of the 20th century: Its assumptions included:

- a "modernist" view of the city as a unified whole, i.e., a coherent regional system in which the center organizes its hinterland;
- an individual-centered understanding of the urban condition; urban process in *The City* is typically grounded in the individual subjectivities of urbanites, their personal choices ultimately explaining the overall urban condition, including spatial structure, crime, poverty, and racism; and
- a linear evolutionist paradigm, in which processes lead from tradition to modernity, from primitive to advanced, from community to society, and so on.

There may be other important assumptions of the Chicago School, as represented in *The City*, that are not listed here. Finding them and identifying what is right or wrong about them is one of the tasks at hand, rather than excoriating the book's contributors for not accurately foreseeing some distant future.

The most enduring of the Chicago School models was the zonal or *concentric ring theory*, an account of the evolution of differentiated urban social areas by E. W. Burgess (1925). Based on assumptions that included a uniform land surface, universal access to a single-centered city, free competition for space, and the notion that development would take place outward from a central core, Burgess concluded that the city would tend to form a series of concentric zones.

[. . .]

Other urbanists subsequently noted the tendency for cities to grow in star-shaped rather than concentric form, along highways that radiate from a center with contrasting land uses in the interstices. This observation gave rise to a *sector theory* of urban structure, an idea advanced in the late 1930s by Homer Hoyt (1933, 1939), who observed that once variations arose in land uses near the city center, they tended to persist as the city expanded. Distinctive sectors thus grew out from the CBD, often organized along major highways.

Hoyt emphasized that "non-rational" factors could alter urban form, as when skillful promotion influenced the direction of speculative development. He also understood that the age of the buildings could still reflect a concentric ring structure, and that sectors may not be internally homogeneous at one point in time.

The complexities of real-world urbanism were further taken up in the multiple nuclei theory of C. D. Harris and E. Ullman (1945). They proposed that cities have a cellular structure in which land-uses develop around multiple growth-nuclei within the metropolis – a consequence of accessibility-induced variations in the land-rent surface and agglomeration (dis)economics. Harris and Ullman also allow that real-world urban structure is determined by broader social and economic forces, the influence of history, and international influences. But whatever the precise reasons for their origin, once nuclei have been established, general growth forces reinforce their pre-existing patterns.

Much of the urban research agenda of the 20th century has been predicted on the precepts of the concentric zone, sector, and multiple-nuclei theories of urban structure. Their influences can be seen directly in factorial ecologies of intra-urban structure, land-rent models, studies of urban economies and diseconomies of scale, and designs for ideal cities and neighborhoods. The specific and persistent popularity of the Chicago concentric ring model is harder to explain, however, given the proliferation of evidence in support of alternative theories. The most likely reasons for its endurance (as I have mentioned) are related to its beguiling simplicity and the enormous volume of publications produced by adherents of the Chicago School. . . .

In the final chapter of *The City*, the same Louis Wirth (1925) had already provided a magisterial review of the field of urban sociology, entitled (with deceptive simplicity and astonishing self-effacement) "A Bibliography of the Urban Community." But what Wirth does in this chapter, in a remarkably prescient way, is to summarize the fundamental premises of the Chicago School and to isolate two fundamental features of the urban condition that was to rise to prominence at the beginning of the 21st century. Specifically, Wirth establishes that the city lies at the center of, and provides the organizational logic for, a complex regional hinterland based on trade. But he also notes that the development of "satellite cities" is characteristic of the "latest phases" of city growth and that the location of such satellites can exert a "determining influence" on the direction of growth (1925, p. 185). He further observes that modern communications have transformed the world into a "single mechanism," where the global and the local intersect decisively and continuously (Wirth, 1923, p. 186).

And there, in a sense, you have it. In a few short paragraphs, Wirth anticipates the pivotal moments that characterize Chicago-style urbanism, those primitives that eventually will separate it from an L.A.-style urbanism. He effectively foreshadowed *avant la lettre* the shift from what I term a "modernist" to a "postmodern" city, and, in so doing, the necessity of the transition from the Chicago to the Los Angeles School. *For it is no longer the center that organizes the urban hinterlands, but the hinterlands that determine what remains of the center.* The imperatives toward decentralization (including suburbanization) have become the principal dynamic in contemporary cities; and the 21st century's emerging world cities (including L.A.) are ground-zero loci in a communications-driven globalizing political economy. From a few, relatively humble first steps, we gaze out over the abyss – the yawning gap of an intellectual fault line separating Chicago from Los Angeles.

CONTEMPORARY URBANISMS IN SOUTHERN CALIFORNIA

I turn now to review the empirical evidence of recent urban developments in Southern California. In this task, I take my lead from what exists rather than what may be considered as a normative taxonomy of urban research. From this, I move quickly to a synthesis that is prefigurative of a proto-postmodern urbanism that serves as a basis for a distinctive L.A. school of urbanism.

Edge cities

Joel Garreau noted the central significance of Los Angeles in understanding contemporary metropolitan growth in the United States. He refers to L.A. as the "great-granddaddy" of edge cities, claiming there are 26 of them within a five-county area in Southern California (Garreau, 1991, p. 9). For Garreau, edge cities represent the crucible of America's urban future. . . . One essential feature of the edge city is that politics is not yet established there. Into the political vacuum moves a "shadow government" – a privatized protogovernment that is essentially a plutocratic alternative to normal politics. Shadow governments can tax, legislate for, and police their communities, but they are rarely accountable, are responsive primarily to wealth (as opposed to numbers of voters), and subject to few constitutional constraints.

Privatopia

Privatopia, perhaps the quintessential edge city residential form, is a private housing development based in common-interest developments (CIDs) and administered by homeowner associations. There were fewer than 500 such associations in 1964; by 1992, there were 130,000 associations privately governing approximately 32 million Americans. . . . In her futuristic novel of L.A. wars between walled-community dwellers and those beyond the walls, Octavia Butler (1993) envisioned a dystopian privatopian future. It includes a balkanized nation of defended neighborhoods at odds with one another, where entire communities are wiped out for a handful of fresh lemons or a few cups of potable water; where torture and murder of one's enemies is common; and where company-town slavery is attractive to those who are fortunate enough to sell their services to the hyper-defended enclaves of the very rich.

Cultures of heteropolis

One of the most prominent sociocultural tendencies in contemporary Southern California is the rise of minority populations. Provoked to comprehend the causes and implications of the 1992 civil disturbances in Los Angeles Charles Jencks zeroes in on the city's diversity as the key to L.A.'s emergent urbanism: "Los Angeles is a combination of enclaves with high identity, and multienclaves with mixed identity, and, taken as a whole, it is perhaps the most heterogenous city in the world" (Jencks, 1993, p. 32).

City as theme park

California in general, and Los Angeles in particular, have often been promoted as places where the American (suburban) Dream is most easily realized. Its oft-noted qualities of optimism and tolerance coupled with a balmy climate have given rise to an architecture and society fostered by a spirit of experimentation, risk-taking, and hope. Many writers have used the "theme park" metaphor to describe the emergence of such variegated cityscapes. . . . Disneyland is the archetype, described by Sorkin (1992, p. 227) as a place of "Taylorized fun," the "Holy See of Creative Geography." What is missing in this new cybernetic suburbia is not a particular building or place, but the spaces between, i.e., the connections that make sense of forms. What is missing, then, is connectivity and community. . . .

Fortified city

The downside of the Southern Californian dream has, of course, been the subject of countless dystopian visions in histories, movies, and novels. In one powerful account, Mike Davis (1992a) noted how Southern Californians' obsession with security has transformed the region into a fortress. This shift is accurately manifested in the physical form of the city, which is divided into fortified cells of affluence and places of terror where police battle the criminalized poor. These urban phenomena, according to Davis (1992a, p. 155), have placed Los Angeles "on the hard edge of postmodernity." The dynamics of fortification involve the omnipresent application of high-tech policing methods to protect the security of gated residential developments and panopticon malls. It extends to space policing, including a proposed satellite observation capacity that would create an invisible Haussmannization of Los Angeles. In the consequent carceral city, the working poor and destitute are spatially sequestered on the mean streets, and excluded from the affluent forbidden cities through security by design.

Interdictory spaces

Elaborating upon Davis' fortress urbanism, Steven Flusty (1994) observed how various types of fortification have extended a canopy of suppression and surveillance across the entire city. His taxonomy identifies how spaces are designed to exclude by a combination of their function and cognitive sensibilities. . . . One consequence of the socio-spatial differentiation described by Davis and Flusty is an acute fragmentation of the urban landscape. Commentators who remark upon the strict division of residential neighborhoods along race and class lines miss the fact that L.A's microgeography is incredibly volatile and varied. In many neighborhoods, simply turning a street corner will lead the pedestrian/driver into totally different social and physical configurations. . . .

Historical geographies of restructuring

. . . In his history of Los Angeles between 1965 and 1992, Soja (1996) attempts to link the emergent patterns of urban form with underlying social processes. He identified six kinds of restructuring, which together define the region's contemporary urban process. In addition to *Exopolis* (noted earlier), Soja lists: *Flexcities*, associated with the transition to post-Fordism, especially deindustrialization and the rise of the information economy; and *Cosmopolis*, referring to the globalization of Los Angeles both in terms of its emergent world city status and its internal multicultural diversification. According to Soja, peripheralization, post-Fordism, and globalization together define the experience of urban restructuring in Los Angeles. Three specific geographies are consequent upon these dynamics: *Splintered Labyrinth*, which describes the extreme forms of social, economic, and political polarization characteristic of the postmodern city; *Carceral City*, referring to the new "incendiary urban geography" brought about by the amalgam of violence and police surveillance; and *Simcities*, the term Soja uses to describe the new ways of seeing the city that are emerging from the study of Los Angeles – a kind of epistemological restructuring that foregrounds a postmodern perspective.

Fordist versus Post-Fordist regimes of accumulation and regulation

. . . In a series of important books, Allen Scott (1988a, 1988b, 1993, 2000) has portrayed the burgeoning urbanism of Southern California as a consequence of this deep-seated structural change in the capitalist political economy. Scott's basic argument is that there have been two major phases of urbanization in the United States. The first related to an era of Fordist mass production, during which the paradigmatic cities of industrial capitalism (Detroit, Chicago, Pittsburgh, etc.) coalesced around industries that were themselves based on ideas of mass production. The second phase is associated with the decline of the Fordist era and the rise of a post-Fordist "flexible production" (what some refer to as "flexible accumulation"). This is a form of industrial activity based on small-size, small-batch units of (typically sub-contracted) production that are nevertheless integrated into clusters of economic activity. Such clusters have been observed in two manifestations: labor-intensive craft forms (in Los Angeles, typically garments and jewelry); and high technology (especially the defense and aerospace industries). According to Scott, these so-called "technopoles" until recently constituted the principal geographical loci of contemporary (sub)urbanization in Southern California. An equally important facet of post-Fordism is the significant informal sector that mirrors the gloss of the high-tech sectors. Post-Fordist regimes of accumulation are associated with analogous regimes of regulation, or social control. . . .

Globalization

Needless to say, any consideration of the changing nature of industrial production sooner or later must encompass the globalization question. In his reference to the global context of L.A's localisms, Mike Davis (1992b) claims that if L.A. is in any sense paradigmatic, it is because the city condenses the intended and unintended spatial consequences of a global post-Fordism. He insists that there is no simple master-logic of restructuring, focusing instead on two key localized macroprocesses: the overaccumulation in Southern California of bank and real estate capital principally from the East Asian trade surplus of the 1980s; and the reflux of low-wage manufacturing and labor-intensive service industries following upon immigration from Mexico and Central America. For instance, Davis (1992b, p. 26) noted how the City of Los Angeles used tax dollars gleaned from international capital investments to subsidize its downtown (Bunker Hill) urban renewal, a process he refers to as "municipalized land speculation." Through such connections, what happens today in Asia and Central America will tomorrow have an effect in Los Angeles. This global/local dialectic has already become an important (if somewhat imprecise) leitmotif of contemporary urban theory, most especially via notions of "world cities" and global "city-regions" (Scott, 1998, 2001).

Politics of nature

The natural environment of Southern California has been under constant assault since the first colonial settlements. Human habitation on a metropolitan scale has only been possible through a widespread manipulation of nature, especially the control of water resources in the American West. On the one hand, Southern Californians tend to hold a grudging respect for nature, living as they do adjacent to one of the earth's major geological hazards, and in a desert environment that is prone to flood, landslide, and fire. On the other hand, its inhabitants have been energetically, ceaselessly, and often carelessly unrolling the carpet of urbanization over the natural landscape for more than a century. This uninhibited occupation has engendered its own range of environmental problems, most notoriously air pollution, but also issues related to habitat loss and encounters between humans and other animals. . . .

LOS ANGELES AS POSTMODERN URBANISM

If all these observers of the Southern California scene could talk with each other, how might they synthesize their visions? At the risk of misrepresenting their work, I can suggest a synthesis that outlines a "proto-postmodern" urban process. It is driven by a global restructuring that is permeated and balkanized by a series of interdictory networks; whose populations are socially and culturally heterogeneous, but politically and economically polarized; whose residents are educated and persuaded to the consumption of dreamscapes even as the poorest are consigned to carceral cities; whose built environment, reflective of these processes consists of edge cities, privatopias, and the like; and whose natural environment is being erased to the point of unlivability while at the same time providing a focus for political action.

[. . .]

The Los Angeles School is distinguishable from the Chicago precepts (as noted above) by the following counter-propositions:

- Traditional concepts of urban form imagine the city organized around a central core; in a revised theory, the urban peripheries are organizing what remains of the center.
- A global, corporate-dominated connectivity is balancing, even offsetting, individual-centered agency in urban processes.
- A linear evolutionist urban paradigm has been usurped by a nonlinear, chaotic process that includes pathological forms such as transnational criminal organizations, common-interest developments (CIDs), and life-threatening environmental degradation (e.g., global warming).

[. . .]

"Keno capitalism" is the synoptic term that Steven Flusty and I have adopted to describe the spatial manifestations that are consequent upon the (postmodern) urban condition implied by these assumptions (see Figure 1). Urbanization is occurring on a quasi-random field of opportunities, in which each space is (in principle) equally available through its connection with the information superhighway (Dear and Flusty, 1998). Capital touches down as if by chance on a parcel of land, ignoring the opportunities on intervening lots, thus sparking the development process. The relationship between development of one parcel and non-development of another is a disjointed, seemingly unrelated affair. While not truly a random process, it is evident that the traditional, center-driven agglomeration economies that have guided urban

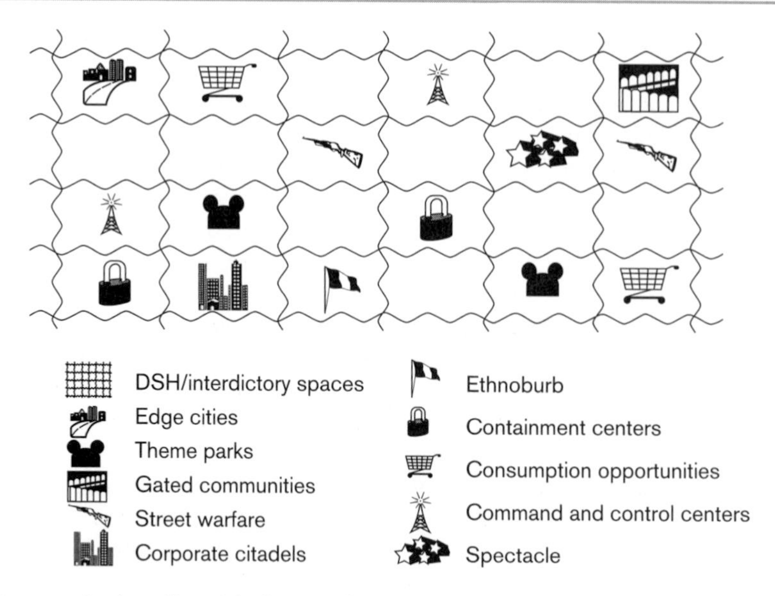

Figure 1 The "keno gaming board" model of postmodern Los Angeles

development in the past no longer generally apply. Conventional city form, Chicago-style, is sacrificed in favor of a non-contiguous collage of parcelized, consumption-oriented landscapes devoid of conventional centers yet wired into electronic propinquity and nominally unified by the mythologies of the (dis)information superhighway.

[...]

INVITATION TO A DEBATE

In these postmodern times, the gesture to a Los Angeles School might appear to be a deeply contradictory intellectual strategy: A "school" has semantic overtones of codification and hegemony; it has structure and authority. Modernists and postmodernists alike might shudder at the irony implied by these associations. And vet, ultimately, I am comfortable in proclaiming the existence of an L.A. school of urbanism for two reasons. First, the Los Angeles School exists as a body of literature, as this essay attests. It exhibits an evolution through history, beginning with analysis of Los Angeles as an aberrant curiosity distinct from other forms of urbanism. The tone of that history has shifted gradually to the point that the city is now commonly represented as indicative of a new form of urbanism supplanting the older forms against which Los Angeles was once judged deviant. Second, the Los Angeles School exists as a discursive strategy demarcating a space both for the exploration of new realities and for resistance to old hegemonies. It is proving to be far more successful than its detractors at explaining the form and function of the urban.

[...]

The fragmented and globally oriented nature of the Los Angeles School counters any potential for a new hegemony. The avowal of a Los Angeles School can become a decolonizing, postcolonial impulse, even as it alerts us to new colonialisms lurking along the historical path. Those who worry about the hegemonic intent of a Los Angeles School may rest assured that its adherents are in fact pathologically anti-leadership. Nor will everyone who writes on L.A. readily identify as a member of the Los Angeles School; some adamantly reject such a notion (e.g., Ethington and Meeker, 2001). The programmatic intent of the Los Angeles School remains fractured, incoherent, and idiosyncratic even to its constituent scholars, who most often perceive themselves as occupying a place on the periphery rather than at the center. The Los Angeles School promotes inclusiveness by inviting as members all those who take Los Angeles as a worthy object of study and a source of insight into the nature of contemporary urbanism. Such a school evades dogma by including divergent empirical and theoretical approaches rooted in philosophies, both modern and postmodern, ranging from Marxist to Libertarian. Admittedly; such a school will be a fragmentary and loosely connected entity, always on the verge of disintegration – but, then again, so is Los Angeles itself.

A unified, consensual description of Los Angeles is equally unlikely, since it would necessitate excluding a plethora of valuable readings on the region. For instance, numerous discursive battles have been fought in L.A. since the events of April 1992 to decide what term best describes them or, more cynically, which term most effectively recasts them as a weapon adaptable to a particular rhetorical arsenal. Those who read the events as a spontaneous, visceral, opportunistic reaction to the acquittal of Rodney King employ the term *riot*. For those who read the events within the context of economic evisceration and social polarization, the term *uprising* is preferred. And those who see in them a more conscious political intentionality apply the term *rebellion*. For its part, civic authority skirts these issues by relying on the supposedly depoliticized term, *civil unrest*. But those concerned with the perspective of Korean participants, literally caught in the middle of the turmoil itself as well as the subsequent rhetoric wars, deploy the Korean tradition of naming an occurrence by its principal date and so make use of the term, *Sa-I-Gu*. Which name is definitive? The polyvocality of the Los Angeles School permits us to replace the question, "Which is it?" with, "Which is it, at which stage of events, at which location in the region, and from whose perspective?" Such an approach may well entail a loss of clarity and certitude, but in exchange it offers a richness of description and interpretation that would otherwise be forfeited in the name of achieving an "official" narrative.

Finally, the temptation to adopt L.A. as a world city template is avoidable because the urban landscapes of Los Angeles are not necessarily original to L.A The luxury compound atop a matrix of impoverished misery, and self-contained communities of fortified homes can also he found in places like Manila and São Paulo. Indeed, Anthony King has suggested that all things ascribed to postmodern urbanism can be seen decades earlier in the principal cities of the colonial world. The Los Angeles School justifies a presentation of L.A. not as *the* model of contemporary urbanism, nor as the privileged locale whence a cabal of regal theoreticians issue proclamations about the way things really are, but as one of a number of space-time geographical prisms through which current processes of urban (re)formation may be advantageously viewed. Hence, the literature of the Los Angeles School largely (although not exclusively) shows itself to be less concerned with looking to L.A. for models of the urban, and more about looking for contemporary expressions of the urban in L.A. Thus, the school and its concepts of contemporary Angeleno urbanism do not represent an emerging vision of contemporary urbanism in total; instead they are but one component in a new comparative urban studies working out of Los Angeles but inviting the participation of (and placing equal importance upon) the on-going experiences and voices of Tijuana, São Paulo, Hong Kong, and the like (cf. Sassen, 1991).

[. . .]

REFERENCES

Burgess, E. W. (1925) "The Growth of the City," in R. E. Park, E. W. Burgess, and R. McKenzie, *The City: Suggestions of Investigation of Human Behavior in the Urban Environment*, pp. 47–62. Chicago, IL: University of Chicago Press.

Butler, O. E. (1993) *Parable of the Sower*. New York: Four Walls Eight Windows.

Davis, M. (1989) "Homeowners and Homeboys: Urban Restructuring in LA," *Enclitic*, Summer, 9–16.

Davis, M. (1990) *City of Quartz: Excavating the Future in Los Angeles*. New York: Verso.

Davis, M. (1992a) "Fortress Los Angeles: The Militarization of Urban Space," in M. Sorkin (ed.), *Variations on a Theme Park*, p. 155. New York: Noonday Press.

Davis, M. (1992b) "*Chinatown* Revisited? The 'Internationalization' of Downtown Los Angeles," in Reid, D. (ed.), *Sex, God and Death in L.A.*, pp. 19–53. New York: Pantheon Books.

Dear, M. (2000) *The Postmodern Urban Condition*. Oxford: Blackwell Publishers.

Dear, M. (ed.) (2001) *From Chicago to LA: Making Sense of Urban Theory*. Thousand Oaks, CA: Sage Publications.

Dear, M. and Flusty S. (1998) "Postmodern Urbanism," *Annals, Association of American Geographers*, 88 (1), 50–72.

Dear, M. and Flusty, S. (eds) (2002) *The Spaces of Postmodernity: A Reader in Human Geography*. Oxford: Blackwell Publishers.

Dear, M. J., Schockman, H. E., and Hise, G. (eds) (1996) *Rethinking Los Angeles*. Thousand Oaks, CA: Sage Publications.

Ethington, P. and Meeker, M. (2001) "'*Saber y Conocer*': The Metropolis of Urban Inquiry," in M. Dear (ed.), *From Chicago to L.A.: Making Sense of Urban Theory*, pp. 403–420. Thousand Oaks, CA: Sage Publications.

Flusty, S. (1994) *Building Paranoia. The Proliferation of Interdictory Space and the Erosion of Spatial Justice*. West Hollywood, CA: Los Angeles Forum for Architecture and Urban Design.

Garreau, J. (1991) *Edge City: Life on the New Frontier*. New York: Anchor Books.

Hoyt, H. (1933) *One Hundred Years of Land Values in Chicago*. Chicago, IL: University of Chicago Press.

Hoyt, H. (1939) *The Structure and Growth of Residential Neighborhoods in American Cities*. Washington, DC: U.S. Federal Housing Administration.

Jencks, C. (1993) *Heteropolis: Los Angeles, the Riots and the Strange Beauty of Hetero-Architecture*. New York: St. Martin's Press.

Sassen, S. (1991) *The Global City*, Princeton, NJ: Princeton University Press.

Scott, A. J. (1988a) *New Industrial Spaces: Flexible Production Organization and Regional Development in North America and Western Europe*. London: Pion.

Scott, A. J. (1988b) *Metropolis: From the Division of Labor to Urban Form*. Berkeley, CA: University of California Press.

Scott, A. J. (1993) *Technopolis: High-Technology Industry and Regional Development in Southern California*. Berkeley, CA: University of California Press.

Scott, A. J. (1998) *Regions and the World Economy: The Coming Shape of Global Production, Competition, and Political Order*. Oxford: Oxford University Press.

Scott, A. J. (2000) *The Cultural Economy of Cities*. London: Sage Publications.

Scott, A. J. (ed.) (2001) *Global City-Regions: Trends, Theory, Policy*. Oxford: Oxford University Press.

Scott, A. J. and Soja, E. W. (1986) "Los Angeles: Capital of the Late 20th Century," *Society and Space*, 4, 249–254.

Scott, A. J. and Soja, E. W. (eds) (1996) *The City: Los Angeles and Urban Theory at the End of the Twentieth Century*. Berkeley, CA: University of California Press.

Soja, E. W. (1989) *Postmodern Geographies: The Reassertion of Space in Critical Social Theory*. New York: Verso.

Soja, E. (1992) "Inside Exopolis: Scenes from Orange County," in M. Sorkin (ed.), *Variations on a Theme Park*. New York: Noonday Press.

Soja, E. (1996) "Los Angeles 1965–1992: The Six Geographies of Urban Restructuring," in A. J. Scott and E. Soja (eds), *The City: Los Angeles & Urban Theory at the End of the Twentieth Century*, pp. 426–462. Los Angeles, CA: University of California Press.

Sorkin, M. (ed.) (1992) *Variations on a Theme Park. The New American City and the End of Public Space*. New York: Hill and Wang.

Wirth, L. (1925) "A Bibliography of the Urban Community" in R. E. Park, E. W. Burgess, and R. McKenzie, *The City: Suggestions of Investigation of Human. Behavior in the Urban Environment*, pp. 161–228. Chicago, IL: University of Chicago Press.

Wirth, L. (1938) "Urbanism as a Way of Life," *American Journal of Sociology*, XLIV (1), 1–24.

PART THREE

Inequality and Social Difference

Plate 6 East 100th Street, New York City, 1966. Children in Spanish Harlem tenement by Bruce Davidson (reproduced by permission of Magnum Photos).

Plate 7 Homeless in São Paulo, 2003 by Ferdinando Scianna (reproduced by permission of Magnum Photos).

INTRODUCTION TO PART THREE

The city is a landscape of inequality and social difference. Our urban society is divided by differences of socio-economic class, race/ethnicity, and social status. The dominant elite members of our society occupy the prime spaces of the central business district and affluent districts and neighborhoods such as Chicago's Gold Coast, New York's Upper East Side "silk stocking district" and the Hollywood hills of Los Angeles. The poor and the homeless inhabit the marginal spaces of the ghetto, the barrio, and skid row in places such as South Central Los Angeles, East Los Angeles, and New York's Bowery. These marginal spaces proliferate in the "inner city" and the "zone in transition" in the interstitial spaces surrounding the dominant spaces of the central city. The presence of slums, skid rows, and other urban "badlands" of the cities are not just symptoms or containers of social inequality. There is a dialectic interaction between space and society. Segregation in poor and deteriorated places disadvantages these people from access to good jobs, housing, schools, and hospitals. The poor and the homeless do not have opportunities for good life chances in the marginal spaces of the city. This social isolation reproduces poverty through the generations. Spatial immobility acts as a barrier to social mobility.

African American and Latino underclass people face a stigmatization and social exclusion very similar to that experienced by the homeless and other socially marginalized people. They are labeled as gangsters, drug dealers, hustlers, and criminals by the media and general public. They inhabit similar marginal spaces of isolation and containment. Some Asian spaces like Chinatown or Little Saigon are also mapped in the public consciousness as places of vice, corruption, and moral depravity. The underclass, homeless and other social minorities inhabit the no man's lands of the city. The Gypsies of Europe, and witches of colonial New England historically generated similar fears as a threat to public health and social order. The inner city is now the site that is labeled as that of the "other" in our contemporary capitalist societies. Racial minority gangsters are our modern folk devils. Trumped up threats of the drug trade, prostitution and AIDS create contemporary moral panics. The poor are feared as a source of disease and contagion, moral defilement, and criminality. Inner city slums are seen as discrepant places.

The problems of American inner cities became an increasing focus of urban sociology beginning in the 1960s. The outmovement of industry and residences to the suburbs had exposed many central city neighborhoods to disinvestments, abandonment and decay. The emergence of an urban "underclass" suffering dilapidated living conditions and crime became associated with neighborhoods such as the South Bronx and Spanish Harlem in New York City and the South Side of Chicago. The sense of urban crisis was further inflamed with the eruption of physically destructive "race riots" in inner city "ghettos" such as Watts and South Central Los Angeles in the 1960s and 1990s and the Overtown district of Miami in the 1980s. These conflagrations of anger and violence among poor minorities gave voice to a growing sense of racial injustice and class inequality in urban America. The urban crisis stimulated "white flight" into new suburban and exurban areas, including gated communities. People in the suburbs have a generalized perception of the inner city as a place of crime, poverty, and racial/social conflict.

Within the field of urban sociology, these trends have stimulated lively research on the concept of the "ghetto underclass," the population of persons and places that are mired in a social world of recurring poverty. Ghetto individuals and households are challenged by chronic unemployment or underemployment, poor access to housing, education and health care, and tendencies toward street crime, substance abuse, welfare dependency, out of wedlock births, and female-headed families. Attention has focused on how the spatial isolation of the underclass fosters their economic marginalization and confronts them with social problems such as crime, substance abuse, and poor living conditions. Living in the ghetto severely reduces the access of residents to good health, education, jobs, and overall life chances. There is a seminal debate about the relative impact of racial versus class factors in the spatial segregation of the urban poor.

William Julius Wilson presented a thesis in *The Declining Significance of Race* (University of Chicago Press, 1976) that social class factors were supplanting racial factors as the main problems facing contemporary Black America. In *The Truly Disadvantaged* (University of Chicago Press, 1984), Wilson articulates the emergence of the urban underclass in late twentieth century America. He explains that an unintended consequence of the successful integration of African Americans into middle-class suburbs is the contemporaneous emergence of a black underclass in the central cities. He recognizes the exodus of the middle class from inner city Black America has resulted in a loss of the "social buffer" of stable working families and community leaders, such as preachers, teachers, and small business owners. They have left the underclass in a condition of "social isolation" with declining life opportunities.

In their co-written chapter, Wacquant and Wilson collaborate on a new analysis that draws central attention to the phenomenon of "hyperghettoization" that confronts Black America as a "crisis" of racial and class exclusion. They emphasize that massive job losses have beset the inner city through branch plant relocations and urban disinvestment and reinvestment in suburban locations, the Sunbelt, and overseas. This deindustrialization process in the metropolitan economy has involved a shift from traditional manufacturing industries to service employment or knowledge-intensive industries. Inner city residents can no longer rely on relatively high-wage factory employment that does not require higher education, in which they acquire on-the-job training. They are relegated to low-paying service sector employment in such arenas as fast-food restaurants and building security. They suffer a jobs–skills mismatch in the postindustrial economy. They cannot acquire financial capital and homeownership has declined. The loss of social capital in the hyperghetto leaves them in social isolation, without access to good educational resources or job networks. The loss of social capital correlates with the decline of local businesses, good schools, and stable neighborhoods of working families.

The persistence of the ghetto contradicts some basic predictions of human ecology theory, namely that racial/ethnic enclaves and colonies of the inner city would eventually dissipate with the upward social mobility and cultural assimilation of minorities and immigrants into American cities and society. Douglas Massey later articulates this phenomenon as the "spatial assimilation" thesis in writings such as "Ethnic Residential Segregation: a Theoretical and Empirical Review," *Sociology and Social Research* 69 (1985): 315–350. Upward social mobility into better jobs and social status is connected with spatial mobility into better homes and neighborhoods. The continuing spatial entrapment of minorities in the inner city is a main factor in their social immobility.

Douglas Massey and Nancy Denton argue that the black ghetto was a conscious creation of white people seeking to isolate and contain poor African Americans during the Jim Crow era of the early twentieth century. Despite the efforts to end residential segregation and integrate blacks, a continuing more serious form of "hypersegregation" has emerged in American cities. The prosecution of racial discrimination via civil rights legislation is not effective enough in housing and financial markets to put a stop to racial steering by realtors, redlining practices by some banks and mortgage companies, and predatory lending by others. The Fair Housing Acts address individual but not the structural roots of discrimination in our financial and housing markets. Massey and Denton find that, especially in older Midwestern and northern cities, underclass communities grew in the 1970s and 1980s and segregation increased. White flight has increasingly alienated white suburban voters from the concerns of the inner cities, and they oppose public spending that would assist the urban underclass. Segregation has undermined the potential for interracial coalitions, and America is fragmented into "white suburbs and chocolate cities." In the era of hypersegregation, the ghetto is politically isolated.

The social and spatial isolation intensifies the creation of an alternative "oppositional culture" or ghetto social world. Residents reject mainstream values from a system that marginalizes them and they embrace unconventional attitudes and operate in informal and underground economic activities. The sense of an alternative culture is intensified through the growth of Ebonics or Black English vernacular. This is a fascinating and rich cultural phenomenon but may impede the successful mainstreaming of underclass children into our society. There is a correlation between spatial and cultural/linguistic isolation.

Loïc Wacquant compares the French Northern African underclass and the African American underclass and finds an ensemble of historical and structural differences in the way that the poor have become marginalized in Paris and Chicago. They have very different experiences of stigmatization relating to both racial characteristics and the public perception of their communities, which are the Parisian "Red Belt" city of La Courneuve and the "Black Belt" neighborhood on the Southside of Chicago. Wacquant found his French informants to have a stronger sense of political and social inclusion, while the Americans felt more marginalized. This expresses some differences between the political culture of socialism and collectivism in Europe versus the ethos of free enterprise and individualism in America.

The homeless tramp is the focus of the following article by James S. Duncan. The homeless and other social minorities face a condition of spatial and social exclusion very similar to that experienced by the underclass. Their presence is a threat to established conventions of behavior and social order, and people fear them as a health or crime hazard. Our vagrancy laws restrict their movement because an address or property ownership is a prerequisite to the rights and entitlements of full citizenship. Thus they are segregated into marginal spaces such as alleyways and back roads and jurisdictional voids commonly known as "skid rows." Their presence may be tolerated in some prime public spaces if they operate under cover or with acceptable etiquette. The segregation of the homeless and related minorities such as hustlers, drug addicts, and runaways in the no man's lands of the city permits their containment by charitable institutions and the state. The recent gentrification and revival of central cities threatens to displace the homeless from marginal spaces under threat of redevelopment.

Immigrant colonies are another phenomenon of racial/ethnic residential segregation. They have grown increasingly since the 1960s, especially among Asian and Latino immigrants, and are an outgrowth of changes in immigration law and the globalization of the U.S. economy. These immigrant enclaves include Chinatown, Koreatown, Little Tokyo, Little Saigon, Little Havana, and Little Haiti. They offer an alternative sub-economy separate from the segmented mainstream economy, which is split into an upper tier of jobs with good mobility ladders, and a lower tier of dead-end jobs in which minorities predominate. These ethnic enclaves offer a protected sector for immigrants newly arrived without English language skills, good education, or official papers. Co-ethnic bosses profit through their ability to self-exploit co-ethnics in arduous sweatshop conditions with marginal benefits and labor rights. Immigrant enclaves create jobs and revenue through the phenomenon of the economic multiplier.

Some immigrant entrepreneurs, however, focus on certain occupational niches as economic and social middlemen between dominant white groups and poor minorities. Chinese, Korean, and Indian immigrants commonly operate small businesses, groceries, liquor stores, and motels. They fulfill a function undesired by white elites. They act as a social buffer between the dominant and oppressed groups of a society and, in situations of crisis, may bear the brunt of underclass anger, as seen in the Black/Korean violence that followed the Rodney King disturbances of 1992 in Los Angeles. Korean/Black conflict has been represented in popular cultural texts such as Ice Cube's song "Black Korea" (from his 1991 album, "Death Certificate") and the Spike Lee movie, *Do the Right Thing* (1989). A comparable portrayal of intergroup conflict between South Asian Indians and African Americans in the U.S. South can be seen in Mina Nair's film, *Mississippi Masala* (1992).

"The Cost of Racial and Class Exclusion in the Inner City"

from *Annals of the American Academy of Political and Social Science* (1989)

Loïc J. D. Wacquant and William Julius Wilson

Editors' Introduction

William Julius Wilson is one of the leading black sociologists and one of the most influential thinkers on issues of urban poverty, race, and social policy in America. His first major contribution to the national debate on the status of African Americans in the U.S. was *The Declining Significance of Race* (University of Chicago Press, 1976) in which he argued that socioeconomic issues were superseding racial issues as the main problems confronting black urban America. He applied his ideas more specifically to the conditions of the urban black poor with his second book, *The Truly Disadvantaged: The Inner City, the Underclass, and Public Policy* (University of Chicago Press, 1987).

In *The Truly Disadvantaged*, Wilson argues that the black ghetto has become a much more dangerous, deprived and socially disorganized place across the course of the twentieth century. He begins with a discussion of the problem of labeling; the term "underclass" like the phrase "culture of poverty," has been used by political conservatives since the 1980s to blame the victims of urban poverty for their own plight. Wilson repudiates the arguments of political conservatives while challenging liberals to reestablish control of public discourse concerning the underclass. He analyzes the effect of structural economic change and the suburbanization of the black middle class in concentrating the problems of the black poor in the inner cities. He asserts that the urban black poor suffer from a "tangle of pathologies" and live in "social isolation" from the mainstream of social life in America. He also discusses the merits of social policies of universalism versus targeted income-tested or race-based programs to address the urban underclass.

In their co-written selection, Loïc Wacquant and William Wilson reiterate and reformulate some of the issues that Wilson initially addressed. They emphasize the dual importance of both class and racial dynamics in the exclusion of blacks in Chicago as a case study of national trends. The mass exodus of jobs and working families from the inner city, coupled with the growth of neoliberal policies of government privatization and reduction of public spending has triggered a process of "hyperghettoization," concentrating blacks in a crisis of joblessness and extreme poverty. They draw attention also to deindustrialization or structural shift in the economy, notably the decentralization of manufacturing employment from the inner city to the suburbs, Sunbelt states, and offshore locations in developing nations.

The decline of institutional structures in the ghetto, what Wilson in *The Truly Disadvantaged* called "social buffers," is described in this selection as the loss of the "pulpit and the press." The loss of the black leadership (such as teachers, clergy, journalists, lawyers, and businessmen) into the suburbs has left the inner city bereft of stable working families and resources for upward social mobility. Wacquant and Wilson describe the loss of educational resources in the hyperghetto, a situation that is all the more stark because of the loss of manufacturing employment from the inner city. These factory jobs were often available for the previous generation without formal education, as work skills could often be acquired on-the-job. They also paid a living wage, unlike the service sector jobs that have replaced factory jobs, with the "runaway plant," and deindustrialization process in American cities. Contemporary residents of the hyperghetto are also poorly suited for employment in the new information and technology-based sectors of the postindustrial economy. John Kasarda has described this problem as "jobs–skills mismatch" in a variety of writings, including a chapter titled "Urban Industrial Transition and the Underclass," in William Wilson, editor, *The Ghetto Underclass: Social Science Perspectives* (Newbury Park, CA: Sage Publications, 1993).

Wacquant and Wilson also consider the growing feminization of poverty in the hyperghetto, as poor households are increasingly headed by single-women. They note the continuing erosion of financial resources for ghetto households, and the decline in homeownership. They note that the households left in the hyperghetto are bereft of links to solidarity groups, networks, and organizations, what the French sociologist Pierre Bourdieu calls "social capital" ("The Forms of Capital," in *Handbook of Theory and Research for the Sociology of Education*, edited by J. G. Richardson (New York: Greenwood Press, 1986). The political scientist Robert Putnam has recently received national attention for his writings on the general decline of social capital and community networks as a general process in postwar U.S. society (*Bowling Alone: The Collapse and Revival of American Community* (New York: Simon and Schuster, 2000).

William Wilson's *The Declining Significance of Race* was winner of the American Sociological Association's Sydney Spivack Award. *The Truly Disadvantaged* was selected by the editors of the *New York Times Book Review* as one of the 16 best books of 1987, and it also received the *Washington Monthly* Annual Book Award and the Society for the Study of Social Problems C. Wright Mills Award. *When Work Disappears: The World of the New Urban Poor* (New York: Alfred A. Knopf) was chosen as one of the notable books of 1996 by the editors of the *New York Times Book Review* and received the Sidney Hillman Foundation Award. He published *The Bridge over the Racial Divide: Rising Inequality and Coalition Politics* in 1999 (Berkeley, CA: University of California Press).

William Julius Wilson received his Ph.D. from Washington State University in 1996. He taught at the University of Massachusetts at Amherst before joining the University of Chicago faculty in 1972. In 1990 he became the director of the Center for the Study of Urban Inequality at the University of Chicago. In 1996, he moved to become the Lewis P. and Linda L. Geyser University Professor at Harvard University. Wilson is a past president of the American Sociological Association. He was a MacArthur Prize fellow from 1987 to 1992 and has been elected to the National Academy of Sciences, the American Academy of Arts and Sciences, the National Academy of Education, and the American Philosophical Society. In June 1996 he was selected by *Time* magazine as one of America's 25 Most Influential People. In 1998, he received the National Medal of Science, the highest scientific honor in the U.S.

Loïc Wacquant was a doctoral student at the University of Chicago, working as a research assistant at the Urban Poverty and Family Structure Project, when he and Wilson began the collaboration that led to this selection. Further biographical background on Wacquant is provided in the introduction to his selection on "Urban Outcasts: Stigma and Division in the Black American Ghetto and the French Urban Periphery."

After a long eclipse, the ghetto has made a stunning comeback into the collective consciousness of America. Not since the riots of the hot summers of 1966–68 have the black poor received so much attention in academic, activist, and policymaking quarters alike. Persistent and rising poverty, especially among children, mounting social disruptions, the continuing degradation of public housing and public schools, concern over the eroding tax base of cities plagued by large ghettos and by the dilemmas of gentrification, the disillusions of liberals over welfare have all combined to put the black inner-city poor back in the spotlight. Owing in large part to the pervasive and ascendant influence of conservative ideology in the United States, however, recent discussions of the plight of ghetto blacks have typically been cast in individualistic and moralistic terms. The poor are presented as a mere aggregation of personal cases, each with its own logic and self-contained causes. Severed from the struggles and structural changes in the society, economy, and polity that in fact determine them, inner-city dislocations are then portrayed as a self-imposed, self-sustaining phenomenon. This vision of poverty has found perhaps its most vivid expression in the lurid descriptions of ghetto residents that have flourished in the pages of popular magazines and on televised programs devoted to the emerging underclass. Descriptions and explanations of the current predicament of inner-city blacks put the emphasis on individual attributes and the alleged grip of the so-called culture of poverty.

This chapter, in sharp contrast, draws attention to the specific features of the proximate social structure in which ghetto residents evolve and strive, against formidable odds, to survive and, whenever they can, escape its poverty and degradation. We provide this different perspective by profiling blacks who live in Chicago's inner city, contrasting the situation of those who dwell in low-poverty areas with residents of the city's ghetto neighborhoods. Beyond its sociographic focus, the central argument running through this article is that the interrelated set of phenomena captured by the term "underclass" is primarily social-structural and that the ghetto is experiencing a "crisis" not because a "welfare ethos" has mysteriously taken over its residents but because joblessness and economic exclusion, having reached dramatic proportions, have triggered a process of hyperghettoization.

Indeed, the urban black poor of today differ both from their counterparts of earlier years and from the white poor in that they are becoming increasingly concentrated in dilapidated territorial enclaves that epitomize acute social and economic marginalization.

[...]

This growing social and spatial concentration of poverty creates a formidable and unprecedented set of obstacles for ghetto blacks. As we shall see, the social structure of today's inner city has been radically altered by the mass exodus of jobs and working families and by the rapid deterioration of housing, schools, businesses, recreational facilities, and other community organizations, further exacerbated by government policies of industrial and urban laissez-faire that have channeled a disproportionate share of federal, state, and municipal resources to the more affluent. The economic and social buffer provided by a stable black working class and a visible, if small, black middle class that cushioned the impact of downswings in the economy and tied ghetto residents to the world of work has all but disappeared. Moreover, the social networks of parents, friends, and associates, as well as the nexus of local institutions, have seen their resources for economic stability progressively depleted. In sum, today's ghetto residents face a closed opportunity structure.

[...]

DEINDUSTRIALIZATION AND HYPERGHETTOIZATION

Social conditions in the ghettos of Northern metropolises have never been enviable, but today they are scaling new heights in deprivation, oppression, and hardship. The situation of Chicago's black inner city is emblematic of the social changes that have sown despair and exclusion in these communities. An unprecedented tangle of social woes is gripping the black communities of the city's South Side and West Side. These racial enclaves have experienced rapid increases in the number and percentage of poor families, extensive out-migration of working- and middle-class households, stagnation – if not real regression – of income, and record levels of unemployment. . . .

The single largest force behind this increasing social and economic marginalization of large numbers of inner-city blacks has been a set of mutually reinforcing spatial and industrial changes in the country's urban political economy that have converged to undermine the material foundations of the traditional ghetto. Among these structural shifts are the decentralization of industrial plants, which commenced at the time of World War I but accelerated sharply after 1950, and the flight of manufacturing jobs abroad, to the Sunbelt states, or to the suburbs and exurbs at a time when blacks were continuing to migrate en masse to Rustbelt central cities; the general deconcentration of metropolitan economies and the turn toward service industries and occupations, promoted by the growing separation of banks and industry; and the emergence of post-Taylorist, so-called flexible forms of organizations and generalized corporate attacks on unions – expressed by, among other things, wage cutbacks and the spread of two-tier wage systems and labor contracting – which has intensified job competition and triggered an explosion of low-pay, part-time work. This means that even mild forms of racial discrimination – mild by historical standards – have a bigger impact on those at the bottom of the American class order. In the labor-surplus environment of the 1970s, the weakness of unions and the retrenchment of civil rights enforcement aggravated the structuring of unskilled labor markets along racial lines, marking large numbers of inner-city blacks with the stamp of economic redundancy.

In 1954, Chicago was still near the height of its industrial power. Over 10,000 manufacturing establishments operated within the city limits, employing a total of 616,000, including nearly half a million production workers. By 1982, the number of plants had been cut by half, providing a mere 277,000 jobs for fewer than 162,000 blue-collar employees – a loss of 63 percent, in sharp contrast with the overall growth of manufacturing employment in the country, which added almost 1 million production jobs in the quarter century starting in 1958. This crumbling of the city's industrial base was accompanied by substantial cuts in trade employment, with over 120,000 jobs lost in retail and wholesale from 1963 to 1982. The mild growth of services – which created an additional 57,000 jobs during the same period, excluding health, financial, and social services – came nowhere near to compensating for this collapse of Chicago's low-skilled employment pool. Because, traditionally, blacks have relied heavily on manufacturing and blue-collar employment for economic sustenance, the upshot of these structural economic changes for the inhabitants of the inner city has been a steep and accelerating rise in labor market exclusion. In the 1950s, ghetto blacks had roughly the same rate of employment as the average Chicagoan, with some 6 adults in 10 working. While this ratio has not changed citywide over the ensuing three decades, nowadays most residents of the Black Belt cannot find gainful employment and must resort to welfare, to participation in the second economy, or to illegal activities in order to survive. . . .

As the metropolitan economy moved away from smokestack industries and expanded outside of Chicago, emptying the Black Belt of most of its manufacturing jobs and employed residents, the gap between the ghetto and the rest of the city, not to mention its suburbs, widened dramatically. By 1980, median family income on the South and West sides had dropped to around one-third and one-half of the city average, respectively, compared with two-thirds and near parity thirty years earlier. Meanwhile, some of the city's white bourgeois neighborhoods and upper-class suburbs had reached over twice the citywide figure. Thus in 1980, half of the families of Oakland had to make do with less than $5,500 a year, while half of the families of Highland Park incurred incomes in excess of $43,000.

A recent ethnographic account by Arne Duncan on changes in North Kenwood, one of the poorest black sections on the city's South Side, vividly encapsulates the accelerated physical and social decay of the ghetto and is worth quoting at some length:

> In the 1960's, 47th Street was still the social hub of the South Side black community. Sue's eyes light up when she describes how the street used to be filled with stores, theaters and nightclubs in which one could listen to jazz bands well into the evening. Sue remembers the street as "soulful." Today the street might be better characterized as soulless. Some stores, currency exchanges, bars and liquor stores continue to exist on 47th. Yet, as one walks down the street, one is struck more by the death of the street than by its life. Quite literally, the destruction of human life occurs frequently on 47th. In terms of physical structures, many stores are boarded up and abandoned. A few buildings have bars across the front and are closed to the public, but they are not empty. They are used, not so secretly, by people involved in illegal activities. Other stretches of the street are

simply barren, empty lots. Whatever buildings once stood on the lots are long gone. Nothing gets built on 47th. . . . Over the years one apartment building after another has been condemned by the city and torn down. Today many blocks have the bombed-out look of Berlin after World War II. There are huge, barren areas of Kenwood, covered by weeds, bricks, and broken bottles.

Duncan reports how this disappearance of businesses and loss of housing have stimulated the influx of drugs and criminal activities to undermine the strong sense of solidarity that once permeated the community. With no activities or organizations left to bring them together or to represent them as a collectivity, with half the population gone in 15 years, the remaining residents, some of whom now refer to North Kenwood as the "Wild West," seem to be engaged in a perpetual *bellum omnium contra omnes* for sheer survival. One informant expresses this succinctly: " 'It's gotten worse. They tore down all the buildings, deterioratin' the neighborhood. All your friends have to leave. They are just spreading out your mellahs [close friends]. It's not no neighborhood anymore.' " With the ever present threat of gentrification – much of the area is prime lake-front property that would bring in huge profits if it could be turned over to upper-class condominiums and apartment complexes to cater to the needs of the higher-income clientele of Hyde Park, which lies just to the south – the future of the community appears gloomy. One resident explains: " 'They want to put all the blacks in the projects. They want to build buildings for the rich, and not us poor people. They are trying to move us all out. In four or five years we will all be gone.' "

Fundamental changes in the organization of America's advanced economy have thus unleashed irresistible centrifugal pressures that have broken down the previous structure of the ghetto and set off a process of hyperghettoization. By this, we mean that the ghetto has lost much of its organizational strength – the "pulpit and the press," for instance, have virtually collapsed as collective agencies – as it has become increasingly marginal economically; its activities are no longer structured around an internal and relatively autonomous social space that duplicates the institutional structure of the larger society and provides basic minimal resources for social mobility, if only within a truncated black class structure. And the social ills that have long been associated with segregated poverty – violent crime, drugs, housing deterioration, family disruption, commercial blight, and educational failure – have reached qualitatively different proportions and have become articulated into a new configuration that endows each with a more deadly impact than before.

If the "organized," or institutional, ghetto of forty years ago described so graphically by Drake and Cayton imposed an enormous cost on blacks collectively, the "disorganized" ghetto, or hyper-ghetto, of today carries an even larger price. For, now, not only are ghetto residents, as before, dependent on the will and decisions of outside forces that rule the field of power – the mostly white dominant class, corporations, realtors, politicians, and welfare agencies – they have no control over and are forced to rely on services and institutions that are massively inferior to those of the wider society. Today's ghetto inhabitants comprise almost exclusively the most marginal and oppressed sections of the black community. Having lost the economic underpinnings and much of the fine texture of organizations and patterned activities that allowed previous generations of urban blacks to sustain family, community, and collectivity even in the face of continued economic hardship and unflinching racial subordination, the inner-city now presents a picture of radical class and racial exclusion. It is to a sociographic assessment of the latter that we now turn.

THE COST OF LIVING IN THE GHETTO

Let us contrast the social structure of ghetto neighborhoods with that of low-poverty black areas of the city of Chicago. For purposes of this comparison, we have classified as low-poverty neighborhoods all those tracts with rates of poverty – as measured by the number of persons below the official poverty line between 20 and 30 percent as of the 1980 census. Given that the overall poverty rate among black families in the city is about one-third, these low-poverty areas can be considered as roughly representative of the average non-ghetto, non-middle-class, black neighborhood of Chicago. In point of fact, nearly all – 97 percent – of the respondents in this category reside outside traditional ghetto areas. Extreme-poverty neighborhoods comprise tracts with at least 40 percent of their residents in poverty in 1980. These tracts make up the historic heart of Chicago's black ghetto: over 82 percent of the respondents in this category inhabit the West and South sides of the city, in areas most of which have been all black for half a century and more, and an additional 13 percent live in immediately adjacent tracts. Thus when we counterpose extreme-poverty areas with low-poverty areas, we are in effect comparing ghetto neighborhoods with other black areas, most of which are moderately poor, that are not part of Chicago's traditional Black Belt. Even though this comparison involves a truncated spectrum of types of neighborhoods, the contrasts it reveals between low-poverty and ghetto tracts are quite pronounced.

It should be noted that this distinction between low-poverty and ghetto neighborhoods is not merely analytical but captures differences that are clearly perceived by social agents themselves. First, the folk category of ghetto does, in Chicago, refer to the South Side and West Side, not just to any black area of the city; mundane usages of the term entail a social-historical and spatial referent rather than simply a racial dimension. Furthermore, blacks who live in extreme-poverty areas have a noticeably more negative opinion of their neighborhood. Only 16 percent rate it as a "good" to "very good" place to live in, compared to 41 percent among inhabitants of low-poverty tracts; almost 1 in 4 find their neighborhood "bad or very bad" compared to fewer than 1 in 10 among the latter. In short, the contrast between ghetto and non-ghetto poor areas is one that is socially meaningful to their residents.

The black class structure in and out of the ghetto

The first major difference between low- and extreme-poverty areas has to do with their class structure. A sizable majority of blacks in low-poverty tracts are gainfully employed: two-thirds hold a job, including 11 percent with middle-class occupations and 55 percent with working-class jobs, while one-third do not work. These proportions are exactly opposite in the ghetto, where fully 61 percent of adult residents do not work, one-third have working-class jobs and a mere 6 percent enjoy middle-class status. For those who reside in the urban core, then, being without a job is by far the most likely occurrence, while being employed is the exception. Controlling for gender does not affect this contrast, though it does reveal the greater economic vulnerability of women, who are twice as likely as men to be jobless. Men in both types of neighborhoods have a more favorable class mix resulting from their better rates of employment: 78 percent in low-poverty areas and 66 percent in the ghetto. If women are much less frequently employed – 42 percent in low-poverty areas and 69 percent in the ghetto do not work – they have comparable, that is, severely limited, overall access to middle-class status: in both types of neighborhood, only about 10 percent hold credentialed salaried positions or better.

These data are hardly surprising. They stand as a brutal reminder that joblessness and poverty are two sides of the same coin. The poorer the neighborhood, the more prevalent joblessness and the lower the class recruitment of its residents. But these results also reveal that the degree of economic exclusion observed in ghetto neighborhoods during the period of sluggish economic growth of the late 1970s is still very much with us nearly a decade later, in the midst of the most rapid expansion in recent American economic history.

As we would expect, there is a close association between class and educational credentials. Virtually every member of the middle class has at least graduated from high school; nearly two-thirds of working-class blacks have also completed secondary education; but less than half – 44 percent – of the jobless have a high school diploma or more. Looked at from another angle, 15 percent of our educated respondents – that is, high school graduates or better – have made it into the salaried middle class, half have become white-collar or blue-collar wage earners, and 36 percent are without a job. By comparison, those without a high school education are distributed as follows: 1.6 percent in the middle class, 37.9 percent in the working class, and a substantial majority of 60.5 percent in the jobless category. In other words, a high school degree is *a conditio sine qua non* for blacks for entering the world of work, let alone that of the middle class. Not finishing secondary education is synonymous with economic redundancy.

Ghetto residents are, on the whole, less educated than the inhabitants of other black neighborhoods. This results in part from their lower class composition but also from the much more modest academic background of the jobless: fewer than 4 in 10 jobless persons on the city's South Side and West Side have graduated from high school, compared to nearly 6 in 10 in low-poverty areas. It should be pointed out that education is one of the few areas in which women do not fare worse than men: females are as likely to hold a high school diploma as males in the ghetto – 50 percent – and more likely to do so in low-poverty areas – 69 percent versus 62 percent.

Moreover, ghetto residents have lower class origins, if one judges from the economic assets of their family of orientation. Fewer than 4 ghetto dwellers in 10 come from a family that owned its home and 6 in 10 have parents who owned nothing, that is, no home, business, or land. In low-poverty areas, 55 percent of the inhabitants are from a home-owning family while only 40 percent had no assets at all a generation ago. Women, both in and out of the ghetto, are least likely to come from a family with a home or any other asset – 46 percent and 37 percent, respectively. This difference in class origins is also captured by differential rates of welfare receipt during childhood: the proportion of respondents whose parents were on public aid at some time when they were growing up is 30 percent in low-poverty tracts and 41 percent in the ghetto. Women in extreme-poverty areas are by far the most likely to come from a family with a welfare record.

Class, gender, and welfare trajectories in low- and extreme-poverty areas

If they are more likely to have been raised in a household that drew public assistance in the past, ghetto dwellers are also much more likely to have been or to be currently on welfare themselves. Differences in class, gender, and neighborhood cumulate at each juncture of the welfare trajectory to produce much higher levels of welfare attachments among the ghetto population.

In low-poverty areas, only one resident in four are currently on aid while almost half have never personally received assistance. In the ghetto, by contrast, over half the residents are current welfare recipients, and only one in five have never been on aid. These differences are consistent with what we know from censuses and other studies: in 1980, about half of the black population of most community areas on the South Side and West Side was officially receiving public assistance, while working- and middle-class black neighborhoods of the far South Side, such as South Shore, Chatham, or Roseland, had rates of welfare receipt ranging between one-fifth and one-fourth.

None of the middle-class respondents who live in low-poverty tracts were on welfare at the time they were interviewed, and only one in five had ever been on aid in their lives. Among working-class residents, a mere 7 percent were on welfare and just over one-half had never had any welfare experience. This same relationship between class and welfare receipt is found among residents of extreme-poverty tracts, but with significantly higher rates of welfare receipt at all class levels: there, 12 percent of working-class residents are presently on aid and 39 percent received welfare before; even a few middle-class blacks – 9 percent – are drawing public assistance and only one-third of them have never received any aid, instead of three-quarters in low-poverty tracts. But it is among the jobless that the difference between low- and extreme-poverty areas is the largest: fully 86 percent of those in ghetto tracts are currently on welfare and only 7 percent have never had recourse to public aid, compared with 62 percent and 20 percent, respectively, among those who live outside the ghetto.

Neighborhood differences in patterns of welfare receipt are robust across genders, with women exhibiting noticeably higher rates than men in both types of areas and at all class levels. The handful of black middle-class women who reside in the ghetto are much more likely to admit to having received aid in the past than their male counterparts: one-third versus one-tenth. Among working-class respondents, levels of current welfare receipt are similar for both sexes – 5.0 percent and 8.5 percent, respectively – while levels of past receipt again display the greater economic vulnerability of women: one in two received aid before as against one male in five. This gender differential is somewhat attenuated in extreme-poverty areas by the general prevalence of welfare receipt, with two-thirds of all jobless males and 9 in 10 jobless women presently receiving public assistance.

The high incidence and persistence of joblessness and welfare in ghetto neighborhoods, reflecting the paucity of viable options for stable employment, take a heavy toll on those who are on aid by significantly depressing their expectations of finding a route to economic self-sufficiency. While a slim majority of welfare recipients living in low-poverty tracts expect to be self-supportive within a year and only a small minority anticipate receiving aid for longer than five years, in ghetto neighborhoods, by contrast, fewer than 1 in 3 public-aid recipients expect to be welfare-free within a year and fully 1 in 5 anticipate needing assistance for more than five years. This difference of expectations increases among the jobless of both genders. For instance, unemployed women in the ghetto are twice as likely as unemployed women in low-poverty areas to think that they will remain on aid for more than five years and half as likely to anticipate getting off the rolls within a year.

Thus if the likelihood of being on welfare increases sharply as one crosses the line between the employed and the jobless, it remains that, at each level of the class structure, welfare receipt is notably more frequent in extreme-poverty neighborhoods, especially among the unemployed, and among women.

[...]

Differences in economic and financial capital

A quick survey of the economic and financial assets of the residents of Chicago's poor black neighborhoods reveals the appalling degree of economic hardship, insecurity, and deprivation that they must confront day in and day out. The picture in low-poverty areas is grim; that in the ghetto is one of near-total destitution.

In 1986, the median family income for blacks nationally was pegged at $18,000, compared to $31,000 for white families. Black households in Chicago's low-poverty areas have roughly equivalent incomes, with 52 percent declaring over $20,000 annually. Those living in Chicago's ghetto, by contrast, command but a fraction of this figure: half of all ghetto respondents live in households that dispose of less than $7500 annually, twice the rate among residents of low-poverty neighborhoods. Women assign their households to much lower income brackets in both areas, with fewer than 1 in 3 in low-poverty areas and 1 in 10 in extreme-poverty areas enjoying more than $25,000 annually. Even those who work report smaller incomes in the ghetto: the proportion of working-class and middle-class households falling under the $7500 mark on the South and West sides – 12.5 percent and 6.5 percent, respectively – is double that of other black neighborhoods, while fully one-half of jobless respondents in extreme-poverty tracts do not reach the $5000 line. It is not surprising that ghetto dwellers also less frequently report an improvement of the financial situation of their household, with women again in the least enviable position. This reflects sharp class differences: 42 percent of our middle-class respondents and 36 percent of working-class blacks register a financial amelioration as against 13 percent of the jobless.

Due to meager and irregular income, those financial and banking services that most members of the larger society take for granted are, to put it mildly, not of obvious access to the black poor. Barely one-third of the residents of low-poverty areas maintain a personal checking account; only one in nine manage to do so in the ghetto, where nearly three of every four persons report no financial asset whatsoever from a possible list of six and only 8 percent have at least three of those six assets. Here, again, class and neighborhood lines are sharply drawn: in low-poverty areas, 10 percent of the jobless and 48 percent of working-class blacks have a personal checking account compared to 3 percent and 37 percent, respectively, in the ghetto; the proportion for members of the middle class is similar – 63 percent – in both areas.

The American dream of owning one's home remains well out of reach for a large majority of our black respondents, especially those in the ghetto, where barely 1 person in 10 belong to a home-owning household, compared to over 4 in 10 in low-poverty areas, a difference that is just as pronounced within each gender. The considerably more modest dream of owning an automobile is likewise one that has yet to materialize for ghetto residents, of which only one-third live in households with a car that runs. Again, this is due to a cumulation of sharp class and neighborhood differences: 79 percent of middle-class respondents and 62 percent of working-class blacks have an automobile in their household, contrasted with merely 28 percent of the jobless. But, in ghetto tracts, only 18 percent of the jobless have domestic access to a car – 34 percent for men and 13 percent for women.

The social consequences of such a paucity of income and assets as suffered by ghetto blacks cannot be overemphasized. For just as the lack of financial resources or possession of a home represents a critical handicap when one can only find low-paying and casual employment or when one loses one's job, in that it literally forces one to go on the welfare rolls, not owning a car severely curtails one's chances of competing for available jobs that are not located nearby or that are not readily accessible by public transportation.

Social capital and poverty concentration

Among the resources that individuals can draw upon to implement strategies of social mobility are those potentially provided by their lovers, kin, and friends and by the contacts they develop within the formal associations to which they belong – in sum, the resources they have access to by virtue of being socially integrated into solidarity groups, networks, or organizations, what Bourdieu calls "social capital." Our data indicate that not only do residents of extreme-poverty areas have fewer social ties but also that they tend to have ties of lesser social worth, as measured by the social position of their partners, parents, siblings, and best friends, for instance. In short, they possess lower volumes of social capital.

Living in the ghetto means being more socially isolated: nearly half of the residents of extreme-poverty tracts have no current partner – defined here as a person they are married to, live with, or are dating steadily – and one in five admit to having no one who would qualify as a best friend, compared to 32 percent and 12 percent, respectively, in low-poverty areas. It also means that intact marriages are less frequent. Jobless men are much less likely than working males to have current partners in both types of neighborhoods: 62 percent in low-poverty neighborhoods and 44 percent in extreme-poverty areas. Black women have a slightly better chance of having a partner if they live in a low-poverty area, and this partner is also more likely to have completed high school and to work steadily; for ghetto residence further affects the labor-market standing of the latter. The partners of women living in extreme-poverty areas are less stably employed than those of female respondents from low-poverty neighborhoods: 62 percent in extreme-poverty areas work regularly as compared to 84 percent in low-poverty areas.

Friends often play a crucial role in life in that they provide emotional and material support, help construct one's identity, and often open up opportunities that one would not have without them – particularly in the area of jobs. We have seen that ghetto residents are more likely than other black Chicagoans to have no close friend. If they have a best friend, furthermore, he or she is less likely to work, is less educated, and twice as likely to be on aid. Because friendships tend to develop primarily within genders and women have much higher rates of economic exclusion, female respondents are much more likely than men to have a best friend who does not work and who receives welfare assistance. Both of these characteristics, in turn, tend to be more prevalent among ghetto females.

Such differences in social capital are also evidenced by different rates and patterns of organizational participation. While being part of a formal organization, such as a block club or a community organization, a political party, a school-related association, or a sports, fraternal, or other social group, is a rare occurrence as a rule – with the notable exception of middle-class blacks, two-thirds of whom belong to at least one such group – it is more common for ghetto residents – 64 percent, versus 50 percent in low-poverty tracts – especially females – 64 percent, versus 46 percent in low-poverty areas – to belong to no organization. As for church membership, the small minority who profess to be, in Weber's felicitous expression, "religiously unmusical" is twice as large in the ghetto as outside: 12 percent versus 5 percent. For those with a religion, ghetto residence tends to depress church attendance slightly – 29 percent of ghetto inhabitants attend service at least once a week compared to 37 percent of respondents from low-poverty tracts – even though women tend to attend more regularly than men in both types of areas. Finally, black women who inhabit the ghetto are also slightly less likely to know most of their neighbors than their counterparts from low-poverty areas. All in all, then, poverty concentration has the effect of devaluing the social capital of those who live in its midst.

CONCLUSION: THE SOCIAL STRUCTURING OF GHETTO POVERTY

The extraordinary levels of economic hardship plaguing Chicago's inner city in the 1970s have not abated, and the ghetto seems to have gone unaffected by the economic boom of the past five years. If anything, conditions have continued to worsen. This points to the asymmetric causality between the economy and ghetto poverty and to the urgent need to study the social and political structures that mediate their relationship. The significant differences we have uncovered between low-poverty and extreme-poverty areas in Chicago are essentially a reflection of their different class mix and of the prevalence of economic exclusion in the ghetto.

Our conclusion, then, is that social analysts must pay more attention to the extreme levels of economic deprivation and social marginalization as uncovered in this article before they further entertain and spread so-called theories about the potency of a ghetto culture of poverty that has yet to receive rigorous empirical elaboration. Those who have been pushing moral–cultural or individualistic –behavioral explanations of the social dislocations that have swept through the inner city in recent years have created a fictitious normative divide between urban blacks that, no matter its reality – which has yet to be ascertained – cannot but pale when compared to the objective structural cleavage that separates ghetto residents from the larger society and to the collective material constraints that bear on them. It is the cumulative structural entrapment and forcible socioeconomic marginalization resulting from the historically evolving interplay of class, racial, and gender domination, together with sea changes in the organization of American capitalism and failed urban and social policies, not a "welfare ethos," that explain the plight of today's ghetto blacks. Thus, if the concept of underclass is used, it must be a structural concept: it must denote a new sociospatial patterning of class and racial domination, recognizable by the unprecedented concentration of the most socially excluded and economically marginal members of the dominated racial and economic group. It should not be used as a label to designate a new breed of individuals molded freely by a mythical and all-powerful culture of poverty.

"Segregation and the Making of the Underclass"

from *American Apartheid* (1993)

Douglas S. Massey and Nancy A. Denton

Editors' Introduction

One hundred years after W. E. DuBois first decried that "the problem of the Twentieth Century is the problem of the color-line," Douglas Massey and Nancy Denton warn that racial segregation is still an imposing obstacle in American society. The "hypersegregation" of blacks and Latinos in urban ghettos at the turn of the new millennium is unlike the residential segregation experienced by the white ethnic minorities that preceded them. The "dark ghetto" has "invisible walls" that may be social, political, educational or economic. Contrary to popular belief, segregation is not at its worst in the South, but in the North. In 1966, Martin Luther King came to Chicago and declared it "The most segregated city in America." Despite King's and all the civil rights leaders' efforts, little has changed. Decades after the passage of the last of the great civil rights acts of the 1960s, blacks remain almost as segregated in American cities as they were in 1968.

In *American Apartheid*, Massey and Denton argue that white people created the underclass ghetto during the first half of the twentieth century in order to isolate growing urban black populations. Despite the Fair Housing Act of 1968, segregation is perpetuated today through an interlocking set of individual actions, institutional practices, and governmental policies. In some urban areas the degree of black segregation is so intense and occurs in so many dimensions simultaneously that it amounts to "hypersegregation." The authors demonstrate that this systematic segregation of African Americans leads inexorably to the creation of underclass communities during periods of economic downturn. Under conditions of extreme segregation, any increase in the overall rate of black poverty yields a marked increase in the geographic concentration of indigence and the deterioration of social and economic conditions in black communities. As ghetto residents adapt to this increasingly harsh environment under a climate of racial isolation, they evolve attitudes, behaviors, and practices that further marginalize their neighborhoods and undermine their chances of success in mainstream American society. Their book is a sober challenge to those who argue that race is of declining significance in the United States today.

This represents perhaps the greatest failure in the national effort to equalize the condition of American blacks. Compared with the substantial changes in employment and political representation and education – the growth of the black middle class, the great increase in black college attendance, the surge in the number of black mayors, state legislators, members of Congress – the indices of residential segregation show almost no change. Despite the overturning of *de jure* segregation, *de facto* segregation continues in America, through practices such as bank and insurance redlining and prejudicial real estate steering. Despite the Mortgage Disclosure Act and the Community Reinvestment Act, Banks still discriminate against Blacks in the home loan market. There is "Segregation with a smile," where realtors actively steer blacks away from white neighborhoods, a practice which is revealed by housing audit studies.

White flight and continuing segregation have isolated racial minorities in central cities, undermining political coalitions and fragmenting the political landscape and the tax base between "white suburbs" and "chocolate cities." White ethnic immigrants of the early twentieth century were able to create pan-ethnic coalitions in urban patronage machines to allocate spending in their neighborhoods and maintain the quality of their schools and infrastructure. But segregation has undermined the ability of blacks to advance their interests, form coalitions, and establish common interests with white voters. The spatial and political isolation of the ghetto makes it easier for racists to act on their own prejudices.

The emergence and persistence of the urban ghetto attests to the condition of spatial apartheid that confronts racial minorities at the turn of the millennium in American cities. Using the index of dissimilarity as a measure of residential segregation, they have collaborated in numerous exhaustive studies that confirm that while ethnics are becoming more spatially assimilated, blacks experience significant continuing segregation. Especially in older cities of the U.S. Northeast and Midwest, there was a growth of underclass communities in the 1970s and 1980s in cities experiencing greater residential segregation. Barriers to spatial mobility are also barriers to social mobility. They liken this condition of spatial and social immobility to a condition of racial apartheid.

Under conditions of extreme segregation, any increase in the overall rate of black poverty yields a marked increase in the geographic concentration of indigence and the deterioration of social and economic conditions in black communities. As ghetto residents adapt to this increasingly harsh environment under a climate of racial isolation, they evolve attitudes, behaviors, and practices that further marginalize their neighborhoods and undermine their chances of success in mainstream American society. Ghetto residents live in a very limited social world that intensifies the growth of Ebonics or Black English vernacular and reinforces the sense of an "oppositional culture" to the American cultural mainstream. Ebonics is a rich cultural phenomenon, but the oppositional culture creates peer pressure against school attendance and social mobility. There can be a climate of intimidation, violence, and fear among the youth of underclass black communities. Spatial isolation feeds cultural and linguistic isolation.

The authors suggest a more vigorous prosecution of realtors and bankers that discriminate against African Americans, stricter enforcement of the Fair Housing Law of 1968, and rental vouchers to African Americans in order to ease segregation. Segregation is a problem without any easy answers. If Massey and Denton's proposals were enacted, would that stop the problem of whites moving away? Perhaps the only real solution requires whites learning how to get along with and not fear African Americans.

Douglas Massey received his Ph.D. in 1978 from Princeton University and has served on the faculties of the University of Chicago and the University of Pennsylvania. His research focuses on international migration, race and housing, discrimination, education, urban poverty, and Latin America, especially Mexico. He is the author, most recently, of *Beyond Smoke and Mirrors: Mexican Immigration in an Age of Economic Integration*, and *Source of the River: The Social Origins of Freshmen at America's Selective Colleges and Universities*. He is a member of the National Academy of Sciences and the American Academy of Arts and Sciences and Past-President of the American Sociological Association and the Population Association of America.

Nancy Denton received her Ph.D. in Demography from the University of Pennsylvania in 1984. She is an associate professor of sociology and director of graduate studies at the State University of New York at Albany, where she is also a research associate at the Center for Social and Demographic Analysis. She recently edited, with Steward Tolnay, a reader titled *American Diversity: A Demographic Challenge for the 21st Century*. She was chair of the Community and Urban Section of the American Sociological Association from 2001 to 2002.

T H R E E

THE MISSING LINK

> It is quite simple. As soon as there is a group area then all your uncertainties are removed and that is, after all, the primary purpose of this Bill [requiring racial segregation in housing].
>
> (Minister of the Interior, Union of South Africa legislative debate on the Group Areas Act of 1950)

During the 1970s and 1980s a word disappeared from the American vocabulary. It was not in the speeches of politicians decrying the multiple ills besetting American cities. It was not spoken by government officials responsible for administering the nation's social programs. It was not mentioned by journalists reporting on the rising tide of homelessness, drugs, and violence in urban America. It was not discussed by foundation executives and think-tank experts proposing new programs for unemployed parents and unwed mothers. It was not articulated by civil rights leaders speaking out against the persistence of racial inequality; and it was nowhere to be found in the thousands of pages written by social scientists on the urban underclass. The word was segregation.

Most Americans vaguely realize that urban America is still a residentially segregated society, but few appreciate the depth of black segregation or the degree to which it is maintained by ongoing institutional arrangements and contemporary individual actions. They view segregation as an unfortunate holdover from a racist past, one that is fading progressively over time. If racial residential segregation persists, they reason, it is only because civil rights laws passed during the 1960s have not had enough time to work or because many blacks still prefer to live in black neighborhoods. The residential segregation of blacks is viewed charitably as a "natural" outcome of impersonal social and economic forces, the same forces that produced Italian and Polish neighborhoods in the past and that yield Mexican and Korean areas today.

But black segregation is not comparable to the limited and transient segregation experienced by other racial and ethnic groups, now or in the past. No group in the history of the United States has ever experienced the sustained high level of residential segregation that has been imposed on blacks in large American cities for the past fifty years. This extreme racial isolation did not just happen; it was manufactured by whites through a series of self-conscious actions and purposeful institutional arrangements that continue today. Not only is the depth of black segregation unprecedented and utterly unique compared with that of other groups, but it shows little sign of change with the passage of time or improvements in socioeconomic status.

If policymakers, scholars, and the public have been reluctant to acknowledge segregation's persistence, they have likewise been blind to its consequences for American blacks. Residential segregation is not a neutral fact; it systematically undermines the social and economic well-being of blacks in the United States. Because of racial segregation, a significant share of black America is condemned to experience a social environment where poverty and joblessness are the norm, where a majority of children are born out of wedlock, where most families are on welfare, where educational failure prevails, and where social and physical deterioration abound. Through prolonged exposure to such an environment, black chances for social and economic success are drastically reduced.

Deleterious neighborhood conditions are built into the structure of the black community. They occur because segregation concentrates poverty to build a set of mutually reinforcing and self-feeding spirals of decline into black neighborhoods. When economic dislocations deprive a segregated group of employment and increase its rate of poverty, socioeconomic deprivation inevitably becomes more concentrated in neighborhoods where that group lives. The damaging social consequences that follow from increased poverty are spatially concentrated as well, creating uniquely disadvantaged environments that become progressively isolated – geographically, socially, and economically – from the rest of society.

[. . .]

We trace the historical construction of the black ghetto during the nineteenth and twentieth centuries. We show that high levels of black–white segregation were not always characteristic of American urban areas. Until the end of the nineteenth century blacks and whites were relatively integrated in both northern and southern cities; as late as 1900, the typical black urbanite still lived in a neighborhood that was predominantly white. The evolution of segregated, all-black neighborhoods occurred later and was not the result of impersonal market forces. It did not reflect the desires of African Americans themselves. On the contrary, the black ghetto was constructed through a series of well-defined institutional practices, private behaviors, and public policies by which whites sought to contain growing urban black populations.

The manner in which blacks were residentially incorporated into American cities differed fundamentally from the path of spatial assimilation followed by other ethnic groups. Even at the height of immigration from Europe, most Italians, Poles, and Jews lived in neighborhoods where members of their own group did not predominate, and as their socioeconomic status and generations spent in the United States rose, each group was progressively integrated into American society. In contrast, after the construction of the black ghetto the vast majority of blacks were forced to live in neighborhoods that were all black, yielding an extreme level of social isolation.

We show that high levels of black–white segregation had become universal in American cities by 1970, and despite the passage of the Fair Housing Act in 1968, this situation had not changed much in the nation's largest black communities by 1980. In these large urban areas black–white segregation persisted at very high level, and the extent of black suburbanization lagged far behind that of other groups. Even within suburbs, levels of racial segregation remained exceptionally high, and in many urban areas the degree of racial separation between blacks and whites was profound. Within sixteen large metropolitan areas – containing one-third of all blacks in the United States – the extent of racial segregation was so intense and occurred on so many dimensions simultaneously that we label the pattern "hypersegregation."

We examine why black segregation continues to be so extreme. One possibility that we rule out is that high levels of racial segregation reflect socioeconomic differences between blacks and whites. Segregation cannot be attributed to income differences, because blacks are equally highly segregated at all levels of income. Whereas segregation declines steadily for most minority groups as socioeconomic status rises, levels of black–white segregation do not vary significantly by social class. Because segregation reflects the effects of white prejudice rather than objective market forces, blacks are segregated no matter how much money they earn.

Although whites now accept open housing in principle, they remain prejudiced against black neighbors in practice. Despite whites' endorsement of the ideal that people should be able to live wherever they can afford to regardless of race, a majority still feel uncomfortable in any neighborhood that contains more than a few black residents; and as the percentage of blacks rises, the number of whites who say they would refuse to enter or would try to move out increases sharply.

These patterns of white prejudice fuel a pattern of neighborhood resegregation because racially mixed neighborhoods are strongly desired by blacks. As the percentage of blacks in a neighborhood rises, white demand for homes within it falls sharply while black demand rises. The surge in black demand and the withering of white demand yield a process of racial turnover. As a result, the only urban areas where significant desegregation occurred during the 1970s were those where the black population was so small that integration could take place without threatening white preferences for limited contact with blacks.

Prejudice alone cannot account for high levels of black segregation, however, because whites seeking to avoid contact with blacks must have somewhere to go. That is, some all-white neighborhoods must be perpetuated and maintained, which requires the erection of systematic barriers to black residential mobility. In most urban housing markets, therefore, the effects of white prejudice are typically reinforced by direct discrimination against black homeseekers. Housing audits carried out over the past two decades have documented the persistence of widespread discrimination against black renters and homebuyers, and a recent comprehensive study carried out by the U.S. Department of Housing and Urban Development suggests that prior work has understated both the incidence and the severity of this racial bias. Evidence also suggests that blacks can expect to experience significant discrimination in the allocation of home mortgages as well.

We demonstrate theoretically how segregation creates underclass communities and systematically builds deprivation into the residential structure of black communities. We show how any increase in the poverty rate of a residentially segregated group leads to an immediate and automatic increase in the geographic concentration of poverty. When the rate of minority poverty is increased under conditions of high segregation, all of the increase is absorbed by a small number of neighborhoods. When the same increase in poverty occurs in an integrated group, the added poverty is spread evenly throughout the urban area, and the neighborhood environment that group members face does not change much.

During the 1970s and 1980s, therefore, when urban economic restructuring and inflation drove up rates of black and Hispanic poverty in many urban areas, underclass communities were created only where increased minority poverty coincided with a high degree of segregation – principally in older metropolitan areas of the northeast and the Midwest. Among Hispanics, only Puerto Ricans developed underclass communities, because only they were highly segregated; and this high degree of segregation is directly attributable to the fact that a large proportion of Puerto Ricans are of African origin.

The interaction of intense segregation and high poverty leaves black neighborhoods extremely vulnerable to fluctuations in the urban economy, because any dislocation that causes an upward shift in black poverty rates will also produce a rapid change in the concentration of poverty and, hence, a dramatic shift in the social and economic composition of black neighborhoods. The concentration of poverty, for example, is associated with the wholesale withdrawal of commercial institutions and the deterioration or elimination of goods and services distributed through the market.

Neighborhoods, of course, are dynamic and constantly changing, and given the high rates of residential turnover characteristic of contemporary American cities, their well-being depends to a great extent on the characteristics and actions of their residents. Decisions taken by one actor affect the subsequent decisions of others in the neighborhood. In this way isolated actions affect the well-being of the community and alter the stability of the neighborhood.

Because of this feedback between individual and collective behavior, neighborhood stability is characterized by a series of thresholds, beyond which various self-perpetuating processes of decay take hold. Above these thresholds, each actor who makes a decision that undermines neighborhood well-being makes it increasingly likely that other actors will do the same. Each property owner who decides not to invest in upkeep and maintenance, for example, lowers the incentive for others to maintain their properties. Likewise, each new crime promotes psychological and physical withdrawal from public life, which reduces vigilance within the neighborhood and undermines the capacity for collective organization, making additional criminal activity more likely.

Segregation increases the susceptibility of neighborhoods to these spirals of decline. During periods of economic dislocation, a rising concentration of black poverty is associated with the simultaneous concentration of other negative social and economic conditions. Given the high levels of racial segregation characteristic of American urban areas, increases in black poverty such as those observed during the 1970s can only lead to a concentration of housing abandonment, crime, and social disorder, pushing poor black neighborhoods beyond the threshold of stability.

By building physical decay, crime, and social disorder into the residential structure of black communities, segregation creates a harsh and extremely disadvantaged environment to which ghetto blacks must adapt. In concentrating poverty, moreover, segregation also concentrates conditions such as drug use, joblessness, welfare dependency, teenage childbearing, and unwed parenthood, producing a social context where these conditions are not only common but the norm. We argue that in adapting to this social environment, ghetto dwellers evolve a set of behaviors, attitudes, and expectations that are sharply at variance with those common in the rest of American society.

As a direct result of the high degree of racial and class isolation created by segregation, for example, Black English has become progressively more distant from Standard American English, and its speakers are at a clear disadvantage in U.S. schools and labor markets. Moreover, the isolation and intense poverty of the ghetto provides a supportive structural niche for the emergence of an "oppositional culture" that inverts the values of middle-class society. Anthropologists have found that young people in the ghetto experience strong peer pressure not to succeed in school, which severely limits their prospects for social mobility in the larger society. Quantitative research shows that growing up in a ghetto neighborhood increases the likelihood of dropping out of high school, reduces the probability of attending college, lowers the likelihood of employment, reduces income earned as an adult, and increases the risk of teenage childbearing and unwed pregnancy.

[. . .]

THE PERPETUATION OF THE UNDERCLASS

> One notable difference appears between the immigrant and Negro populations. In the case of the former, there is the possibility of escape, with improvement in economic status in the second generation.
>
> (1931 report to President Herbert Hoover by the Committee on Negro Housing)

If the black ghetto was deliberately constructed by whites through a series of private decisions and institutional practices, if racial discrimination persists at remarkably high levels in U.S. housing markets, if intensive residential segregation continues to be imposed on blacks by virtue of their skin color, and if segregation concentrates poverty to build a self-perpetuating spiral of decay into black neighborhoods, then a variety of deleterious consequences automatically follow for individual African Americans. A racially segregated society cannot be a race-blind society; as long as U.S. cities remain segregated – indeed, hypersegregated – the United States cannot claim to have equalized opportunities for blacks and whites. In a segregated world, the deck is stacked against black socioeconomic progress, political empowerment, and full participation in the mainstream of American life.

In considering how individuals fare in the world, social scientists make a fundamental distinction between individual, family, and structural characteristics. To a great extent, of course, a person's success depends on individual traits such as motivation, intelligence, and especially, education. Other things equal, those who are more highly motivated, smarter, and better educated will be rewarded more highly in the labor market and will achieve greater socioeconomic success.

Other things generally are not equal, however, because individual traits such as motivation and education are strongly affected by family background. Parents who are themselves educated, motivated, and economically successful tend to pass these traits on to their children. Children who enter the middle and upper classes through the accident of birth are more likely than other, equally intelligent children from other classes to acquire the schooling, motivation, and cultural knowledge required for socioeconomic success in contemporary society. Other aspects of family background, moreover, such as wealth and social connections, open the doors of opportunity irrespective of education or motivation.

Yet even when one adjusts for family background, other things are still not equal, because the structural organization of society also plays a profound role in shaping the life chances of individuals. Structural variables are elements of social and economic organization that lie beyond individual control, that are built into the way society is organized. Structural characteristics affect the fate of large numbers of people and families who share common locations in the social order.

Among the most important structural variables are those that are geographically defined. Where one lives – especially, where one grows up – exerts a profound effect on one's life chances. Identical individuals with similar family backgrounds and personal characteristics will lead very different lives and achieve different rates of socioeconomic success depending on where they reside. Because racial segregation confines blacks to a circumscribed and disadvantaged niche in the urban spatial order, it has profound consequences for individual and family well-being.

Social and spatial mobility

In a market society such as the United States, opportunities, resources, and benefits are not distributed evenly across the urban landscape. Rather, certain residential areas have more prestige, greater affluence, higher home values, better services, and safer streets than others. Marketing consultants have grown rich by taking advantage of this "clustering of America" to target specific groups of consumers for wealthy corporate clients. The geographic differentiation of American cities by socioeconomic status does more than conveniently rank neighborhoods for the benefit of the demographer, however; it also creates a crucial connection between social and spatial mobility.

As people get ahead, they not only move up the economic ladder, they move up the residential ladder as well. As early as the 1920s, sociologists at the University of Chicago noted this close connection between social and spatial mobility, a link that has been verified many times since. As socioeconomic status improves, families relocate to take advantage of opportunities and resources that are available in greater abundance elsewhere. By drawing on benefits acquired through residential mobility, aspiring parents not only consolidate their own class position but enhance their and their children's prospects for additional social mobility.

In a very real way, therefore, barriers to spatial mobility are barriers to social mobility, and where one lives determines a variety of salient factors that affect individual well-being: the quality of schooling, the value of housing, exposure to crime, the quality of public services, and the character of children's peers. As a result, residential integration has been a crucial component in the broader process of socioeconomic advancement among immigrants and their children. By moving to successively better neighborhoods, other racial and ethnic groups have gradually become integrated into American society. Although rates of spatial assimilation have varied, levels of segregation have fallen for each immigrant group as socioeconomic status and generations in the United States have increased.

The residential integration of most ethnic groups has been achieved as a by-product of broader processes of socioeconomic attainment, not because group members sought to live among native whites per se. The desire for integration is only one of a larger set of motivations, and not necessarily the most important. Some minorities may even be antagonistic to the idea of integration, but for spatial assimilation to occur, they need only be willing to put up with integration in order to gain access to socioeconomic resources that are more abundant in areas in which white families predominate.

To the extent that white prejudice and discrimination restrict the residential mobility of blacks and confine them to areas with poor schools, low home values, inferior services, high crime, and low educational aspirations, segregation undermines their social and economic well-being. The persistence of racial segregation makes it difficult for aspiring black families to escape the concentrated poverty of the ghetto and puts them at a distinct disadvantage in the larger competition for education, jobs, wealth, and power. The central issue is not whether African Americans "prefer" to live near white people or whether integration is a desirable social goal, but how the restrictions on individual liberty implied by severe segregation undermine the social and economic well-being of individuals.

Extensive research demonstrates that blacks face strong barriers to spatial assimilation within American society. Compared with other minority groups, they are markedly less able to convert their socioeconomic attainments into residential contact with whites, and because of this fact they are unable to gain access to crucial resources and benefits that are distributed through housing markets. Dollar for dollar, blacks are able to buy fewer neighborhood amenities with their income than other groups.

Among all groups in the United States, only Puerto Ricans share blacks' relative inability to assimilate spatially; but this disadvantage stems from the fact that many are of African origin. Although white Puerto Ricans achieve rates of spatial assimilation that are comparable with those found among other ethnic groups, those of African or racially mixed origins experience markedly lower abilities to convert socioeconomic attainments into contact with whites. Once race is controlled, the "paradox of Puerto Rican segregation" disappears.

Given the close connection between social and spatial mobility, the persistence of racial barriers implies the systematic exclusion of blacks from benefits and resources that are distributed through housing markets. We illustrate the severity of this black disadvantage with data specially compiled for the city of Philadelphia in 1980. The data allow us to consider the socioeconomic character of neighborhoods that poor, middle-income, and affluent blacks and whites can be expected to inhabit, holding education and occupational status constant.

In Philadelphia, poor blacks and poor whites both experience very bleak neighborhood environments; both groups live in areas where about 40 percent of the births are to unwed mothers, where median home values are under $30,000, and where nearly 40 percent of high school students score under the 15th percentile on a standardized achievement test. Families in such an environment would be unlikely to build wealth through home equity, and children growing up in such an environment would be exposed to a peer environment where unwed parenthood was common and where educational performance and aspirations were low.

[...]

For blacks, in other words, high incomes do not buy entree to residential circumstances that can serve as springboards for future socioeconomic mobility; in particular, blacks are unable to achieve a school environment conducive to later academic success. In Philadelphia, children from an affluent black family are likely to attend a public school where the percentage of low-achieving students is three times greater than the percentage in schools attended by affluent white children. Small wonder, then, that controlling for income in no way erases the large racial gap in SAT scores. Because of segregation, the same income buys black and white families educational environments that are of vastly different quality.

Given these limitations on the ability of black families to gain access to neighborhood resources, it is hardly surprising that government surveys reveal blacks to be less satisfied with their residential circumstances than socioeconomically equivalent whites. This negative evaluation reflects an accurate appraisal of their circumstances rather than different values or ideals on the part of blacks. Both races want the same things in homes and neighborhoods; blacks are just less able to achieve then. Compared with whites, blacks are less likely to be homeowners, and the homes they do own are of poorer quality, in poorer neighborhoods, and of lower value. Moreover, given the close connection between home equity and family wealth, the net worth of blacks is a small fraction of that of whites, even though their incomes have converged over the years. Finally, blacks tend to occupy older, more crowded dwellings that are structurally inadequate compared to those inhabited by whites; and because these racial differentials stem from segregation rather than income, adjusting for socioeconomic status does not erase them.

[...]

THE FUTURE OF THE GHETTO

> The isolation of Negro from white communities is increasing rather than decreasing . . . Negro poverty is not white poverty. Many of its causes . . . are the same. But there are differences – deep, corrosive, obstinate differences – radiating painful roots into the community, the family, and the nature of the individual.
>
> (President Lyndon Johnson, address to Howard University, June 4, 1965)

After persisting for more than fifty years, the black ghetto will not be dismantled by passing a few amendments to existing laws or by implementing a smattering of bureaucratic reforms. The ghetto is part and parcel of modern American society; it was manufactured by whites earlier in the century to isolate and control growing urban black populations, and it is maintained today by a set of institutions, attitudes, and practices that are deeply embedded in the structure of American life. Indeed, as conditions in the ghetto have worsened and as poor blacks have adapted socially and culturally to this deteriorating environment, the ghetto has assumed even greater importance as an institutional tool for isolating the by-products of racial oppression: crime, drugs, violence, illiteracy, poverty, despair, and their growing social and economic costs.

For the walls of the ghetto to be breached at this point will require an unprecedented commitment by the public and a fundamental change in leadership at the highest levels. Residential segregation will only be eliminated from American society when federal authorities, backed by the American people, become directly involved in guaranteeing open housing markets and eliminating discrimination from public life. Rather than relying on private individuals to identify and prosecute those who break the law, the U.S. Department of Housing and Urban Development and the Office of the Attorney General must throw their full institutional weight into locating instances of housing discrimination and bringing those who violate the Fair Housing Act to justice; they must vigorously prosecute white racists who harass and intimidate blacks seeking to exercise their rights of residential freedom; and they must establish new bureaucratic mechanisms to counterbalance the forces that continue to sustain the residential color line.

Given the fact that black poverty is exacerbated, reinforced, and perpetuated by racial segregation, that black–white segregation has not moderated despite the federal policies tried so far, and that the social costs of segregation inevitably cannot be contained in the ghetto, we argue that the nation has no choice but to launch a bold new initiative to eradicate the ghetto and eliminate segregation from American life. To do otherwise is to condemn the United States and the American people to a future of economic stagnation, social fragmentation, and political paralysis.

Race, class, and public policy

In the United States today, public policy discussions regarding the urban underclass frequently devolve into debates on the importance of race versus class. However one defines the underclass, it is clear that African Americans are overrepresented within in it. People who trace their ancestry to Africa are at greater risk than others of falling into poverty, remaining there for a long time, and residing in very poor neighborhoods. On almost any measure of social and economic well-being, blacks and Puerto Ricans come out near the bottom.

The complex of social and economic problems that beset people of African origin has led many observers to emphasize race over class in developing remedies for the urban underclass. According to these theories, institutional racism is pervasive, denying blacks equal access to the resources and benefits of American society, notably in education and employment. Given this assessment, these observers urge the adoption of racial remedies to assist urban minorities; proposals include everything from special preference in education to affirmative action in employment.

Other observers emphasize class over race. The liberal variant of the class argument holds that blacks have been caught in a web of institutional and industrial change. Like other migrants, they arrived in cities to take low-skilled jobs in manufacturing, but they had the bad fortune to become established in this sector just as rising energy costs, changing technologies, and increased foreign competition brought a wave of plant closings and layoffs. The service economy that arose to replace manufacturing industries generated high-paying jobs for those with education, but poorly paid jobs for those without it.

Just as this transformation was undermining the economic foundations of the black working class, the class theorists argue, the civil rights revolution opened up new opportunities for educated minorities. After the passage of the 1964 Civil Rights Act, well-educated blacks were recruited into positions of responsibility in government, academia, and business, and thus provided the basis for a new black middle class. But civil rights laws could not provide high-paying jobs to poorly educated minorities when there were no jobs to give out. As a result, the class structure of the black community bifurcated into an affluent class whose fortunes were improving and a poverty class whose position was deteriorating.

The conservative variant of the class argument focuses on the deleterious consequences of government policies intended to improve the economic position of the poor. According to conservative reasoning, federal antipoverty programs implemented during the 1960s – notably the increases in Aid to Families with Dependent Children – altered the incentives governing the behavior of poor men and women. The accessibility and generosity of federal welfare programs reduced the attractiveness of marriage to poor women, increased the benefits of out-of-wedlock childbearing, and reduced the appeal of low-wage labor for poor men. As a result, female-headed families proliferated, rates of unwed childbearing rose, and male labor force participation rates fell. These trends drove poverty rates upward and created a population of persistently poor, welfare-dependent families.

Race- and class-based explanations for the underclass are frequently discussed as if they were mutually exclusive. Although liberal and conservative class theorists may differ with respect to the specific explanations they propose, both agree that white racism plays a minor role as a continuing cause of urban poverty; except for acknowledging the historical legacy of racism, their accounts are essentially race-neutral. Race theorists, in contrast, insist on the primacy of race in American society and emphasize its continuing role in perpetuating urban poverty; they view class-based explanations suspiciously, seeing them as self-serving ideologies that blame the victim.

By presenting the case for segregation's present role as a central cause of urban poverty, we seek to end the specious opposition of race and class. The issue is not whether race *or* class perpetuates the urban underclass, but how race *and* class *interact* to undermine the social and economic well-being of black Americans. We argue that race operates powerfully through urban housing markets, and that racial segregation interacts with black class structure to produce a uniquely disadvantaged neighborhood environment for African Americans.

If the decline of manufacturing, the suburbanization of employment, and the proliferation of unskilled service jobs brought rising rates of poverty and income inequality to blacks, the negative consequences of these trends were exacerbated and magnified by segregation. Segregation concentrated the deprivation created during the 1970s and 1980s to yield intense levels of social and economic isolation. As poverty was concentrated, moreover, so were all social traits associated with it, producing a structural niche within which welfare dependency and joblessness could flourish and become normative. The expectations of the urban poor were changed not so much by generous AFDC payments as by the spatial concentration of welfare recipients, a condition that was structurally built into the black experience by segregation.

If our viewpoint is correct, then public policies must address both race and class issues if they are to be successful. Race-conscious steps need to be taken to dismantle the institutional apparatus of segregation, and class-specific policies must be implemented to improve the socioeconomic status of minorities. By themselves, programs targeted to low-income minorities will fail because they will be swamped by powerful environmental influences arising from the disastrous neighborhood conditions that blacks experience because of segregation. Likewise, efforts to reduce segregation will falter unless blacks acquire the socioeconomic resources that enable them to take full advantage of urban housing markets and the benefits they distribute.

Although we focus in this chapter on how to end racial segregation in American cities, the policies we advocate cannot be pursued to the exclusion of broader efforts to raise the class standing of urban minorities. Programs to dismantle the ghetto must be accompanied by vigorous efforts to end discrimination in other spheres of American life and by class-specific policies designed to raise educational levels, improve the quality of public schools, create employment, reduce crime, and strengthen the family. Only a simultaneous attack along all fronts has any hope of breaking the cycle of poverty that has become deeply rooted.

"Urban Outcasts: Stigma and Division in the Black American Ghetto and the French Urban Periphery"

from *International Journal of Urban and Regional Research* (1993)

Loïc J. D. Wacquant

Editors' Introduction

Loïc Wacquant promotes a comparative perspective on the phenomenon of the urban underclass, through his comparative fieldwork among Northern Africans in the Parisian "Red Belt" city La Courneuve and its infamous public housing concentration called the Quatre mille, and African Americans in the classic "Black Belt" district known as the Southside of Chicago. Through compelling ethnography, he finds similarities in how the American inner city and the French urban periphery are stigmatized as territories of deprivation, dereliction, and danger. He detects commonality in the way both his informant groups experience demonization as undesirable and degraded social outsiders. He finds important differences, however, in the ensemble of historical and social processes that have marginalized these two groups as reservoirs of low-skilled labor in a de-industrializing economy. Race, space, and place are experienced in a similar fashion, but constituted differently in the two locations.

Wacquant draws a sharp contrast between French minorities and American minorities in the sense of political inclusion and cultural identity. His French North African informants carry an outlook of collectivism and social class-consciousness from the factory shop floor, neighborhood politics, and participation in street marches. He found, by contrast, that African Americans of the Southside of Chicago feel more marginalized by the individualistic ethos of U.S. society. They have a different experience of racial minority status; the French minorities are more adept at a kind of cultural code switching, while African Americans experience racial difference as a more permanent status of exclusion. This debunks the claim that European poverty is being "Americanized."

Loïc Wacquant was born and raised in Southern France, and educated in Montpellier and Paris. He received his Ph.D. in Sociology at the University of Chicago in 1994 after earlier graduate studies in industrial economics and fieldwork on the island of New Caledonia in the South Pacific. His research interests include urban poverty, criminal justice, race relations, and social theory. Wacquant has conducted fieldwork on the South Side of Chicago, and in the jails of big cities in the United States, France and Brazil. He has acted as consultant on issues of urban poverty, violence, ethnicity, and crime to central and local governments, unions, and the courts in France, Argentina, Brazil, Norway, Sweden, and to the OECD.

For the past several years, he has been Professor of Sociology and Research Associate at the Earl Warren Legal Institute, University of California at Berkeley, where he was also affiliated with the Program in Medical Anthropology and the Center for Urban Ethnography, and Researcher at the Centre de Sociologie Européenne du Collège de France. He is now on the faculty of the Graduate Faculty at the New School University in New York. He has been a member of the Society of Fellows at Harvard University and a MacArthur Foundation Fellow. His books include *An Invitation to Reflexive Sociology* (with Pierre Bourdieu; 1992, translated into 18 languages); *Prisons of Poverty* (1999, translated into 15 languages); *Body and Soul: Notebooks of an Apprentice Boxer* (2000, translated into nine languages); *Parias Urbanos* (2001); *Punir les Pauvres* (2004); *Deadly Symbiosis: Race and the Rise of Neoliberal Penality* (2004); *The Mystery of Ministry: Pierre Bourdieu and Democratic Politics* (forthcoming in 2005).

Wacquant recently completed an anthology of the works of Marcel Mauss on *Belief, Exchange, and Social Transformation* (forthcoming with the University of Chicago Press), a special thematic issue of *Ethnography* on "Dissecting the Prison" (vol. 3, no 4, December 2002), and the English translation of his first book on prizefighting, *Body and Soul* (Oxford: Oxford University Press, 2004). He is currently working on the sequel, *The Passion of the Pugilist*, an analysis of the dialectic of desire and domination in which he puts forth an argument for "carnal sociology." He is also preparing an anti-reader on the works of Pierre Bourdieu (with whom he worked closely for over a decade) entitled *Practice, Power, Knowledge: The Essential Bourdieu*; and writing up an ethnography of life strategies and "institutional stretching" on Chicago's South Side, *In the Zone: Making Do in the Dark Ghetto at Century's End*. His next project is a comparative historical sociology of forms and mechanisms of racial domination provisionally entitled *Peculiar Institutions*. He is co-founder and co-editor of the international and interdisciplinary journal *Ethnography*, director of the new book series "Practice and Symbolic Power" published by Verso (with W. W. Norton in the U.S.A.), and a regular contributor to *Le Monde diplomatique*.

Two interconnected trends have reshaped the visage of western European cities over the past decade or so. The first is the pronounced rise of multifarious urban inequalities and the crystallization of novel forms of socio-economic marginality, some of which appear to have a distinctly "ethnic" component and to feed (off) processes of spatial segregation and public unrest. The second is the surge and spread of ethno-racial or xenophobic ideologies and tensions consequent upon the simultaneous increase in persistent unemployment and the settlement of immigrant populations formerly thought of as guest workers.

[. . .]

The coincidence of new forms of urban exclusion with ethno-racial strife and segregation has given *prima facie* plausibility to the notion that European poverty is being "Americanized." Hence many European analysts (though by no means all) have turned to the United States for analytic assistance in an effort to puzzle out the current degradation of urban conditions and relations in their respective countries: thus the transatlantic diffusion of concepts, models, and sometimes ready-made theories from recent (and not-so-recent) American social science. This is visible in the worried and confused public discussion in France – and in other countries such as Belgium, Germany and Italy – about the presumed formation of immigrant "ghettos" in degraded working-class neighborhoods harboring large low-income housing tracts known as *cités*. It can be detected also in the spread of the notion of "underclass" in Britain and in its smuggling into the Netherlands to address the strain put on citizenship by the emerging concatenation of joblessness, ethnic discrimination and neighborhood decline. Such conceptual borrowings, however, stand on shaky analytical grounds inasmuch as they presume exactly that which needs to be established: namely, that the American conceptual idiom of "race relations" has purchase on the urban realities of Europe – leaving aside the question of whether conventional American categories (or newer concepts such as the largely mythical notion of "underclass") pack any analytical power on their own turf to start with.

The best way to answer, or at least productively reframe, this question is through a systematic, empirically grounded, cross-national comparison of contemporary forms of urban inequality and ethno-racial/class exclusion which (1) does not presuppose that the analytical apparatus forged on one continent should be imposed wholesale on the other, and is sensitive to the fact that all "national" conceptual tools have embedded within them specific social, political and moral assumptions reflective of the particular history of society and state in each country; (2) attends consistently to the meanings and lived experiences of social immobility and marginality; and (3) strives to embed individual strategies and collective trajectories firmly into the local social structure as well as within the broader national framework of market and state.

This chapter is part of a broader attempt to contribute to such a comparative sociology through an analysis of the social and mental structures of urban exclusion in the American "Black Belt" and the French "Red Belt." "Black Belt" is used here to denote the remnants of the historic "dark ghetto" of the large metropolis of the North-East and Mid-West of the United States, that is the dilapidated racial enclaves of the metropolitan core which have dominated recent public and academic discussions of race and poverty in North America. The expression "Red Belt" refers not simply to the townships of the outer ring of Paris that form(ed) the historic stronghold of the French Communist party, but more generally to the traditional mode of organization of "workers' cities" in France, anchored by industrial male employment, a strong workerist culture and solidaristic class consciousness, and civic incorporation of the population through a dense web of union-based and municipal organizations creating a close integration of work, home and public life. It is in such peripheral working-class neighborhoods that urban inequalities and unrest have coalesced, making the question of the *banlieue* perhaps the most pressing public issue in the France of the 1980s.

The analysis that follows uses data from a variety of primary and secondary sources and combines observations drawn from censuses, surveys and field studies of the American ghetto and the French *banlieue*. On the French side, I concentrate on the Red Belt city of La Courneuve and its infamous public housing concentration called the Quatre mille (after the nearly 4,000 units it originally contained). La Courneuve is an older, Communist governed, north-eastern suburb of Paris with a population of 36,000 situated midway between the national capital and the Roissy-Charles de Gaulle airport, in the midst of a densely urbanized, declining industrial landscape. On the American side, I focus on the South Side ghetto of Chicago where I conducted ethnographic fieldwork in 1988–91. The South Side is a sprawling, all-black zone containing some 130,000 inhabitants, the majority of whom are unemployed and live under the official federal "poverty line." I have presented elsewhere a detailed sociography of both sites which highlighted a number of parallel morphological traits and trends. In summary, both locales were found to have a declining population with a skewed age and class structure characterized by a predominance of youths, manual workers and deskilled service personnel, and to harbor large concentrations of "minorities" (North African immigrants on the one side, blacks on the other) which exhibit unusually high levels of unemployment caused by deindustrialization and labor market changes. This comparison also turned up structural and ecological differences suggesting that the declining French working-class, *banlieue* and the black American ghetto constitute two different socio-spatial formations, produced by different institutional logics of segregation and aggregation and resulting in pronouncedly higher levels of blight, poverty and hardship in the ghetto. To simplify greatly: exclusion operates on the basis of color reinforced by class and state in the Black Belt, but mainly on the basis of class and mitigated by the state in the Red Belt, with the result that the former is a racially and culturally homogeneous universe characterized by low organizational density and state penetration, whereas the latter is fundamentally heterogeneous in terms of both class and ethno-national recruitment, with a strong presence of public institutions.

The purpose of this article is to flesh out some of the invariants and variations in the socio-organizational and cognitive structures of urban exclusion by contrasting two dimensions of daily life salient in both the French *banlieue* and the black American ghetto, though, as we shall see, with significantly discrepant inflections, degrees of urgency and socio-political dynamics. The paper addresses the powerful territorial stigma that attaches to residence in an area publicly recognized as a "dumping ground" for poor people, downwardly mobile working-class households and marginal groups and individuals. Poverty is too often (wrongly) equated with material dispossession or insufficient income. But, in addition to being deprived of adequate conditions and means of living, to be poor in a rich society entails having the status of a social anomaly and being deprived of control over one's collective representation and identity: the analysis of public taint in the American ghetto and the French urban periphery serves to stress the *symbolic dispossession* that turns their inhabitants into veritable social outcasts.

[. . .]

TERRITORIAL STIGMATIZATION – ITS EXPERIENCE AND EFFECTS

Any comparative sociology of the "new" urban poverty in advanced societies must begin with the powerful stigma attached to residence in the bounded and segregated spaces, the "neighborhoods of exile" to which the populations marginalized or condemned to redundancy by the postfordist reorganization of the economy and state are increasingly being relegated – not only because it is arguably the single-most protrusive feature of the lived experience of those assigned to, or entrapped in, such areas but also because this stigma helps explain certain similarities in their strategies of coping or escape and thereby many of the surface cross-national commonalities that have given apparent validity to the idea of a transatlantic convergence between the "poverty regimes" of Europe and the United States.

"It's like there's a plague here"

Because they constitute the lowest tier of that nation's public housing complex, have undergone continual material and demographic decline since their erection in the early 1960s and received a strong inflow of foreign families from the mid-1970s onwards, the *cités* of the French urban periphery suffer from a negative public image that instantly associates them with rampant delinquency, immigration and insecurity, so much so that they are almost universally called "little Chicagos," both by their residents and by outsiders. To dwell in a Red Belt low-income estate means to be confined to a branded space, to a blemished setting experienced as a "trap." Thus the media and its inhabitants themselves routinely refer to the Quatre mille as a "dumpster," "the garbage can of Paris," or even a "reservation," a far cry from the official bureaucratic designation of "sensitive neighborhood" used by the public officials in charge of the state's urban renewal program. In recent years, the press of stigmatization has augmented sharply with the explosion of discourses on the alleged formation of so-called *cités-ghettos* widely (mis)represented as growing pockets of "Arab" poverty and disorder symptomatic of the incipient "ethnicization" of France's urban space.

It should be noted however that the Quatre mille does not exist as such in the perceptions of its residents. The indigenous taxonomies the latter use to organize their daily round distinguish numerous subunits within the large estate which in effect has only an administrative and symbolic existence – though it has real consequences. What appears from the outside to be a monolithic ensemble is seen by its members as a finely differentiated congery of "micro-locales": those from the northern cluster of the project, in particular, want nothing to do with their counterparts of the southern section, whom they consider to be "hoodlums" (*racaille* or *caillera* in the local youth slang), and vice versa. "For the residents of the Quatre mille, to change building sometimes means to change lives." Yet it remains that dwellers in *cités* have a vivid awareness of being "exiled" to a degraded space that collectively disqualifies them. Rachid, a former resident of the Quatre mille, gives virulent expression to this sense of indignity when asked about the eventuality of moving back into the project: "For us to return there, it would be to be insulted once again. The Quatre mille are an insult . . . For many people, the Quatre mille are experienced as a shame." When the interviewer inquires about the possibility of salvaging the housing project through renovation, his answer is no less blunt:

> To renovate is to take part in shame. If you agree to play this game, then in a way you're endorsing shame. We've come to a point of no return where you got no solution but to raze the whole thing. Besides the people here agree there's only one solution: "Gotta blow it up." Go and ask them. . . . When you don't feel good inside, when you don't feel good outside, you got no jobs, you got nothing going for you, then you break things, that's the way it is. The shit they're doing trying to fix the garbage disposal and the hallway entrance, the painting, that's no use: it's gonna get ripped right away. It's dumb. It's the whole thing that's the problem. . . . You gotta raze the whole thing.

For Sali, another North African youth from the Quatre mille, the project is "a monstrous universe" seen as an instrument of social confinement by its inhabitants: "It's a jail. They [second-generation residents] are in jail, they got tricked real good, so when they get together, they have karate fights against the mail boxes and bust everything up. It's all quite easy to understand." The verbal violence of these youths, as well as the vandalism they invoke, must be understood as a response to the socio-economic and symbolic violence they feel subjected to by being thus relegated to a defamed place. Not surprisingly, there is great distrust and bitterness among them about the ability of political institutions and the willingness of local leaders to rectify the problem.

It is hardly possible for residents of the *cité* to overlook the scorn of which they are the object since the social taint of living in a low-income housing project that has become closely associated with poverty, crime and moral degradation affects all realms of existence – whether it is searching for employment, pursuing romantic involvements, dealing with agencies of social control such as the police or welfare services or simply talking with acquaintances. Residents of the Quatre mille are quick to impute the ills of their life to the fact of being "stuck" in a "rotten" housing project that they come to perceive through a series of homological oppositions (*cité*/city, us/them, inside/outside, low/high, savage/civilized) that reproduce and endorse the derogatory judgment of outsiders. When asked where they reside, many of "those who work in Paris say vaguely that they live in the northern suburbs" rather than reveal their address in La Courneuve. Some will walk to the nearby police station when they call taxicabs to avoid the humiliation of being picked up at the doorstep of their building. Parents forewarn their daughters against going out with "guys from the Quatre mille."

Residential discrimination hampers the search for jobs and contributes to entrench local unemployment since inhabitants of the Quatre mille encounter additional distrust and reticence among employers as soon as they mention their place of residence. A janitor in the *cité* relates a typical incident in which he helped new tenants contact firms by telephone only to be told that there no longer was any position open whenever he revealed where he was calling from: "It's like there's a plague here," he says in disgust. Territorial stigmatization affects interactions not only with employers but also with the police, the courts and street-level welfare bureaucracies, all of which are especially likely to modify their conduct and procedures based on residence in a degraded *cité*. "All youths recount the change of attitude of policemen when they notice their address during identity checks," for to be from a *cité* carries with it a reflex suspicion of deviance if not of outright guilt. A high school student tells of being stopped by subway controllers in the Paris Métro: "We took out our identity cards. When they saw that we were from the Quatre mille, I swear to you! They went . . . they turned pale."

"People really look down on you"

In America, the dark ghetto stands similarly as the national symbol of urban "pathology," and its accelerating deterioration since the racial uprisings of the mid-1960s is widely regarded as incontrovertible proof of the moral dissolution, cultural depravity and behavioral deficiencies of its inhabitants. The journalistic reports and academic (pseudo) theories that have proliferated to account for the putative emergence of a so-called "underclass" in its midst have accelerated the *demonization of the black urban (sub)proletariat* by symbolically severing it from the "deserving" working class and by obfuscating – and thereby retroactively legitimating – the state policies of urban abandonment and punitive containment responsible for its downward slide.

Today, living in the historic Black Belt of Chicago carries an automatic presumption of social unworthiness and moral inferiority which translates into an acute consciousness of the symbolic degradation associated with being confined to a loathed and despised universe. A student from a vocational high school on the city's South Side voices this sense of being cut off from and cast out of the larger society thus:

> People really look down on you because of where you come from and who you are. People don't want to have anything to do with you . . . You can tell when you go places, people are looking at you like you are crazy or something.

The defamation of the ghetto is inscribed first in the brute facts of its physical dilapidation and of the separateness and massive inferiority of its resident institutions, be they public schools, social agencies, municipal services, neighborhood associations or financial and commercial outlets. It is constantly reaffirmed by the diffident and contemptuous attitudes of outsiders: banks, insurance companies, taxis, delivery trucks and other commercial services avoid the Black Belt or venture into it only gingerly; kin and kith are reluctant to visit. "Friends from other places don't want really to come here. And you yourself, you wouldn't want to invite intelligent people here: there's markings and there's writing on the wall, nasty, whatever," says an unemployed mother of three who lives in a West Side project. Children and women living in public housing in the inner city find it difficult to develop personal ties with outsiders once the latter learn of their place of residence.

Desmond Avery, who lived in both the Cabrini Green project in Chicago and in the Quatre mille, remarks that residential discrimination is at least as prevalent in the Windy City as in the Parisian periphery. Ghetto dwellers are well aware that living in a stigmatized section of town penalizes them on the labor market: "Your address, its impression for jobs." Residing on the South Side, and even more so in a public housing project whose name has become virtually eponymous with "violence and depravity," is yet another hurdle in the arduous quest for employment. A jobless woman who lives in the ill-reputed Cabrini Green housing development remarks:

> It's supposed to be discrimination, but they get away with it, you know. Yes, it's important where you live. Employers notice, they notice addresses, when that application's goin' through personnel, they are lookin' at that address: (worried tone) "Oh, you're from *here*!?"

Over and beyond the scornful gaze of outsiders and the reality of exclusion from participation in society's regular institutions, the thoroughly depressed state of the local economy and ecology exerts a pervasive effect of demoralization upon ghetto residents. Indeed, the words "depressing" and "uninspiring" come up time and again in the description that the latter give of their surroundings. Moreover, two-thirds of the inhabitants of the South Side and West Side of Chicago expect that their neighborhood will either stay in the same state of blight or deteriorate further in the near future; the only route for improvement is to move out, to which nearly all aspire. But the possibility of accumulating resources in preparation for upward mobility is further eroded by the predatory nature of relations between residents and by the pressure towards social uniformity which weighs on those who try to rise above the poverty level common to most inhabitants of the area: "They won't let you get ahead. Stealin' from you and robbin' you and all that kinda thing," laments a 27-year-old machine operator from the far South Side. Given the inordinate incidence of violent crime, living in a ghetto neighborhood also entails significant physical risk and, as a corollary, high levels of psychic stress which tend to "drag you down" and "wear you out." No wonder that life in the Black Belt is suffused with a sense of gloom and fatality, a social *fatum* which obstructs the future from view and seems to doom one to a life of continued failure and rejection.

FROM SOCIAL STIGMATIZATION TO SOCIAL "DISORGANIZATION"

Paradoxically, the experiential burden of territorial stigmatization weighs more heavily on the residents of the French *banlieue* than it does on their counterparts of the American ghetto, in spite of the fact that the latter constitutes a considerably more desolate and oppressive environment. Three factors help account for this apparent disjuncture between objective conditions and the subjective (in)tolerance of those who evolve in them. First, the very idea of relegation into a separate space of *institutionalized social inferiority and immobility* stands in blatant violation of the French ideology of unitarist citizenship and participation in the national community, an ideology fully embraced and forcefully invoked by youths from the Red Belt, especially second-generation immigrants of North African origins in their street protest and marches of the past decade. By contrast, the color line of which the black ghetto is the most visible institutional expression is so ingrained in the make-up of the American urban landscape that it has become part of the order of things: racial division is a thoroughly taken-for granted constituent of the organization of the metropolitan economy, society and polity. Second, residents of the American ghetto are more prone to embracing a highly individualistic ideology of achievement than are their counterparts of the French *cités*. Many, if not most, adhere to the social darwinistic view that social position ultimately reflects one's moral worth and personal strivings, so that no one in the long run can be consistently held back by his or her place of residence. A third and most crucial difference between Red Belt and Black Belt is found in the nature of the stigma they carry: this stigma is only residential in the former, but jointly and inseparably spatial-cum-racial in the latter. The French *banlieue* is but a territorial entity which furthermore contains a mixed, multi-ethnic population; it suffices for inhabitants of the Quatre mille or any other *cité* to hide their address in order to "pass" in the broader society. No readily perceptible physical or cultural marker brands them as members of the Red Belt, and use of simple techniques of "impression management" enable them to shed the stigma, if only temporarily. Thus adolescents from poor Parisian *banlieues* regularly go to "hang out" in the up-market districts of the capital to escape their neighborhood and gain a sense of excitement. By traversing spaces that both symbolize and contain the life of higher classes, they can live for a few hours a fantasy of social inclusion and participate, albeit by proxy, in the wider society. This "consciousness switch" renders more intolerable the idea of permanent exclusion and the outcast status associated with consignment to a degraded *cité*.

Residents of the American Black Belt are not granted the luxury of this dual "awareness context." For the ghetto is not simply a spatial entity, or a mere aggregation of poor families stuck at the bottom of the class structure: it is a uniquely *racial formation* that spawns a society-wide web of material and symbolic associations between color, place and a host of negatively valued social properties. The fact that color is a marker of identity and a principle of vision and division that is immediately available for interpretation and use in public space and interaction makes it nearly impossible for inner-city dwellers to shed the stigma attached to ghetto residence. For instance, they cannot casually cross over into adjacent white neighborhoods, for there the sight of a young black man evokes an image of someone dangerous, destructive, or deviant" so that they will promptly be trailed and stopped, nay systematically harassed, by police. Ghetto blacks in America suffer from *conjugated stigmatization*: they cumulate the negative symbolic capital attached to color and consignment to a specific, reserved and inferior territory itself devalued for being both the repository of the lowest class elements of society and a racial reservation. In a race-divided society such as the United States, where all spheres of life are thoroughly color-coded and given low chances of escaping the ghetto, the best one can do is make a virtue of necessity and learn to live with a stigma that is both illegitimate and unacceptable to French working-class youths of the Red Belt *cités*.

Yet the main effect of territorial stigmatization is similar in both countries: it is to stimulate practices of internal social differentiation and distancing that work to decrease interpersonal trust and undercut local social solidarity. To regain a measure of dignity and reaffirm the legitimacy of their own status in the eyes of society, residents of both *cité* and ghetto typically overstress their moral worth as individuals (or as family members) and join in the dominant discourse of denunciation of those who undeservingly "profit" from social assistance programs, *faux pauvres* and "welfare cheats." It is as if they could gain value only by devaluing their neighborhood and their neighbors. They also engage in a variety of strategies of social distinction and withdrawal which converge to undermine neighborhood cohesion. These take three main forms: mutual avoidance, reconstitution and elaboration of "infra-differences" or micro-hierarchies, and the diversion of public opprobrium onto scapegoats such as notorious "problem families" and foreigners or drug dealers and single mothers. In the French *cité*, residents commonly insist that they are there only "by accident" and carp over the waste of public resources allocated to those who, "contrary to them," do not genuinely need assistance. Similarly, in Chicago's ghetto, residents disclaim belonging to the neighborhood as a network of mutual acquaintance and exchange and strive to set themselves aside from what they know to be a place and population of ill repute. This 41-year-old nurse from the West Side neighborhood of North Lawndale, one of the most destitute in the city, speaks for many of her peers in both Black Belt and Red Belt when she says: "Hell, I don't know what people [around here] do, I guess I'm pretty much on my own. I don't associate with people in the neighborhood; I mean I speak to them, but as far as knowing what they're about, I don't know."

To sum up, residents of the French *cité* and of the American ghetto each form an *impossible community*, perpetually divided against themselves, which cannot but refuse to acknowledge the collective nature of their predicament and who are therefore inclined to deploy strategies of distancing and "exit" that tend to validate negative outside perceptions and feed a deadly self-fulfilling prophecy through which public taint and collective disgrace eventually produce that which they claim merely to record: namely, social atomism, community "disorganization" and cultural anomie.

[. . .]

CONCLUSION

The purpose of this paper has been to uncover some of the similarities and differences between the "new urban poverty" in France and in America as it is locally structured and experienced by those whom the term (or its equivalent) has come to designate in these two countries. Rather than compare national aggregate statistics on income, standards of living or consumption patterns, which often measure little more than properties of the survey bureaucracies and procedures which generate them and take no account of the specific welfare-state and socio-spatial environments in which individuals and groups actually evolve in each society, I have proceeded by way of a contextualized examination of a master aspect of life in a stigmatized neighborhood of concentrated poverty: territorial indignity and its debilitating consequences upon the fabric and form of the local social structure.

Drawing out the organizational and cognitive texture of everyday living in the Parisian Red Belt and in Chicago's Black Belt, how the residents of these blighted areas negotiate and experience social immobility and ostracization in "the ghetto" – as social myth in one case and enduring historic reality in the other – highlights the distinctively racial dimension of inner city poverty in the United States. It also points to the uncertainty in the process of collective identity formation in the Red Belt caused by the demise of traditional agencies of class formation. Whether France and America converge or continue to differ in the future with regard to the social and spatial patterning of inequality in the city, there can be little doubt that racial separation, where it prevails, radicalizes the objective and subjective reality of urban exclusion; and that state support (or tolerance) of segregation and recognition of ethnoracial divisions only serve to intensify the cumulation of urban dispossession and to exacerbate the destructive consequences of socio-economic marginality, not only for those upon whom it is imposed and for their neighborhoods, but for the broader society as well.

"The Immigrant Enclave: Theory and Empirical Examples"

from Susan Olzak and Joane Nagel (eds), *Competitive Ethnic Relations* (1986)

Alejandro Portes and Robert D. Manning

Editors' Introduction

Alejandro Portes and Robert Manning offer an insightful comparative overview of the major schools of literature on immigrant enclaves in the U.S. today. Immigrant enclaves are important phenomena to consider since the 1960s in American cities, during which time the U.S. experienced the growth of an urban underclass and persisting race and class segregation despite the success of social mobility and spatial integration for some of the black middle class. We observe also the onset of globalization and neoliberal economic policies that stimulated the mobility of labor and capital. The Hart–Cellar Immigration Law of 1965 and subsequent immigration, banking, and free trade accords have led to the widespread emergence of Latino and Asian residential and business enclaves in U.S. cities. Chinatown, Koreatown and Little Havana offer an interesting comparison with immigrant enclaves in American cities of the early twentieth century, which were more commonly white ethnic (such as Little Italy and the Jewish enclave), and operated during a different phase of economic growth. Immigrant enclaves challenge us to look beyond the traditional black/white dynamics of American cities, and consider how new immigrants are being inserted into the existing system of race and class stratification.

Portes and Manning first make the point that the proliferation of immigrant enclaves challenges traditional precepts of assimilation theory, as the persistence of ethnic identity and ethnic communities is more permissible and commonplace since the 1960s. Whereas social mobility in American society was in the early twentieth century more predicated on the suppression of ethnic ancestry and the acquisition of American cultural values, we have seen that socioeconomic prosperity in contemporary America can be promoted along with a continuing commitment to ethnicity and ethnic enclaves. Immigrant enclaves are growing as often as they are disappearing, and furthermore, they arise sometimes in the suburbs. Immigrant enclaves are furthermore an alternative for economic incorporation beyond the existing dichotomy of a segmented labor market in which there is an upper tier of jobs that offers good mobility ladders, and a lower tier of dead-end jobs in which minorities predominate.

The contradiction is that immigrant enclaves offer opportunity for some ethnic people through the exploitation of co-ethnic others. Ethnic enclaves offer a kind of protection by hiring immigrants who may be undocumented or lack good English language skills. They commonly offer on-the-job training rather than requiring higher education. While being in this "protective" sector, they may also be subject to severe exploitation, working with low wages, poor benefits, no labor rights, and "sweatshop" conditions that may violate labor law. On the other hand, some co-ethnics benefit, notably bosses and immigrant business owners. Forward, backward, and consumption linkages within immigrant enclaves commonly result in the re-circulating of dollars in the ethnic economy via the phenomenon of the economic multiplier.

Immigrant enclaves commonly employ co-ethnic workers and serve co-ethnic customers. There are, however, a number of ethnic actors that act as "middleman minorities" between white elites and blacks and other minorities. Chinese and Korean immigrants, for instance, commonly operate small businesses in lower-class black communities, filling a niche considered socially undesirable or unprofitable by white businesses and corporations. Middleman minorities operate as a social buffer between the elite and oppressed and underclass groups of a given society. In times of political or economic crisis, the oppressed minorities may lash out with racial violence or anger against the middleman groups. Intergroup conflicts of this sort occurred in Miami in the early 1980s between Blacks and Cubans, and in the 1990s in Los Angeles and New York between Asians and Blacks.

For further reading on middleman minorities, see Edna Bonacich, "A Theory of Middleman Minorities," *American Sociological Review* 38 (October 1973): 583–594, who gives a historical perspective that highlights the experience of global Jewish traders and the Chinese in Southeast Asia. Pyong Gap Min considers Korean/Black American relations in *Caught in the Middle: Korean Communities in New York and Los Angeles* (Berkeley, CA: University of California Press, 1996). Spike Lee offers a provocative interpretation of Black/Korean relations in Brooklyn, New York, in his film, *Do the Right Thing*, especially in the scene where Radio Raheem confronts Korean-American shopkeepers with a request for "D" batteries. The confrontation is reiterated in the Ice Cube song, "Black Korea," in the album *Death Certificate* (Priority Records, 1991).

Alejandro Portes is the Howard Harrison and Gabrielle Snyder Beck Professor of Sociology and director of the Center for Migration and Development at Princeton University. He formerly taught at Johns Hopkins University, where he held the John Dewey Chair in Arts and Sciences, Duke University, and the University of Texas-Austin. He served as president of the American Sociological Association in 1997–99. Born in Havana, Cuba, he came to the United States in 1960. He was educated at the University of Havana, Catholic University of Argentina, and Creighton University. He received his M.A. and Ph.D. from the University of Wisconsin-Madison.

Portes is the author of some 200 articles and chapters on national development, international migration, Latin American and Caribbean urbanization, and economic sociology. His books include *City on the Edge – the Transformation of Miami* (Berkeley, CA: University of California Press, 1993), co-authored with Alex Stepick and winner of the 1995 Robert Park Award for best book in community and urban sociology and the Anthony Leeds Award for best book in urban anthropology in 1995. His current research is on the adaptation process of the immigrant second generation and the rise of transnational immigrant communities in the United States. His most recent books, co-authored with Rubén G. Rumbaut, are *Legacies: The Story of the Immigrant Second Generation* and *Ethnicities: Children of Immigrants in America* (Berkeley, CA: University of California Press, 2001). *Legacies* was the winner of the 2002 Distinguished Scholarship Award from the American Sociological Association and of the 2002 W. I. Thomas and Florian Znaniecki Award for best book from the International Migration Section of ASA.

Robert Manning is currently the Caroline Werner Gannett Professor at the Rochester Institute of Technology. He held previous faculty appointments at American University and Georgetown University. He is a specialist in comparative economic relations, immigration, and minority relations in the United States. He received his Ph.D. from the Johns Hopkins University, his M.A. from Northern Illinois University, and his B.A. from Duke University. His recent book, *Credit Card Nation* (New York: Basic Books, 2000) focuses on the damaging social and political consequences of America's increasing reliance on credit cards.

INTRODUCTION

The purpose of this chapter is to review existing theories about the process of immigrant adaptation to a new society and to recapitulate the empirical findings that have led to an emerging perspective on the topic. This emerging view revolves around the concepts of different modes of structural incorporation and of the immigrant enclave as one of them. These concepts are set in explicit opposition to two previous viewpoints on the adaptation process, generally identified as assimilation theory and the segmented labor markets approach.

The study of immigrant groups in the United States has produced a copious historical and sociological literature, written mostly from the assimilation perspective. Although the experiences of particular groups varied, the common theme of these writings is the unrelenting efforts of immigrant minorities to surmount obstacles impeding their entry into the "mainstream" of American society. From this perspective, the adaptation process of particular immigrant groups followed a sequential path from initial economic hardship and discrimination to eventual socioeconomic mobility arising from increasing knowledge of American culture and acceptance by the host society. The focus on a "core" culture, the emphasis on consensus building, and the assumption of a basic patterned sequence of adaptation represent central elements of assimilation theory.

[. . .]

The second general perspective takes issue with this psychosocial and culturalist orientation as well as with the assumption of a single basic assimilation path. This alternative view begins by noting that immigrants and their descendants do not necessarily "melt" into the mainstream and that many groups seem not to want to do so, preferring instead to preserve their distinct ethnic identities. A number of writers have focused on the resilience of these communities and described their functions as sources of mutual support and collective political power. Others have gone beyond descriptive accounts and attempted to establish the causes of the persistence of ethnicity. Without exception, these writers have identified the roots of the phenomenon in the economic sphere and, more specifically, in the labor-market roles that immigrants have been called on to play.

Within this general perspective, several specific theoretical approaches exist. The first focuses on the situation of the so-called unmeltable ethnics – blacks, Chicanos, and American Indians – and finds the source of their plight in a history of internal colonialism during which these groups have been confined to specific areas and made to work under uniquely unfavorable conditions. In a sense, the role of colonized minorities has been to bypass the free labor market, yielding in the process distinct benefits both to direct employers of their labor and, indirectly, to other members of the dominant racial group. This continuation of colonialist practices to our day explains, according to this view, the spatial isolation and occupational disadvantages of these minorities.

A second approach attempts to explain the persistence of ethnic politics and ethnic mobilization on the basis of the organization of subordinate groups to combat a "cultural division of labor." The latter confined members of specific minorities to a quasi-permanent situation exploitation and social inferiority. Unlike the first view, the second approach does not envision the persistence of ethnicity as a consequence of continuing exploitation, but rather as a "reactive formation" on the part of the minority to reaffirm its identity and its interests. For this reason, ethnic mobilizations are often most common among groups who have already abandoned the bottom of the social ladder and started to compete for positions of advantage with members of the majority.

A final variant focuses on the situation of contemporary immigrants to the United States. Drawing on the dual labor market literature, this approach views recent immigrants as the latest entrants into the lower tier of a segmented labor market where women and other minorities already predominate. Relative to the latter, immigrants possess the advantages of their lack of experience in the new country, their legal vulnerability, and their greater initial motivation. All of these traits translate into higher productivity and lower labor costs for the firms that employ them. Jobs in the secondary labor market are poorly paid, require few skills, and offer limited mobility opportunities. Hence, confinement of immigrants to this sector insures that those who do not return home are relegated to a quasi-permanent status as disadvantaged and discriminated minorities.

What these various structural theories have in common is the view of resilient ethnic communities formed as a result of a consistently disadvantageous economic position and the consequent absence of a smooth path of assimilation. These situations, ranging from slave labor to permanent confinement to the secondary labor market, are not altered easily. They have given rise, in time, either to hopeless communities of "unmeltable" ethnics or to militant minorities, conscious of a common identity and willing to support a collective strategy of self-defense rather than rely on individual assimilation.

These structural theories have provided an effective critique of the excessively benign image of the adaptation process presented by earlier writings. However, while undermining the former, the new structural perspective may have erred in the opposite direction. The basic hypothesis advanced in this chapter is that several identifiable modes of labor-market incorporation exist and that not all of them relegate newcomers to a permanent situation of exploitation and inferiority. Thus, while agreeing with the basic thrust of structural theories, we propose several modifications that are necessary for an adequate understanding of the different types of immigrant flows and their distinct processes of adaptation.

MODES OF INCORPORATION

In the four decades since the end of World War II, immigration to the United States has experienced a vigorous surge reaching levels comparable only to those at the beginning of the century. Even if one restricts attention to this movement, disregarding multiple other migrations elsewhere in the world, it is not the case that the inflow has been of a homogeneous character. Low-wage labor immigration itself has taken different forms, including temporary contract flows, undocumented entries, and legal immigration. More importantly, it is not the case that all immigrants have been directed to the secondary labor market. For example, since the promulgation of the Immigration Act of 1965, thousands of professionals, technicians, and craftsmen have come to the United States, availing themselves of the occupational preference categories of the law. This type of inflow, dubbed "brain drain" in the sending nations, encompasses today sizable contingents of immigrants from such countries as India, South Korea, the Philippines, and Taiwan, each an important contributor to U.S. annual immigration.

The characteristics of this type of migration have been described in detail elsewhere. Two such traits deserve mention, however. First, occupationally skilled immigrants – including doctors, nurses, engineers, technicians, and craftsmen – generally enter the primary labor market; they contribute to alleviate domestic shortages in specific occupations and gain access, after a period of time, to the mobility ladders available to native workers. Second, immigration of this type does not generally give rise to spatially concentrated communities; instead, immigrants are dispersed throughout many cities and regions, following different career paths.

Another sizable contingent of entrants whose occupational future is not easily characterized a priori are political refugees. Large groups of refugees, primarily from Communist-controlled countries, have come to the United States, first after the occupation of Eastern Europe by the Soviet Army, then after the advent of Fidel Castro to power in Cuba, and finally in the aftermath of the Vietnam War. Unlike purely "economic" immigrants, refugees have often received resettlement assistance from various governmental agencies. The economic adaptation process of one of these groups, the Cubans, will be discussed in detail in this chapter. For the moment, it suffices to note that all the available evidence runs contrary to the notion of a uniform entry of political refugees into low-wage secondary occupations; on the contrary, there are indications of their employment in many different lines of work.

A third mode of incorporation has gained the attention of a number of scholars in recent years. It consists of small groups of immigrants who are inserted or insert themselves as commercial intermediaries in a particular country or region. These "middleman minorities" are distinct in nationality, culture, and sometimes race from both the superordinate and subordinate groups to which they relate. They can be used by dominant elites as a buffer to deflect mass frustration and also as an instrument to conduct commercial activities in impoverished areas. Middlemen accept these risks in exchange for the opportunity to share in the commercial and financial benefits gained through such instruments as taxation, higher retail prices, and usury. Jews in feudal and early modern Europe represent the classic instance of a middleman minority. Other examples include Indian merchants in East Africa, and Chinese entrepreneurs in Southeast Asia and throughout the Pacific Basin. Contemporary examples in the United States include Jewish, Korean, and other Oriental merchants in inner-city ghetto areas and Cubans in Puerto Rico.

Primary labor immigration and middleman entrepreneurship represent two modes of incorporation that differ from the image of an homogeneous flow into low-wage employment. Political refugees, in turn, have followed a variety of paths, including both of the above as well as insertion into an ethnic enclave economy. The latter represents a fourth distinct mode. Although frequently confused with middleman minorities, the emergence and structure of an immigrant enclave possess distinct characteristics. The latter have significant theoretical and practical implications, for they set apart groups adopting this entry mode from those following alternative paths. We turn now to several historical and contemporary examples of immigrant enclaves to clarify their internal dynamics and causes of their emergence.

IMMIGRANT ENCLAVES

Immigration to the United States before World War I was, overwhelmingly, an unskilled labor movement. Impoverished peasants from southern Italy, Poland, and the eastern reaches of the Austro-Hungarian Empire settled in dilapidated and crowded areas, often immediately adjacent to their points of debarkation, and took any menial jobs available. From these harsh beginnings, immigrants commenced a slow and often painful process of acculturation and economic mobility. Theirs was the saga captured by innumerable subsequent volumes written from both the assimilation and the structural perspectives.

Two sizable immigrant groups did not follow this pattern, however. Their most apparent characteristic was the economic success of the first generation, even in the absence of extensive acculturation. On the contrary, both groups struggled fiercely to preserve their cultural identity and internal solidarity. Their approach to adaptation thus directly contradicted subsequent assimilation predictions concerning the causal priority of acculturation to economic mobility. Economic success and "clannishness" also earned for each minority the hostility of the surrounding population. These two immigrant groups did not have a language, religion, or even race in common and they never overlapped in significant numbers in any part of the United States. Yet, arriving at opposite ends of the continent, Jews and Japanese pursued patterns of economic adaptation that were quite similar both in content and in their eventual consequences.

Jews in Manhattan

The first major wave of Jewish immigration to the United States consisted of approximately 50,000 newcomers of German origin, arriving between 1840 and 1870. These immigrants went primarily into commerce and achieved, in the course of a few decades, remarkable success. By 1900, the average income of German-Jewish immigrants surpassed that of the American population. Many individuals who started as street peddlers and small merchants had become, by that time, heads of major industrial, retail, and financial enterprises.

The second wave of Jewish immigration exhibited quite different characteristics. Between 1870 and 1914, over two million Jews left the Pale of Settlement and other Russian-dominated regions, escaping Czarist persecution. Major pogroms occurred before and during this exodus. Thus, unlike most immigrants of the period, the migration of Russian and Eastern Europe Jews was politically motivated and their move was much more permanent. In contrast to German Jews, who were relatively well educated, the Yiddish-speaking newcomers came, for the most part, from modest origins and had only a rudimentary education. Although they viewed the new Russian wave with great apprehension, German Jews promptly realized that their future as an ethnic minority depended on the successful integration of the newcomers. Charitable societies were established to provide food, shelter, and other necessities, and private schools were set up to teach immigrants English, civics, and the customs of the new country.

Aside from its size and rapidity of arrival, turn-of-the-century Jewish immigration had two other distinct characteristics. First was its strong propensity toward commerce and self-employment in general in preference to wage labor; as German

Jews before them, many Russian immigrants moved directly into street peddling and other commercial activities of the most modest sort. Second was its concentration into a single, densely populated urban area – the lower East Side of Manhattan. Within this area, those who did not become storekeepers and peddlers from the start found employment in factories owned by German Jews, learning the necessary rudiments for future self-employment. . . .

The economic success of many of these ventures did not require and did not entail rapid acculturation. Immigrants learned English and those instrumental aspects of the new culture required for economic advancement. For the rest, they preferred to remain with their own and maintained, for the most part, close adherence to their original religion, language, and values. Jewish enclave capitalism depended, for its emergence and advancement, precisely on those resources made available by a solidaristic ethnic community: protected access to labor and markets, informal sources of credit, and business information. It was through these resources that upstart immigrant enterprises could survive and eventually compete effectively with better-established firms in the general economy.

The emergence of a Jewish enclave in East Manhattan helped this group bypass the conventional assimilation path and achieve significant economic mobility in the course of the first generation, well ahead of complete acculturation. Subsequent generations also pursued this path, but the resources accumulated through early immigrant entrepreneurship were dedicated primarily to furthering the education of children and their entry into the professions. It was at this point that outside hostility became most patent, as one university after another established quotas to prevent the onrush of Jewish students. The last of these quotas did not come to an end until after World War II.

Despite these and other obstacles, the movement of Jews into higher education continued. Building on the economic success of the first generation, subsequent ones achieved levels of education, occupation, and income that significantly exceeded the national average. The original enclave is now only a memory, but it provided in its time the necessary platform for furthering the rapid social and economic mobility of the minority. Jews did enter the mainstream of American society, but they did not do so starting uniformly at the bottom, as most immigrant groups had done; instead, they translated resources made available by early ethnic entrepreneurship into rapid access to positions of social prestige and economic advantage.

Japanese on the West Coast

The specific features of Japanese immigration differ significantly from the movement of European Jews, but their subsequent adaptation and mobility patterns are similar. Beginning in 1890 and ending with the enactment of the Gentlemen's Agreement of 1908, approximately 150,000 Japanese men immigrated to the West Coast. They were followed primarily by their spouses until the Immigration Act of 1924 banned any further Asiatic immigration. Although nearly 300,000 Japanese immigrants are documented in this period, less than half of this total remained in the United States. This is due, in contrast to the case of the Jews, to the sojourner character of Japanese immigrants: the intention of many was to accumulate sufficient capital for purchasing farm land or settling debts in Japan. Hence this population movement included commercial and other members of the Japanese middle class who, not incidentally, were explicitly sponsored by their national government.

[. . .]

Japanese immigrants were initially welcomed and recruited as a form of cheap agricultural labor. Their reputation as thrifty and diligent workers made them preferable to other labor sources. Nativist hostilities crystallized, however, when Japanese immigrants shifted from wage labor to independent ownership and smallscale farming. This action not only reduced the supply of laborers but it also increased competition for domestic growers in the fresh-produce market. In 1900, only about 40 Japanese farmers in the entire United States leased or owned a total of 5,000 acres of farmland. By 1909, the number of Japanese farmers had risen to 6000 and their collective holdings exceeded 210,000 acres. Faced with such "unfair" competition, California growers turned to the political means at their disposal. In 1913, the state legislature passed the first Alien Land Law, which restricted land ownership by foreigners. This legislation did not prove sufficient, however, and, in subsequent years, the ever-accommodating legislature passed a series of acts closing other legal loopholes to Japanese farming.

[. . .]

The ability of the first-generation Issei to escape the status of stoop labor in agriculture was based on the social cohesion of their community. Rotating credit associations offered scarce venture capital, while mutual-aid organizations provided assistance in operating farms and urban businesses. Capitalizations as high as $100,000 were financed through ethnic credit networks. Economic success was again accompanied by limited instrumental acculturation and by careful preservation of national identity and values. It was the availability of investment capital, cooperative business associations, and marketing practices (forward and backward economic linkages) within the ethnic enclave that enabled Japanese entrepreneurs to expand beyond its boundaries and compete effectively in the general economy. This is illustrated by the production and marketing of fresh produce. In 1920, the value of Japanese crops was about 10 percent of the total for California, when the Japanese comprised less than 1 percent of the state's population; many retail outlets traded exclusively with a non-Japanese clientele.

During the early 1940s, the Japanese ethnic economy was seriously disrupted but not eliminated by the property confiscations and camp internments accompanying World War II. After the war, economic prosperity and other factors combined to reduce local hostility toward the Japanese. Older Issei and many of their children returned to small business, while other secondgeneration Nisei, like their Jewish predecessors, pursued higher education and entered the whitecollar occupations en masse, This mobility path was completed by the third or Sansei generation, with 88 percent of their members attending college. Other third-generation Japanese have continued, however, the entrepreneurial tradition of their parents. Like Jews before them, Japanese-Americans have made use of the resources made available by early immigrant entrepreneurship to enter the mainstream of society in positions of relative advantage.

[. . .]

CONTEMPORARY EXAMPLES

As a mode of incorporation, the immigrant enclave is not only of historical interest since there are also several contemporary examples. Enclaves continue to be, however, the exception in the post-World War II period, standing in sharp contrast to the more typical pattern of secondary labor immigration. Furthermore, there is no guarantee that the emergence and development of contemporary ethnic enclaves will have the same consequences for their members that they had among turn-of-the-century immigrants.

Koreans in Los Angeles

The Korean community of Los Angeles is a recent product of liberalized U.S. immigration laws and strengthened political and economic ties between the two nations. Since 1965–1968, South Korean immigration to the United States has increased sixfold, swelling the Korean population of Los Angeles from less than 9000 in 1970 to over 65,000 in 1975. Approximately 60 percent of Korean immigrants settle in Los Angeles. In addition to increasing the size of this population flow, U.S. immigration law has altered its class composition. Korean immigrants come predominantly from the highly educated, Westernized, Christian strata of urban Korea.

[. . .]

Korean entrepreneurs, like Jewish and Japanese immigrants before them, are highly dependent on the social and economic resources of their ethnic community. Some immigrants managed to smuggle capital out of Korea, but most rely on individual thrift and ethnic credit systems. For instance, a Korean husband and wife may save their wages from several service and factory jobs until enough capital is accumulated to purchase a small business. This process usually takes 2 or 3 years. Rotating credit systems, which are based on mutual trust and honor, offer another common source of venture capital. This economic institution could not exist without a high degree of social solidarity within the ethnic community. There are more than 500 community social and business associations in Los Angeles, and nearly every Korean is an active member of one or more of them.

In addition, Korean businessmen have utilized public resources from the U.S. Small Business Administration as well as loans and training programs sponsored by the South Korean government.

The ability of the Korean community to generate a self-sustaining entrepreneurial class has had a profound impact on intraethnic labor relations and patterns of ethnic property transfers. For example, labor relations are enmeshed in extended kinship and friendship networks. In this context of "labor paternalism," working in the ethnic economy frequently entails the obligation of accepting low pay and long hours in exchange for on-the-job training and possible future assistance in establishing a small business. Hence, employment in the ethnic economy possesses a potential for advancement entirely absent from comparable low-wage labor in the secondary labor market.

Along the same lines, business practices are fundamentally influenced by cultural patterns. Koreans patronize coethnic businesses and frequently rely on referrals from members of their social networks. Korean-owned businesses, moreover, tend to remain in the community through intraethnic transactions. This is because economic mobility typically proceeds through the rapid turnover of immigrant-owned enterprises. A common pattern of succession, for instance, may begin with a business requiring a relatively small investment, such as a wig shop, and continue with the acquisition of enterprises requiring progressively larger capitalizations: grocery stores, restaurants, gas stations, liquor stores, and finally real estate. This circulation of businesses within the ethnic economy provides a continuous source of economic mobility for aspiring immigrant entrepreneurs.

[. . .]

Cubans in Miami

Over the past 20 years, nearly 900,000 Cubans or about 10 percent of the island's population have emigrated, mostly to the United States. The overwhelming proportion of the Cuban population in America, estimated at roughly 800,000, resides in the metropolitan areas of south Florida and New York. This movement of Cuban emigres has not been a continuous or socially homogeneous flow. Instead, it is more accurately described as a series of "waves," marked by abrupt shifts and sudden discontinuities. This pattern has supported the emergence of an enclave economy through such features as spatial concentration, the initial arrival of a moneyed, entrepreneurial class, and subsequent replenishments of the labor pool with refugees coming from more modest class origins.

In 1959, when Fidel Castro overthrew the regime of Fulgencio Batista, the Cuban community in the United States numbered probably less than 30,000. The political upheavals of the Revolution, however, precipitated a massive emigration from the island. Not surprisingly, the Cuban propertied class, including landowners, industrialists, and former Cuban managers of U.S.-owned corporations, were the first to leave, following close on the heels of leaders of the deposed regime. In the first year of the exodus, approximately 37,000 emigres settled in the United States; most were well-to-do and brought considerable assets with them. After the defeat of the exile force in the Bay of Pigs, in April 1961, Cuban emigration accelerated and its social base expanded to include the middle and urban working classes. By the end of 1962, the first phase of Cuban emigration had concluded and over 215,000 refugees had been admitted to the United States. The emerging Cuban community in south Florida, unlike earlier Japanese and contemporary Korean settlements, was thus fundamentally conditioned by political forces.

Political factors continued to shape the ups and downs of Cuban emigration as well as its reception by American society over the next two decades. In this period, three additional phases can be distinguished: November 1962 to November 1965, December 1965 to April 1973, and May 1973 to November 1980, including 74,000 in the second phase, 340,000 in the third, and 124,769 in the last. This massive influx of refugees to south Florida generated local complaints about the social and economic strains placed in the area. Accordingly, the policy of the Cuban Refugee Program, originally established by the Kennedy Administration, was oriented from the start to resettle Cubans away from Miami. Assistance to the refugees was often made contingent on their willingness to relocate. Although over 469,000 Cubans elected to move by 1978, many subsequently returned to metropolitan Miami. There is evidence that many of these "returnees" made use of their employment in relatively high-wage Northern areas to accumulate savings with which to start new business ventures in Miami. By 1980, the Cuban-born population of the city, composed to a large extent of returnees from the North, was six times greater than the second largest Cuban concentration in New Jersey.

Although a number of Cuban businesses appeared in Miami in the 1960s, they were mostly restaurants and ethnic shops catering to a small exile clientele. An enclave economy only emerged in the 1970s as a result of a combination of factors, including capital availability, access to low-wage labor provided by new refugee cohorts, and the increasingly tenuous hope of returning to Cuba. Cuban-owned firms in Dade County increased from 919 in 1967 to about 8000 in 1976 and approximately 12,000 in 1982.

[. . .]

CONCLUSION: A TYPOLOGY OF THE PROCESS OF INCORPORATION

Having reviewed several historical and contemporary examples, we can now attempt a summary description of the characteristics of immigrant enclaves and how they differ from other paths. The emergence of an ethnic enclave economy has three prerequisites: first, the presence of a substantial number of immigrants with business experience acquired in the sending country; second, the availability of sources of capital; and third, the availability of sources of labor. The latter two conditions are not too difficult to meet. The requisite labor can usually be drawn from family members and, more commonly, from recent arrivals. Surprisingly perhaps, capital is not a major impediment either since the sums initially required are usually small. When immigrants did not bring them from abroad, they could be accumulated through individual savings or pooled resources in the community. It is the first condition that appears critical. The presence of a number of immigrants skilled in the art of buying and selling is common to all four cases reviewed above. Such an entrepreneurial-commercial class among early immigrant cohorts can usually overcome other obstacles; conversely, its absence within an immigrant community will confine the community to wage employment even if sufficient resources of capital and labor are available.

Enclave businesses typically start small and cater exclusively to an ethnic clientele. Their expansion and entry into the broader market requires, as seen above, an effective mobilization of community resources. The social mechanism at work here seems to be a strong sense of reciprocity supported by collective solidarity that transcends the purely contractual character of business transactions. For example, receipt of a loan from a rotating credit association entails the duty of continuing to make contributions so that others can have access to the same source of capital. Although, in principle, it would make sense for the individual to withdraw once his loan is received, such action would cut him off from the very sources of community support on which his future business success depends.

Similarly, relations between enclave employers and employees generally transcend a contractual wage bond. It is understood by both parties that the wage paid is inferior to the value of labor contributed. This is willingly accepted by many immigrant workers because the wage is only one form of compensation. Use of their labor represents often the key advantage making poorly capitalized enclave firms competitive. In reciprocity, employers are expected to respond to emergency needs of their workers and to promote their advancement through such means as on-the-job training, advancement to supervisory positions, and aid when they move into self-employment. These opportunities represent the other part of the "wage" received by enclave workers. The informal mobility ladders thus created are, of course, absent in the secondary labor market where there is no primary bond between owners and workers or no common ethnic community to enforce the norm of reciprocity.

Paternalistic labor relations and strong community solidarity are also characteristic of middleman minorities. Although both modes of incorporation are similar and are thus frequently confused, there are three major structural differences between them. First, immigrant enclaves are not exclusively commercial. Unlike middleman minorities, whose economic role is to mediate commercial and financial transactions between elites and masses, enclave firms include in addition a sizable productive sector. The latter may comprise agriculture, light manufacturing, and construction enterprises; their production, marketed often by coethnic intermediaries, is directed toward the general economy and not exclusively to the immigrant community.

Second, relationships between enclave businesses and established native ones are problematic. Middleman groups tend to occupy positions complementary and subordinate to the local owning class; they fill economic niches either disdained or feared by the latter. Unlike them, enclave enterprises often enter in direct competition with existing domestic firms. There is no evidence, for example, that domestic elites deliberately established or supported the emergence of the Jewish, Japanese, Korean, or Cuban business communities as means to further their own economic interests. There is every indication, on the other hand, that this mode of incorporation was largely self-created by the immigrants, often in opposition to powerful domestic interests. Although it is true that enclave entrepreneurs have been frequently employed as subcontractors by outside firms in such activities as garmentmaking and construction, it is incorrect to characterize this role as the exclusive or dominant one among these enterprises.

Third, the enclave is concentrated and spatially identifiable. By the very nature of their activities, middleman minorities must often be dispersed among the mass of the population. Although the immigrants may live in certain limited areas, their businesses require proximity to their mass clientele and a measure of physical dispersion within it. It is true that middleman activities such as moneylending have been associated in several historical instances with certain streets and neighborhoods, but this is not a necessary or typical pattern, Street peddling and other forms of petty commerce require merchants to go into the areas where demand exists and avoid excessive concentration of the goods and services they offer. This is the typical pattern found today among middleman minorities in American cities.

Enclave businesses, on the other hand, are spatially concentrated, especially in their early stages. This is so for three reasons: first, the need for proximity to the ethnic market which they initially serve; second, proximity to each other, which facilitates exchange of information, access to credit, and other supportive activities; third, proximity to ethnic labor supplies on which they crucially depend. Of the four immigrant groups discussed above, only the Japanese partially depart from the pattern of high physical concentration. This can be attributed to the political persecution to which this group was subjected. Originally, Japanese concentration was a rural phenomenon based on small farms linked together by informal bonds and cooperative associations. Forced removal of this minority from the land compelled their entry into urban businesses and their partial dispersal into multiple activities.

Physical concentration of enclaves underlies their final characteristic. Once an enclave economy has fully developed, it is possible for a newcomer to live his life entirely within the confines of the community. Work, education, and access to health care, recreation, and a variety of other services can be found without leaving the bounds of the ethnic economy. This institutional completeness is what enables new immigrants to move ahead economically, despite very limited knowledge of the host culture and language. Supporting empirical evidence comes from studies showing low levels of English knowledge among enclave minorities and the absence of a net effect of knowledge of English on their average income levels.

Table 1 summarizes this discussion by presenting the different modes of incorporation and their principal characteristics. Two caveats are necessary. First, this typology is not exhaustive, since other forms of adaptation have existed and will undoubtedly emerge in the future, Second, political refugees are not included, since this entry label does not necessarily entail a unique adaptation path. Instead, refugees can select or be channeled in many different directions, including self-employment, access to primary labor markets, or confinement to secondary sector occupations.

Having discussed the characteristics of enclaves and middleman minorities, a final word must be said about the third alternative to employment in the lower tier of a dual labor market. As a mode of incorporation, primary sector immigration also has distinct advantages, although they are of a different order from those pursued by "entrepreneurial" minorities. Dispersal throughout the receiving country and career mobility based on standard promotion criteria makes it imperative for immigrants in this mode to become fluent in the new language and culture. Without a supporting ethnic community, the second generation also

Table 1 Typology of modes of incorporation

Variable	*Primary sector immigration*	*Secondary sector immigration*	*Immigrant enclaves*	*Middleman minorities*
Size of immigrant population	Small	Large	Large	Small
Spatial concentration, national	Dispersed	Dispersed	Concentrated	Concentrated
Spatial concentration, local	Dispersed	Concentrated	Concentrated	Dispersed
Original class composition	Homogeneous: skilled workers and professionals	Homogeneous: manual laborers	Heterogeneous: entrepreneurs, professionals, and workers	Homogeneous: merchants and some professionals
Percent occupational status distribution	High mean status/ low variance	Low mean status/low variance	Mean status/ high variance	Mean status/ low variance
Mobility opportunities	High: formal promotion ladders	Low	High: informal ethnic ladders	Average: informal ethnic ladders
Institutional diversification of ethnic community	None	Low: weak social institutions	High: institutional completeness	Medium: strong social and economic institutions
Participation in ethnic organizations	Little or none	Low	High	High
Resilience of ethnic culture	Low	Average	High	High
Knowledge of host country language	High	Low	Low	High
Knowledge of host country institutions	High	Low	Average	High
Modal reaction of host community	Acceptance	Discrimination	Hostility	Mixed: elite acceptance/ mass hostility

becomes thoroughly steeped in the ways of the host society. Primary sector immigration thus tends to lead to very rapid social and cultural integration. It represents the path that approximates most closely the predictions of assimilation theory with regard to (1) the necessity of acculturation for social and economic progress and (2) the subsequent rewards received by immigrants and their descendants for shedding their ethnic identities.

Clearly, however, this mode of incorporation is open only to a minority of immigrant groups. In addition, acculturation of professionals and other primary sector immigrants is qualitatively different from that undergone by others. Regardless of their differences, immigrants in other modes tend to learn the new language and culture with a heavy "local" content. Although acculturation may be slow, especially in the case of enclave groups, it carries with it elements unique to the surrounding community – its language inflections, particular traditions, and loyalties. On the contrary, acculturation of primary sector immigrants is of a more cosmopolitan sort. Because career requirements often entail physical mobility, the new language and culture are learned more rapidly and more generally, without strong attachments to a particular community. Thus, while minorities entering menial labor, enclave, or middleman enterprise in the United States have eventually become identified with a certain city or region, the same is not true for immigrant professionals, who tend to "disappear," in a cultural sense, soon after their arrival.

Awareness of patterned differences among immigrant groups in their forms of entry and labor market incorporation represents a significant advance, in our view, from earlier undifferentiated descriptions of the adaptation process. This typology is, however, a provisional effort. Just as detailed research on the condition of particular minorities modified or replaced earlier broad generalizations, the propositions advanced here will require revision. New groups arriving in the United States at present and a revived interest in immigration should provide the required incentive for empirical studies and theoretical advances in the future.

"Men Without Property: The Tramp's Classification and Use of Urban Space"

from *Antipode* (1983) [1978]

James S. Duncan

Editors' Introduction

James S. Duncan provides a fascinating account of the tactics and strategies employed by homeless people to find secure spaces for rest and residence spaces while eluding the police, imprisonment and the moral authority of the state. The public fear of the tramp, Duncan asserts, comes from their status as a willful negation of established social order. Vagrancy laws restrict their movement because they reflect the prevailing ideology concerning private places and freedom of movement. As long as the tramp adopts a low profile, he can occupy marginal spaces such as alleyways, spaces under bridges, dumps, railroad yards, and other no man's lands of the city. Sometimes urban districts appear that are "jurisdictional voids" completely ceded to the homeless. As a way of containing the homeless, police accept these "skid rows" and charitable or state agencies may congregate their shelters and transitional facilities there.

Tramps are generally driven out of prime public and private places as eyesores, public nuisances and threats to the social order. Full citizenship rights are not extended to those without property. Their presence may be tolerated in some public spaces, however, such as parks, libraries and transportation stations. By employing "props" such as newspapers and books casually draped over their faces, they may catch a snooze or read under the tolerant surveillance of the policeman. These props help legitimize the tramp while submerging his stigmata. Prostitutes and gay hustlers may cruise in the park, as "the coppers will let you whistle low, but not loud." Tramps, hustlers, drug dealers and other socially marginalized people may perform a kind of jurisdictional etiquette or interaction ritual with the forces of law and social order. This negotiation process helps reproduce middle class moral order in the city.

Duncan alludes to human ecologist Robert Park in his references to the moral order of the city. Duncan also alludes to critical legal theory when he notes that Vagrancy Laws emerged as early as the fourteenth century as an attempt on the part of landowners to expropriate labor at a time when it was in short supply and its value was high. Twentieth century policies to regulate the homeless, on the other hand, control people with no labor value. Skid row deprives people of their "natural rights" rather than being an outgrowth of urban "natural selection." Tramps would not be regulated in a truly free society. Duncan has a view of symbolic interaction that recognizes inequalities of power and access to property and movement.

The recent gentrification and redevelopment process in many American cities has given the plight of the homeless new visibility. Marginal spaces of the inner city that were previously ceded to tramps are increasingly subject to police surveillance, and interest on the part of developers, planners, and public officials. There has been new outcry over the homeless as a problem of mental illness and a threat to public health and order. With the return of jobs and people to the central city, there are fears about a link between tramps and criminality. The police do "sweeps" through skid row under trumped-up charges of parole violation and other illegal activity.

There has been a series of classic urban ethnographic studies on tramps and the homeless. They include: Elliot Liebow, *Talley's Corner: A Study of Negro Streetcorner Men* (Boston: Little, Brown, 1967), Ulf Hannerz, *Soulside: Inquiries into Ghetto Culture and Community* (New York: Columbia University Press, 1969), David A. Snow and Leon Anderson, *Down on Their Luck: A Study of Homeless Street People* (Berkeley: University of California Press, 1993), Mitchell Duneier, *Sidewalk* (New York: Farrar, Straus, and Giroux, 1999).

James Duncan is currently a University Lecturer in Geography and Fellow of Emmanuel College, University of Cambridge. He previously taught at the University of British Columbia and Syracuse University. He obtained his B.A. at Dartmouth College and his M.A. and Ph.D. at Syracuse University. His books include *Housing and Identity: Cross-Cultural Perspectives* (New York: Holmes and Meier Publishers, 1982), *Place/Culture/Representation*, edited with David Ley (London: Routledge, 1993), *Landscapes of Privilege: The Politics of the Aesthetic in an American Suburb*, written with Nancy G. Duncan (New York: Routledge, 2004) and *A Companion to Cultural Geography*, edited with Nuala C. Johnson and Richard H. Schein (Oxford: Blackwell, 2004).

[...]

Differential mobility, access to space and inequalities in power to influence others' use of space reflect the interrelationships between social groups in the city. In fact, land is divided up and access to space is limited in such ways that land can be said to constitute a relationship between men. Power and other social relationships between men [*sic*] are in various ways enacted in the use of land and in restrictions of others' use of land.

This paper is about the difficulty that tramps, being members of an extremely marginal group, encounter when they try to carve out a niche for themselves in the American city whose moral orders have little place for them. It concerns the strategies they employ to exist in the nooks and crannies of the urban world whose moral order denies the legitimacy of their nomadic existence. The tramp roams from city to city by freight train, on foot, or by broken down car. He attempts to adapt as best he can to the hostile urban environment. He wanders the skid rows, hides in alleyways, sleeps under bridges, on the sidewalks, in garbage dumps and in parks. He is perpetually on the move to avoid arrest, to look for a job or a handout, and to find a place to sleep.

In the first half of this essay I will discuss the relationship between the two groups under consideration, the tramp and the host population, the latter being those who have control over the various areas of the city through which the tramp wanders. Also mentioned will be the dominant ideology which provides a framework within which the host group operates. Next I consider the strategies the tramp employs to make a place for himself in the spatial order of the city. In particular, I will discuss the way the tramp takes the role of the host group in order to derive a classification of the city pertinent to his attempts to survive there.

THE MORAL ORDER OF THE LANDSCAPE

The city is composed of more or less well defined social areas each of which is controlled by one or more groups who sustain a moral order there. The moral order of an area is the public order. I use the term moral order here to capture the feeling of a group that the way it organizes its world is inherently correct. It is defined here as the set of customary relations in an area and the etiquette governing its landscape; it constitutes what is believed by the dominant group to be the proper arrangement and use of artifacts and the proper form that interaction in that landscape should take. It stipulates what people under what circumstances are allowed to engage in what activity in what places.

This moral order is not arrived at, however, without some negotiation on the part of the participants. The negotiation process is not only of a formal political nature but also arises out of routine social interaction. Superimposed upon local moral orders is the largely middle class moral order of the city as a whole. This is codified in laws and enforced by the police and other official agencies. Indirect attempts at control are also made by architects and planners who design buildings and outdoor spaces to keep order. The local moral orders also have their guardians in the form of residents (peer pressure), shopkeepers, gangs, and others who enforce the local etiquette. There is a certain tension between the local and city-wide moral orders that arises from the differences between them.

The police are expected to keep this tension under control. They accept the fact that local moral orders do exist and often take precedence over the city-wide order as long as they do not grossly contradict it or spill over into other areas.

[. . .]

The police tolerate certain deviant behavior as normal on skid row, which is one of the few areas the tramp may occupy with a minimum of risk. In this sense they respect the moral order of the row; they find, however, that other behavior there so flagrantly violates the larger moral order that they must intervene.

THE MARKET PLACE IDEOLOGY OF THE HOST GROUP

In a society such as ours, whose organization is based on individual property rights, a poor person will be viewed as a problem for the group controlling the area in which he lives. He possesses little property and hence has little stake in the existing order which functions primarily to protect property and ensure that orderly market relations take place. The tramp poses all of the same problems for the controlling group that the poor local person does but these problems are magnified as he, being an itinerant, has even less stake in the area. Not only does he lack property but he rarely has any ties with local residents. Thus the tramp feels no obligation to maintain the moral order and furthermore his mobility makes it difficult to force him to comply with it. The locals view the tramp as a threat and do their best to drive him away.

Social value in our society is based primarily upon property, real property or labor which one is willing and able to sell. While the tramp occasionally has a few possessions, a watch, a ring, a little money made on a temporary job or panhandled, perhaps a radio purchased in good times to be sold when the times get rough, these possessions come and go. The only property a tramp has on a fairly regular basis is his labor power which he can sell if he is willing, though often he is not because the wages offered are often less than can be made panhandling. Furthermore, his lack of skill or his frequent inability to hold a job because of alcoholism makes his labor worth very little to society.

[. . .]

According to the dominant ideology in society one must have sufficient property to be able to own or rent a place to sleep if he is not to be charged with vagrancy. Vagrancy laws, which restrict the movement of individuals from place to place, have been in effect in England since the fourteenth century and were adopted by the colonies and subsequently the United States. The social problem that they were originally intended to alleviate has long since disappeared while other totally unrelated problems have become their target. Vagrancy laws should be of special interest to geographers because they reflect the prevailing ideology concerning private places and freedom of movement. For example, it is illegal not to own the right of access to some private place. Every person who wanders about without access to a private place to sleep, who can show no means of support ". . . shall be deemed a tramp, and shall be subject to imprisonment . . ." It is such access that provides ". . . evidence that a man has a legitimate relationship to the social structure." The vagrancy laws are selectively applied to contain tramps within skid row or to banish them entirely from the city. Thus, those who are penniless or without normal occupation in the broad sense of the term may be severely restricted in their spatial mobility and even declared criminal.

STRATEGIES FOR GETTING BY

Let us now turn to a consideration of the strategies that are adopted by the tramp who finds that the moral order of the host group has no place for him. There are a variety of possible strategies, e.g., mobility, and eschewing cumbersome possessions and attachments to members of the local population. I will restrict myself here to a discussion of those strategies that make use of the landscape.

In order to avoid the guardians of the host landscape, the tramp must adopt a low profile. Often this is not easy to accomplish because the tramp wears old, worn clothes that are easily recognizable. It is not easy to change this because it involves an outlay of resources that is usually not available. He would also severely diminish his chances of being able to panhandle successfully because his clothes are a sign, often the only visible one, that he needs a handout.

> I believe that the dirtier the man is the better for the street make. If I am dirty, a man will give me a coin rather than have me walk down the street and have people think that I might be one of his friends.

Since passing as a member of the established moral order is not a viable method of adopting a low profile, the tramp must resort to another tactic, that of using the landscape as a cover. This method requires a large amount of environmental knowledge. He must know for example where he can find a warm place to spend a winter's night unobserved.

> You've got to have ingenuity. You've got to know New York, its people, how to get around. I sleep in the subways nowadays. It works out fine. . . . You can't sleep there when an officer's on duty. That's from eight p.m. to four a.m. So I go there about ten minutes till four; sleep to noon, usually . . . I like the Eighth Avenue line. Less stops. I sleep on one of the front or back cars; never the middle. Too many disturbances.

At times he must out of necessity sleep on the street and is therefore potentially open to harassment from the police and jackrollers (muggers). He finds ingenious ways to hide himself in this kind of exposed landscape. "There's garbage cans back there . . . I'd lay there with half my body in a garbage can and the upper half in a pasteboard box . . . that way no one knows." Once a tramp finds a place that provides good cover he may wish to protect it from others. Often some ingenious tactics are employed to attempt to claim a bit of public space.

> Snead, a dope addict, slightly past middle age, has been living for some time in the doorway of an unoccupied building. This doorway consists of a sort of vestibule and has plenty of room for a person to lie down. His only furniture is a broom and a box of broken glass. When he leaves he scatters the broken glass over the floor to keep others away. And when he returns he uses the broom to sweep up the glass.

Most of this environmental knowledge is very specialized. It is of interest to the tramp alone although certain members of the local population, notably the police, may also be tangentially interested. He knows certain ethnic streets where begging is especially good, specific street corners where truckers stop to pick up cheap labor, and that public libraries are good places to sleep during the day. He knows many different spots where he can sleep and keep warm on a cold winter night, such as in bus depot toilet stalls, near city steam pipes, in heated sandboxes in railroad yards, in warm stacks of bricks in brickyards, and in open churches. In fact Spradley discovered that tramps could name over one hundred categories of sleeping places.

Whereas most tramps are aware of these relatively stable elements of the urban environment, the more astute quickly capitalize on changes in the environment. For example, shortly after a chapter of Alcoholics Anonymous started holding meetings in a church near the Bowery in New York, a number of tramps congregated about the church during meetings taking advantage of the fact that members of A.A. are a "soft touch" for a handout.

[. . .]

Alongside the marketplace ideology (which guides relations between most strangers within the city) struggles the tramp's non-market ideology, which is based on reciprocity and is termed "brotherhood of the road" by its adherents. The maintenance of this alternative ideology may be considered a strategy for survival. A tramp shares with others whether he knows them or not, knowing that he may be taken care of when he is in need. He will share his bottle, flop (any of a wide variety of places to sleep), clothes, food, and information about local police practices and availability of spot jobs, etc. A tramp may give his jacket to another inmate when he leaves jail and then go out into the street to "cut in" on someone's bottle. According to informal rules one must never refuse to share his bottle.

A tramp's environmental knowledge must include all the intricacies of how the host group classifies space and specifically the social value that it attaches to different landscapes. Gaining such knowledge is the most important of all his strategies.

PRIME AND MARGINAL SPACE

Most citizens find public versus private the primary constraining classification in their use of space: one ought to be able to use any public space by virtue of his being a citizen provided he behaves in a fairly "normal" and legal fashion for the time and place. Private spaces, of course, are far more exclusive. The distinction between public and private is usually clear although it can be a little fuzzy around the edges. The propertyless are generally excluded from private places (they can sometimes stay in flophouses or go into bars or missions but not on a regular basis). What is more, they are also driven out of public places, for full citizenship rights are apparently not extended to the propertyless, a notion which has survived from the eighteenth century when citizenship was legally extended only to those who had property. Public property apparently, then, belongs to the citizenry as a whole which has the right to exclude the tramp.

[...]

In contrast to the propertied population and the relatively simple public versus private distinction which restricts its spatial movements, tramps are forced to use a different system of determining usable space within the city. This system is based on an unstated scale of social value which the host groups apply to different areas. The scale ranges from what I would term prime to marginal space. This implies no inherent value in the space itself; on the contrary the value is assigned to the space by a given group on the basis of how it uses the space. The social value of space, furthermore, is an ever-evolving phenomenon which is based upon varying degrees of consensus. Thus the tramp must learn the social value that the host group assigns to different types of urban space and to the regular temporal variations in these values; he must then resign himself to spending as much time as possible in marginal space.

Marginal space includes alleys, dumps, space under bridges, behind hedgerows, on the roofs of buildings, and in other no man's lands such as around railroad yards, which are not considered worth the cost of patrolling. As one tramp put it,

> If you are under the bridge down below Pike Street Market you are safe – the police don't walk down there, just too lazy.

Skid row and a few other very poor residential or commercial areas are also considered marginal. Here the tramp can achieve a low profile because the shabbiness of his dress does not stand out as it does elsewhere. Skid row is ceded to the tramps because the authorities realize that tramps must stay somewhere and there are definite advantages from the point of view of social control to keeping them together in one place.

[...]

The terms prime and marginal refer not to a dichotomy but a continuum. It must also be noted that they are relative terms. By looking at the city as a whole and grossly dividing it into space which is prime and marginal in the eyes of the host group I have in effect lumped all the citizens of the city except the tramps into one group, which is an obvious oversimplification. Whether a place is prime or marginal depends upon the perspective from which one views the situation. The tramp is aware of the diversity of perspectives within the host population. In day-to-day decisions he must see things from their perspective. He must take into consideration the fact that what one person considers prime space another considers marginal. He takes the perspective of some into account more than others because he realizes the relative inequalities of power among these groups to enforce their perspective. It must be remembered, however, that there is some general agreement as to prime and marginal in the gross sense for the lower class citizen is forced to a certain extent to accept the perspective of the middle class citizen and to realize that effectively his area is considered marginal even if it is of prime importance to him.

Likewise the tramp must "take the role of" the police officers in their subdivision of marginal space into more or less marginal. For example, the police categorize abandoned buildings as extremely marginal space.

> The patrolman knows these buildings are in use, for he constantly sees evidence . . . (but) he does not examine these places on his route. . . . They are no longer buildings in the formal sense. They are like the alleys the patrolman does not go into because he does not see them as being either public or private places . . .

These places, which are neither public nor private, have been classified by the patrolman as places with no property value and hence as jurisdictional voids. The tramp is quick to see the personal benefits to be derived from the patrolman's classification of marginal space. He also learns to divide marginal space on the basis of accessibility and inaccessibility to the police, who, for matters of safety and convenience like to patrol their beat in cars. In effect the police have given up jurisdiction over certain areas in which they cannot drive a car.

[. . .]

Similarly tramps take advantage of the policeman's classification of alleys as inaccessible space and use them for drinking and other illicit activities.

THE USE OF PRIME SPACE

The tramp is faced with a problem that really has no solution. In order to survive without being arrested he must occupy marginal areas, while in order to secure the wherewithal to survive he must often venture into high risk, prime areas. The vast majority of the city is prime space and therefore space which the tramp finds dangerous. Prime space, however, does not have a uniformly high social value. The tramp knowing the relative value assigned to different spaces by the host group minimizes his risk by using prime space which has moderate social value.

> . . . the central business district, and certain main thoroughfares, are the only places in the metropolis (with the exception of skid row) where the bum can successfully ply his trade. If he tries door-to-door begging . . . householders call the police.

It is extremely difficult to adopt a low profile in any prime area while one is trying to panhandle. On most city streets there are few if any places that provide cover. Furthermore, interaction with strangers violates the norms of conduct on such public streets, and panhandling is therefore easily noticed. "Normal" and hence acceptable conduct on such a street consists of walking purposefully and not interacting with others unless they are walking with you.

> In Chicago, an individual in the uniform of a hobo can loll on "the stem," but once off this preserve he is required to look as if he were intent on getting to some business destination.

The practice of picking up cigarette butts from the street, "snipe shooting," illustrates some of the problems inherent in using various types of prime space. To compete for these snipes one must have a detailed knowledge of the good snipe hunting areas.

> He told me the best place for snipes was outside a church – the Sunday folk always left long ones before they entered. On a good day you could pick up enough to make a pack of almost new ready-rolls.

However, although one can find good snipes on the Loop one must be careful of the police there.

> I wouldn't advise you to hunt snipes in the Loop, for while the butts are as long as this (measuring with his hands the length of a lead pencil), you are almost sure to get picked up by the police.

The advantages to be gained from using prime space, therefore, must be weighted against the risks.

There are certain other prime spaces that the tramp uses upon occasion in order to keep warm or to catch some sleep where he may not be noticed by the police. Two such places are public libraries, and train stations.

> Thus in some urban libraries, the staff and the local bums may reach a tacit understanding that dozing is permissible as long as the dozer first draws out a book and props it up in front of his head.

Likewise it appears that

> . . . newspaper readers never seem to attract the attention (of the station guards) and even the seediest vagrant can sit in Grand Central all night without being molested if he continues to read a newspaper.

In both cases the tramp adopts the strategy of using a prop, in this case reading material, in order to legitimize his presence in this prime space. These props appear to submerge his stigmatized identity sufficiently to stop it from spreading to others in the place. This can only be accomplished in impersonal, low interaction places where the stigmatized person can easily be disattended by others. These are places and props which isolate individuals, "remove them" as it were from others. It is as if they were required to build a wall around themselves in order that their presence might be tolerated in prime space.

For matters of convenience I have been speaking of the host group as if it were a single monolithic group which of course it is not. Within it there are many social worlds and factions with more or less conflicting interests. Some are reasonably friendly to tramps and occasionally will form temporary alliances and make deals with them against other segments of the host population. For example, tramps have a certain working relationship with homosexuals who exchange food, shelter, a chance to get cleaned up, clothes and occasionally money, for sexual relations. A third party to this relationship is often the police who tolerate such activity if it remains reasonably unobtrusive. Tramps whistle to get the attention of homosexuals who are "cruising" the park.

> They whistle low. They cannot whistle loud. The coppers will hear them. The coppers will let you whistle low, but not loud.

Such agreements are the product of implicit bargaining, or negotiation, between the police and tramps. The tramps continually try to make gains in freedom to use certain prime space to make contact with members of the host population while the police attempt to hold them in check without expending too much extra effort.

The fact that the tramp is forced to enter prime space on occasion should not obscure the fact that his principal strategy is to occupy marginal space. Excursions into prime space constitute a small but dangerous part of the tramp's existence.

ON THE SEPARATION OF NORMAL AND SPOILED IDENTITIES

There is a belief in our society that the social value of an individual must be roughly commensurate with the social value of the place he frequents. In a society which is socially mobile the objects and settings that one surrounds oneself with are an inseparable part of one's self. This argument logically leads to the notion that if one shares a setting with another then, to a degree, he shares his identity thus allowing stigma to spread by spatial association. A tramp does not have a "normal," that is propertied, identity but to put it in Goffman's terms, one that is "spoiled." Such people must be separated out for their spoiled identities can spoil the setting and by extension the "normal" people who are a part of it. Hence the hierarchical division of space according to social value is of critical importance to the host population.

Private spaces are the most closely associated with an individual's identity. The home in particular is an important part of one's self. Impersonal, highly public places do not usually constitute an integral aspect of an individual's identity. Sharing such a place does not entail sharing a part of one's social identity to the extent it does in a private place. Thus the separation of spoiled and normal identities is more easily accomplished. If the tramp walks briskly and purposefully through such an area he runs a fairly low risk of being arrested. By his rapid movement he makes little claim on space, thereby reducing the stigma he might otherwise spread. Sitting or standing, however, constitute much more of an involvement of one's identity with a place, i.e., it is an implicit claim that one somehow "belongs" there. The tramp therefore is rarely allowed to remain stationary in such an area for very long unless, as I have mentioned, he restricts himself to a few places such as the library and uses a prop to disassociate himself from others, so as to minimize the spread of stigma.

One could say that the prime areas with the highest social value are those residential sections with which middle to upper class individuals associate their identity. These areas convey positive information about such people. Other prime space includes the highly public places which are fairly neutral in terms of association with individuals' identities. In contrast, marginal areas are negative in the sense that they convey stigmatizing information about their inhabitants.

It is interesting to note that not all highly public areas are neutral in the above sense. There appears to be a generalized association of the identity of many propertied citizens with certain key places which are thought to represent the city. This association is generally termed civic pride. These key places are thought to be an important part of the city's presentation of self to outsiders. This is especially true when large numbers of tourists come to the city for special events or during the tourist season. At these times the city is on display and must be cleared of its unaesthetic elements.

> Drunks are also arrested in significant numbers before any large conventions, celebrations, or fairs. This is done primarily to "clean up" the area so it will not be unnecessarily offensive to visitors.

Thus the social value of prime space ebbs and flows at different times. The tramp cannot afford to be unaware of the temporal nature of the value of space.

Certain areas of the central business district of large cities, such as the Loop in Chicago, are thought to be continuously on display and therefore tramps are discouraged from ever going there.

> The main emphasis of the court is stay out of the Loop. That statement was made again and again, in fact in some instances a judge even stated, if you have to beg do it anywhere but in the Loop.

Skid row is thought of as an area which is morally bankrupt. The characterization of its inhabitants as morally defective leads to the conception of the row as an open asylum. People with "normal" propertied identities therefore should be kept out. The police have little sympathy for tourists who want to see the row or locals who want to take a "moral holiday" thereby engaging in illicit activity. These people cause trouble for the police and as such are discouraged from entering this marginal space.

CONCLUSION

Classification and use of space are intimately tied. One can describe the tramp's use of urban space; one can map it; however, one cannot begin to explain it without reference to the tramp's classification of various areas within the city. To understand his classification, one must examine the process by which he derives it. His classification of areas represents a plan of action, a plan which takes into account the action of others. Reconciling these lines of action requires negotiation which must be viewed in the context of power relationships and the ideology or concept of society held by the more powerful parties to the relationship. Occupying a very marginal place in the prevailing concept of society, tramps are in a very poor position from which to negotiate for rights to use space. Their classification of areas within the city is largely shaped by the prime/marginal distinction of the host group. Similarly their strategy of occupying marginal space is a direct result of the host's strategy of containment. The tramp, however, pays a price for using what is defined by the host group as marginal space. To quote sociologist Erving Goffman:

> A status, a position, a social place is not a material thing, to be possessed, and then displayed, it is a pattern of appropriate conduct, coherent, embellished, and well articulated . . . it is . . . something that must be enacted and portrayed.

By occupying marginal space, the tramp acts out and reconfirms his social marginality in the eyes of the host group. His strategy merely reaffirms the host's perspective and causes only minimal adjustments to be made in the host's moral order.

The division of the city into prime and marginal space as I have outlined it here is not in itself as important as the idea that the classification and use of urban areas by any group must be viewed in the context of that group's relation to other groups. Any group, including the most powerful, must negotiate with others, and this inevitably leads to compromise in their perspective. Thus the classification of prime and marginal, or relatively unsafe and safe, to look at the same distinction from the perspective of the tramp, should serve to indicate the relative or interactional nature of any regionalization of the city by any individual or social group.

PART FOUR

Gender and Sexuality

Plate 8 Grandmother, Brooklyn, New York, 1993 by Eugene Richards (reproduced by permission of Magnum Photos).

Plate 9 Gay Pride Parade, New York City, 1998 by Nikos Economopoulos (reproduced by permission of Magnum Photos).

INTRODUCTION TO PART FOUR

Cities are not just landscapes of socioeconomic and racial/ethnic inequality; they are also terrains of inequality by gender and sexuality. While the urban underclass and homeless experience spatial isolation and containment in the marginal spaces of the inner city, women have since the postwar era experienced spatial entrapment in the suburbs. Suburbanization may be seen as a form of segregation. Only more recently have women found greater spatial mobility with their entry into the labor force and the widespread gentrification movements occurring in many central cities. The "return to the city" has been correlated with a set of demographic transitions surrounding a decline in traditional male-led households with children, and the growth of double-income no-children households, and double-income gay and lesbian households, as well as single households. The growing visibility of gays and lesbians in American society is partly correlated with the growing visibility of gay and lesbian communities in the central city.

Susan Saegart, in an essay titled, "Masculine Cities and Feminine Suburbs," *Signs* (Spring 1981): S92-S111, establishes the structural and symbolic dichotomy of women in the suburbs and men in the cities. The symbolism of suburban women has been promoted in a range of cultural representations from advertising, literature, to film and television. This kind of symbolism reinforces inequalities of power in economic and urban life. While the suburbs are the domain of women, family and private life, cities are the domain of men, work, and public life.

In her essay, published in the same 1981 issue of *Signs*, Ann Markusen elaborates further on the patriarchal structure of our cities. The separation of residential life from the central city in single-family households in the suburbs is the creation of a patriarchal capitalist society that differentiates and reproduces an understanding of men's work as waged work and household work as unpaid women's work. Economists define households as places of buying power and consumption, not production. Even the notion of the "journey to work" implies the household is not an area of production values.

Markusen blames patriarchy for the spatial segregation of women in the suburbs, not just the benign operation of the real estate market, the growth of automobile transportation, and the activity of the Federal Housing Administration in the postwar era. Suburban life leads to inefficiency and waste through increased commuting time and energy consumption, and alienation in household life. Suburbs erode extended family and community networks and other forms of social capital. They militate against the collectivization of household work and replace parks and collective play space with private yards. Patriarchy further serves capitalism by dividing the workforce and blunting class-consciousness. Patriarchy does reinforce the power of men as heads of households and primary decision makers, and eaters of better home-cooked meals. The suburbs reinforce the power of American individualism and privatism in the social and political culture.

Markusen points out that new urban trends are disrupting the old dichotomy of masculine cities and feminine suburbs, including: (a) the onset of gentrification along with the breakdown of the traditional patriarchal household, (b) the growth of retired households and retirement communities, and (c) the growth of small and medium-sized towns as households flee large congested cities. She supports a range of urban social policies to promote the interests of women and families, including: (a) fostering child care in extended family arrangements or neighborhood cooperatives, (b) fostering the development of non-patriarchal collectivized affordable housing, and (c) enhancing the efficiency of the household production sector, through combined live/work spaces, as seen in the "loft spaces" of urban artistic colonies. We may speculate also on the impact of the Internet on urban space and participation of women in our society. Electronic connectivity through the World Wide Web breaks down distances of time and space and permits economic participation from places far from urban central places through telecommuting and the growth of non-place based Internet enterprises. The Internet affords some opportunity for women to "cocoon" in residential space while participating in economic and cultural life.

Melissa Gilbert offers a revision of the "spatial entrapment" thesis through her concept of "rootedness" which suggests that spatial limitations on women can confer resources as much as creating constraints. Where white feminists are typically more focused on improving the social and spatial mobility of women, Gilbert's perspective may be associated with the "womanism" of Alice Walker and the writings of bell hooks on home places. Gilbert finds that for low-income women of color living in the inner city, there is a certain power of place that they procure due to the density of social capital networks. Mobility is normally associated with empowerment for women, but relative "place-based" immobility may have certain advantages. In her fieldwork with low-income women of color in Worcester, Massachusetts, Gilbert finds that being close to a dense network of kin and friends improves their ability to procure child-care. These social networks provide access to or information about a variety of other social capital resources, including employment, housing, education, health-care, and church. She finds that race intersects with gender to shape the spatiality of daily life and determine economic security.

Sy Adler and Johanna Brenner address some related issues in their article on lesbians and gay men in the city. Whereas gay men have recently raised their spatial and political visibility in communities such as Greenwich Village, New York, and the Castro district of San Francisco, lesbians have been relatively invisible, with less of a presence in terms of bars, businesses, festivals, and politics. This lesser visibility is not a symptom of absence, Adler and Brenner point out, but an expression of a different form of social networking and identity politics. They point out that women have historically played an important role in community politics and urban social movements. This activity may involve non-territorially based interpersonal networks and non-place-based political issues.

Finally, Donald Donham explores the transformation of gay urban life in Soweto, South Africa with the end of the apartheid system of racial segregation. During the apartheid period, gay South African black laborers commuted from the townships in order to work in the white-dominated economy, and took up *skesanas*, sexually submissive males, as "wives." While racial segregation fostered a sexual culture in which these black laborers closeted their homosexuality, they have become liberated in the post-apartheid period to publically assume their gay identity, bringing what Donham calls a "modernization" of sexuality.

"City Spatial Structure, Women's Household Work, and National Urban Policy"

from *Signs* (1981)

Ann R. Markusen

Editors' Introduction

The Women's Room is Marilyn French's 1977 classic tale of suburban isolation, in which the main character, Mira Ward, questions the accepted social norms regarding women's limited place in society. "The school had been planned for men, and there were places, she had been told, where women were simply not permitted to go. It was odd. Why? she wondered. Women were so unimportant anyway, why would anyone bother to keep them out?" After many twists and turns, Mira Ward's journey ends up embracing the relevance of women's work – unappreciated and unnoticed as it is – to the world at large.

In this selection, Ann Markusen directs our attention to work within the (mostly suburban) home – the "household reproduction of labor" – which has largely been ignored by classical urban scholars. She argues that the single-family home, geographically isolated in suburbs and from workplaces, is a product of the patriarchal (male-dominated) organization of household production. Hence, she leads us to the analysis of gender relations, not as *shaped by* urban form, but as *shaping how* cities and suburbs are built and ordered.

As Markusen points out, social reproduction (the production of the workforce) within the household should be considered an economic activity because it contributes to the overall reproduction of society. The cooking of meals and the provision of personal services, education, and child care are all labor activities that contribute to the maintenance of society. However, the compensation for labor time of the household workers – primarily women – is quite different from the terms under which other forms of labor are hired and organized. Household reproduction rests on the persistence of patriarchical gender relationships. More than half of all adult women under age sixty-five work for wages but they still bear the primary responsibility for household work. The economic nature of this activity, Markusen argues, is largely hidden by the informal economic contract involved (marriage). Patriarchy is the organizing principle of household labor and social production, hence urban spatial structure is a product of patriarchy as much as it is of capitalism.

Identity-based social movements, such as the women's movement, have challenged patriarchy and the power dynamic between men and women it entails. And in doing so, the isolated household domain akin to suburbanization has given way to alternative spatial developments. Women have mobilized to change land-use laws and homeownership financing and have been instrumental in the planning of urban mass transportation. Nonetheless and despite such efforts, Markusen concludes, patriarchy continues as does the dominant cultural preference for the suburban cul-de-sac.

While it may be commonplace for us now to think of household labor as productive labor (if not hard work), we should recognize Ann Markusen's work as an early call to urbanists to include gender as an important variable in their research. Her work inspired a generation of urbanists, especially feminist geographers, to take up the task of critically interrogating the interplay between gender and spatial forms. Other early key texts on gender and urban space include Dolores Hayden's "What Would a Non-Sexist City Be Like? Speculations on Housing, Urban Design, and Human Work;" Gerda Wekerle's "Women in the Urban Environment;" and Susan Saegert's "Masculine Cities and Feminine Suburbs: Polarized Ideas, Contradictory Realities" – all of which appear in *Women and the American City*, edited by Catherine R. Stimpson, Elsa Dixler, Martha J. Nelson, and Kathryn B. Yatrakis (Chicago: University of Chicago Press, 1981).

Ann Markusen is Professor of Planning and Public Policy at the University of Minnesota. She specializes in the areas of regional economics and planning and the high tech and defense industries as stimulants of regional development. Her books include, among others, *From Defense to Development? International Perspectives on Realizing the Peace* (London: Routledge, 2004) and *Second Tier Cities: Rapid Growth Outside the Metropole in Brazil, Korea, Japan and the United States* (Minneapolis: University of Minnesota Press, 1999).

This chapter investigates the interrelationship between city spatial structure, women's household work, and urban policy. It first differentiates between two types of work in urban space: wage-labor production and household reproduction of labor power, generally and incorrectly ignored in analyses of urban spatial structure and dynamics by neoclassical location theorists and Marxist urbanologists alike. I contend that social reproduction, organized within the patriarchal household where an unequal internal division of labor favors men, profoundly affects and explains the use of urban space. The paper then presents a theoretical argument regarding the evolution of contemporary urban spatial structure. It argues that the dominance of the single-family detached dwelling, its separation from the workplace, and its decentralized urban location are as much the products of the patriarchal organization of household production as of the capitalist organization of wage work. While this arrangement is apparently inefficient and onerous from the point of view of women, it offers advantages to men and poses contradictions for capitalism.

Challenging such patriarchal structuring of urban space are residential choices that certain demographic groups are currently making. New spatial developments – such as retirement communities, gentrification (urban renewal for upper- and middle-income households), and the growth of small towns – suggest that the dominant urban decentralized form of housing and land use may pose major obstacles to efficient household production. A second, and less anarchic, challenge is from the women's movement. Since the 1960s women have organized to change land-use laws, to restructure the patterns of housing ownership and housing-unit structure, to form urban cooperatives for child care and other sharing of household work, and to restructure the urban transportation system. Nevertheless, current patterns of homeownership, real estate construction practices, and the permanency of urban physical structure are formidable barriers to the nonpatriarchal restructuring of urban space.

[...]

PRODUCTION AND SOCIAL REPRODUCTION IN URBAN SPACE

Human energy is largely spent in one of two activities: the production of commodities for market exchange and the reproduction of labor power. The former is organized, at least in capitalist society, within the institution of plants, shops, or offices, where employers hire workers (i.e., buy their labor power) for wages in order to produce commodities such as food, clothing, and shelter, machines, insurance, and so on. These are then sold to consumers or other employers for prices that more than cover wages and costs of production. Both neoclassical and Marxist economic analysts, studying the relationships among employer, worker, and output, seek to explain the conditions under which workers enter the labor market (labor force participation), under which their labor power may not be purchased by employers (unemployment), and by which certain levels of output are forthcoming from the use of certain production processes (productivity).

Similar studies look at the composition of output as a function of both demand (consumers' purchases) and supply (production conditions), and at the savings and spending behavior of workers, who receive wages, and of corporations and capitalists, who receive profits. Such studies contribute to the construction of aggregate Keynesian indicators, such as Gross National Product, and to interpretations of their causation that will guide a government intervention meant to perfect the operation of the production sphere. Marxist accounts emphasize different features of the same production process, particularly the exploitation of labor power and the inherent tendency toward crisis. Marxists consequently derive opposing views of the causes of capitalist economic problems, predicting that a socialist transformation is required to solve production sphere problems.

In the urban setting, these production concerns take on a spatial dimension. As urban economic problems, production problems become those of the transportation of employees to work (the journey to work), of the availability of desirable production space (including rights to pollute), of the fit of skills of a local population to regional industrial structure, and the multiplier process whereby locally generated income and locally stimulated production result in further increases in output and income in secondary and tertiary sectors. Neoclassical urban economic studies, at both the micro and macro level, aim at so characterizing the production process in urban space that government policy can substitute in cases of market failure (pollution control), strengthen or subsidize the market where it is weak (transportation), and stimulate the aggregate level of activity in a local economy when it is recessed (countercyclical revenue sharing). In Marxist versions, urban problems stem from the very structure of capitalism, because both economic crisis and class conflict produce contradictions in urban structure and governance.

While these are generally considered to be the "economic" realities of urban life, the sphere of social reproduction, or reproduction of the production sector workforce, is equally economic in the sense that it, too, requires labor time. Social reproduction involves the activities of both government and households that reproduce both current and future generations of labor power. These include the direct provision of the conditions of physical and mental health, cooked meals, personal services, education, maintenance of living conditions, and child care. However, the organization of social reproduction, meaning the context surrounding the use and compensation for labor time of the household workers involved, is quite different from the terms under which labor is hired and organized in the capitalist wage-labor sector.

Of the two basic institutions involved in social reproduction, government hires workers in much the same way that private businesses do. However, government does not compensate laborers strictly on the basis of productivity, does not aim to make a profit on their labor, and does not derive revenues from the sale of labor-produced output. It instead raises funds via a complex taxation system. Demands for public sector outputs are registered, not by a market process, but by a political system where voter-elected representatives determine levels of output and tax payments. Much more needs to be understood about the way in which the public sector operates, and the ways in which the structuring of urban space conditions the efficiency and equity of the local public sector. In minor ways, the recently proposed national urban policy addresses certain of these conditions by proposing incentives to metropolitan areas to adopt tax-sharing schemes and by requiring neighborhood citizen participation in certain public sector urban programs. Another subject worthy of research is the interrelationship between household production and local public sector structures.

Of greater interest to us now, however, is the structuring of social reproduction in the household realm of urban society. Most economists and other social scientists treat the household as a social and economic unit without disaggregating it to its individual members. Conventional economic analysis treats the household as a consuming and resource-owning unit, whose sole relationships to wage-labor and commodity markets consist of supplying labor to and consuming the output of the commodity production process. Similarly, Marxist analyses of the household under capitalism have emphasized its function as a private sphere absorbing the alienation of the capitalist workplace and even as a triumph of the working-class family preventing complete proletarianization. Both types of analysis obscure the role that household members, chiefly women, perform in the reproduction of labor power and the social relationships of the household in which it takes place.

In urban space, the location and organization of households are generally considered of economic interest only as the locational source of labor power and buying power. Neoclassical studies hypothesize that household-location decisions are primarily a function of the journey to work and secondarily reflect preferences for accessibility to open space, good public services, retail markets, and housing quality. Even the term "journey to work" suggests, incorrectly, that no work is done in the household. The locational outcomes are always presented as utility-maximizing decisions undertaken by the household as a whole. They have not been analyzed as decisions per se about production input required in the reproduction of labor power (e.g., minimizing travel time to schools, to health care, to markets, and to recreation). Nor has household-location analysis considered conflict among the members of the household, particularly between those who are primarily engaged in the capitalist labor market and those primarily responsible for social reproduction activities (especially when working as wage labor also), that is, conflict between men and women or husbands and wives.

The Marxist feminist literature is beginning to correct these omissions. It argues that the household is not a passive consumption unit, but one in which people reproduce their labor power, of both current and future generations, through a process that involves considerable male/female division of labor, extensive expenditure of labor time, and particular composition of output that has its own quality and distributional patterns. Even though more than half of all adult women under age sixty-five work for wages, they still bear the primary responsibility for household work. The products of their labor are meals, clean and mended clothes, home health care, preschool education of children, financial and transportation services, and so on. Yet, the economic nature of this activity is largely hidden by the informal economic contract involved. Marriage, this view argues, is an implicit rather than an explicit contract for the exchange and organizational control of labor power in the household.

HOUSEHOLD PRODUCTION, WOMEN'S ROLES, AND THE STRUCTURING OF U.S. URBAN SPACE

The most striking aspects of modern U.S. city spatial structure are the significant spatial segregation of residence from the capitalist workplace, the increasing low-density settlement, and the predominant single-family form of residential housing. Most contemporary analysts variously ascribe these developments to the rise of mass production techniques, to the automobile, to FHA mortgages, and to class and racial segregation. None of the analyses mentions women's household work, household social relations, or patriarchy as primary determinants of this structure. Feminists critical of urban structure claim that current forms are inefficient for social reproduction and reinforce women's roles as household workers and as members of the secondary labor force. Yet no one has systemically critiqued the myopia of both neoclassical location theory and Marxist urban spatial theory by documenting the centrality of the patriarchal form of household organization as a necessary and causal condition responsible for contemporary urban structure and its problems.

[. . .]

Our more specific concern here is with the spatial form produced by the conjuncture of patriarchy and capitalism in advanced industrial countries. By patriarchal structuring of social reproduction, I follow the definition of Hartmann, in which the labor power of women within a capitalism system is employed partially or wholly in the service of men in the household and where the returns to both women's and men's labor are contained in the family wage, which the man controls. Women's work involves the same basic activities that occur in capitalist production, but organized differently. A woman produces a meal for the household by purchasing raw material inputs at the grocery store or by growing a garden, combining them with her labor time, energy, and machine power provided by kitchen equipment, and serving them to household members. In contrast, her hired counterpart in the restaurant may do the same things, but her service is sold for a price that covers both her own wages and her employer's required return on investment. The productivity and efficiency of household production in the former case are just as important as in the latter, except that in the latter the market test is the willingness of the consumer to buy the restaurant meal at the price covering costs, while the quality and costliness of the household meal are directly "tested" by household members' satisfaction, and the implicit price is represented by the expenditure of household income and time.

The organization of household production has not been without its students and efficiency-promoting studies. Beginning in the middle of the nineteenth century, home economics specialists investigated ways of increasing the efficiency and quality of household production. They introduced mechanized techniques (sewing, washing, and dish-washing machines; vacuum cleaners; etc.), and they rationalized organization of women's time, using methods reminiscent of scientific management in the factory. Such efficiency concerns, however, have accepted the nuclear household as the unit of analysis. Most home economists have confined their prescriptions to changes within the household, rather than to changes in the size, composition, or spatial organization of households in urban places, including their juxtaposition to other institutions that supply inputs into the household production process. A few considered collective kitchens and apartment hotels, but for the most part dramatic changes in household organization were championed by utopians: Charlotte Perkins Gilman, for instance, in *Herland*, her feminist utopia originally published in 1915, and Melusina Fay Peirce in her 1868–71 campaign for cooperative housekeeping in Cambridge, Massachusetts.

Similarly, few treatments of optimal household location in urban space try to account for the maximization of efficiency in household production. Most location models assume that the household chooses between "*the* journey to work" (implying only one wage worker) and various consumption goods such as housing, open space, and public sector services. Gravity models (models that weight the multiple spatial orientations of the firm, e.g., toward sources of inputs and markets for outputs, by the transportation costs between each in order to predict the profit-maximizing location) have not been developed for intraurban household location. The exclusion of household production from the calculus for urban spatial location is damaging, because it may lead to erroneous conclusions about the ability of certain urban-policies to affect spatial form or to a blindness regarding the vulnerability of urban form to changes in household structure.

The fundamental separation between "work" spheres and home corresponds roughly to the division of primary responsibility between adult men and women for household production and wage labor, at least historically. Since patriarchy is the organizing principle of the former sphere, urban spatial structure must be as much a product of patriarchy as it is of capitalism. Patriarchy may thus contribute to and condition urban problems. The recent literature by feminists critiquing urban structure, both the single-family household and its spatial decentralization, argues that current patterns are inefficient for women because they result in wasted labor time and curtailed access to jobs and other facets of urban life. Popular accounts, such as Betty Friedan's critique of women's household experience in *The Feminine Mystique* and Marilyn French's detailing of suburban housework in the bestselling novel *The Women's Room*, also document the waste and alienation inherent in the current structure of urban housing, suburban neighborhoods, and intraurban transportation networks.

[. . .]

If these contentions regarding the inefficient structuring of urban space are true, why do women continue to choose household production roles in the existing urban structure? Primarily because their choices are so limited. In household decisions men and women are conditioned by the limited range of their options, a fact which conventional location analysis obscures. The major force encouraging women to remain in the household sector is the inaccessibility of jobs in the capitalist labor market, and the occupational segregation that keeps women's wages at strikingly low levels and women confined in low-skill, no-advancement jobs. These conditions, in turn, have been traced to patriarchal compromises struck between male workers and capitalist employers, where the primary incentive for male workers to oppose women's incorporation into wage labor was the level of service and power men commanded in the household. Even when women work for wages, and a majority do, their lower pay reinforces their role as supplemental breadwinners for the household, and they continue to bear the primary responsibility for housework.

A second, and tempering, observation must be made. Given the quality and paucity of the alternatives, women's control over their own working conditions may be greater in the household workplace than in a factory or service establishment. Of course, this may not be universally true; some husbands require extraordinary cleanliness and service on command, and some use physical violence to enforce their demands. Household production organized in single-family units permits some variation of tasks and development of a higher level of skills, compared to the specialization and deskilling generally enforced in the lower ranks of the labor market, a station to which most women would be assigned if working as wage labor. This consideration of quality of working conditions cautions us against concluding that household production per se is inefficient. A feminist society, in which oppressive household divisions of labor were eliminated, might still choose to perform many tasks in the reproduction of labor power within a modified form of household.

But, if contemporary urban spatial structure is dysfunctional for most women, then it must be functional or efficient from the point of view of capitalism or patriarchy or both. To distinguish between the two, we must look at the ways in which the conditions governing women's household production in urban settings have evolved with developments in the capitalist workplace and in the patriarchal household. Capitalist structuring of twentieth-century society has profoundly influenced the household, both directly through its requirements for the type of labor power to be reproduced and indirectly through its organization of commodity production. Increasingly, because of its hierarchical elaboration of the labor force, the capitalist production sphere required dramatically different conditions for the rearing of children. This has resulted in class-differentiated neighborhoods with tremendous responsibility for child care devolving upon women in individual households. Women may collaborate in the decision to live in suburbia, because, given the options, suburban locations offer safer, less crime-ridden, and less racially tense environments for both blacks and whites for bringing up children. Furthermore, since the types of household labor involved in caring for children are dissimilar from those involved in reproducing the labor power of adults, it may be that differentials in quality, productivity, and required labor time result in the relatively high incidence of households with small children in metropolitan suburban locations.

Second, capitalist organization of land and housing as commodities brokered by the real estate industry, built by the construction industry, and financed by the banking sector, has found current patterns of residential decentralization and single-family dwellings profitable. However, such profitability need not require this particular spatial form. For instance, in the absence of the patriarchal nuclear family, these same industries might have made just as much profit off the construction, sale, and financing of high-density, urban apartment-like living quarters accommodating various forms of household composition. It is true that the single-family form of housing is more amenable to the type of recycling of urban space that occurs with block-busting, speculation, and continual construction and destruction of housing values. Nevertheless, this argument cannot explain the single-family unit; more accurately, it reflects its popularity. The housing and land use recycling process has, however, indirectly promoted the isolation of the suburban household worker by eroding the extended family and community network that previously helped informally to collectivize household work in urban areas. The dynamics of urban land use increasingly result in the wholesale slummification or renewal of neighborhoods, a process that undermines entire ethnic communities and disperses their residents. This is also a function of the increasing labor mobility required by employers, which produces individual household migration patterns, both intraurban and interregional, that disrupt long-term community ties.

A final factor in the individualization of urban household work has been the restructuring of the income basis for retired people toward social security and pensions and away from inclusion in family support systems. This development has eliminated the need for generations to occupy the same household or to settle near each other. Incomes of the elderly are frequently insufficient to support residence in the same neighborhood as those of their sons and daughters. As a result of age segregation of urban and even regional space (seniors moving to Florida and Arizona), women have lost the propinquity of parents and other older neighbors as child carers.

[. . .]

It is unnecessary to attempt to unravel completely the relationship between patriarchy and capitalism in promoting contemporary labor spatial structure. While eliminating capitalism does not necessarily end patriarchy, as the socialist countries show, it is not clear that we could eliminate patriarchy without transforming capitalism. Some argue that patriarchy indirectly serves capitalism by dividing the workforce and blunting class consciousness. However, acceptance of this view does not require a corollary that *all* forms and products of patriarchal relationships are functional for capitalism. Even if the current urban structure is efficient for capitalism as a whole, one can make a case that it is functional for patriarchy, and that the widespread indifference to women's concerns about its inefficiency can be attributed mainly to patriarchal, rather than capitalist, ideology.

How do men benefit from contemporary urban spatial structure? I can offer several hypotheses. First, the "ideal" single-family, detached, urban or suburban dwelling embraces the contemporary patriarchal form of household organization. Within it, the man (when present) is generally considered the head of household (until the current census, he was officially so), the primary wage earner, and the major support of the household. This form discourages extended family or community sharing of housework; deploys the machinery of housework in individual units, which makes sharing difficult; and replaces collective play space (parks) with individual yards. The journey to work of the husband tends to dominate location decisions, and the distance to jobs, combined with inadequate public transportation, discourages women from working for wages. The publicly subsidized (FHA) propagation of homeownership entails significant commitments on the part of both women and men to maintenance work on the individual structure, toil that is confined and controlled within the household unit.

Second, the apparently wasteful expenditure of women's labor time under such circumstances underscores the nature and maintenance of power and privilege of men in the patriarchal household unit. Thorstein Veblen argued years ago that men receive satisfaction from the conspicuous consumption of women's labor time in the household: the more waste apparent, the more social status accruing to them. If this motive prevails, productivity of women's time in the household is not important, and a concern with it would directly conflict with accrual of status. Feminist historians have observed that working-class women as well as men aspired to replicate the upper-class family, where women did not have to work for a wage, but would instead perform the role of household manager and servant. The lack of interest in household efficiency in the larger spatial context also reinforces the illusion that only men "really" work and "provide" for their families. At any rate, we can conclude that in general, American men have not registered complaints about the waste of women's labor time, or inefficiency, of dispersed, residential, single-family dwelling patterns.

A third, and nonconflicting, hypothesis concerns the quality of output achievable under the patriarchal form of household organization. The advantages of this form of organization include the flexibility of scheduling of certain services (meals, recreation, etc.) and the potentially higher quality of home-cooked meals, homegrown vegetables, and personally tailored services. (Of course, such personally crafted services could be worse, as well!) In this sense, the current form of household production may not be inefficient, that is, wasteful of women's labor time. This argument would hold if women's preferences coincide with men's preferences; given the alternatives, women may indeed enjoy their own home-cooked meals and may prefer taking care of their own children to cheap restaurants or for-profit custodial child care. But it may also benefit men at women's expense. Flexible and quality meals for husbands may require the squandering of women's labor time.

We are, of course, dealing with shifting definitions of efficiency. In the argument regarding quality of household output, the "efficiency" of a home-cooked meal depends on which side of the dinner table you occupy. While it appears that men as a rule gain from the current organization of household production, these gains are qualified by their position as wage-earning workers. It is possible that such personal service could become very expensive with certain changes in capitalist workplace organization, commodity composition, and women's resistance. For instance, if women's wages were to rise enough to make their wage labor attractive; if women were to demand full partnership in household work; and if the quality and price of market-produced childcare, meals, and so on were to improve substantially, men might agree more readily to women's wage work, with a diminution of the quality and privilege associated with full-time household work, and to restructuring of the household in urban space. Nevertheless, the single most significant factor in the structure of the patriarchal household is the derogation of primary responsibility for household labor upon women. If men were to assume their share of responsibility for housework and child care, then we might experience significant changes in household structure and spatial location in urban areas. Such an event might show us just how inefficient contemporary household production patterns really are. Similarly, the question of whether current patterns are functional for capitalism depends upon the net effects on profit produced through such arrangements and, further, on the way in which patriarchal concessions to male workers are necessary to the maintenance of a docile workforce.

Of course factors other than the patriarchal form of household production influence location decisions and housing choices, particularly race and class characteristics and the relationship between households and wage-labor work locations. Furthermore, American individualism has undoubtedly coincided with patriarchal household structure to promote the single-family suburban housing that is more common in U.S. cities than in European ones. Yet, the central theoretical point stands: patriarchy profoundly shapes American urban spatial structure and dynamics. It implies that the dismantling of patriarchal household arrangements might call for dramatic restructuring of cities.

THE RESHAPING OF URBAN SPACE

Patriarchy is not a static system. Currently under widespread attack by the women's movement, it also must accommodate itself to changes in capitalism, particularly the increasing availability of commodity substitutes for household labor (child care, restaurant meals, and so on). In unorganized as well as organized ways, changes are taking place in urban living arrangements that foreshadow the future evaluation of urban spatial structure. For instance, three significant demographic changes have occurred in the 1970s in the United States that involve new forms or locations of household production. Their emergence contributes substance to the argument that decentralized urban single-family households are not the most efficient workplaces for the reproduction of labor power and human life. Each instance involves a different segment of women in the population, but all three suggest significant efficiencies deriving from a more integrated use of urban space or a more collective form of household or neighborhood structure.

The first phenomenon is gentrification, that is, the reverse migration of middle- and upper-income residents to urban centers, bringing housing rehabilitation or rebuilding with them. While no urban analysts have pursued this explanation, it seems clear that gentrification is in large part a result of the breakdown of the patriarchal household. Households of gay people, singles, and professional couples with central business district jobs increasingly find central locations attractive. Particularly important has been the success of both gays and women in the professional and managerial classes in gaining access to decent-paying jobs. Gentrification must also be ascribed to the growth of high-income professional and managerial jobs per se in big cities, where the needs of contemporary large-scale corporations have concentrated jobs associated with the control and management of large-scale economic enterprises and with government-related activity.

Gentrification in large part corresponds to the two-income (or more) professional household that requires both a relatively central urban location to minimize journey-to-work costs of several wage earners and a location that enhances efficiency in household production (stores are nearer) and in the substitution of market-produced commodities (laundries, restaurants, child care) for household production. In some areas, such as the SoHo district in New York City, large networks of women-headed households share child care and other household production activities in a well-organized manner. The flexibility in designing living space out of converted lofts enhances this collectivization and permits variation in household composition.

Second is the movement of retired households to relatively nonurban settings, all the way from southern New Jersey and the Upper Peninsula in Michigan to retirement colonies like Leisure World in Florida and in the Southwest. While these households generally remain nuclear, the division of labor between men and women frequently breaks down. One striking feature of these communities is that they provide relatively easy access to the normal range of inputs in the household production process, eliminating the difficulties posed by urban and suburban traffic, parking, and high taxes. Condominiums reduce the household maintenance tasks. Such communities are experiencing the revival of noncommodity production and barter among residents, women and men alike, with a more explicit emphasis on the quality of household production. Retirement housing is also frequently more collective in shared space and facilities. Buildings or complexes may include group dining facilities, group recreational facilities, and small health-care and therapy centers.

Finally, there is striking evidence that the fastest growing American communities are small and medium-sized towns, not urban areas. Although suburbs continue to be developed, the total suburban population is not rising rapidly, and many urban areas as a whole are experiencing depopulation. While decentralization of employment and the relatively lower cost of living are most often cited as causes, my own informal observations suggest that women are frequently strong proponents of such moves, because they provide greater possibilities for shared child care and greater community involvement in social reproduction. Such places also involve easier access to jobs, even though they may not offer better pay than those available in urban areas. I have noticed, for example, that wives of construction workers who are regionally mobile often choose to live permanently in small towns rather than cities because they make household life easier while their husbands migrate to seasonal jobs.

These latter two examples are not cases of the breakdown of patriarchy per se but of household and urban spatial patterns that have evolved from a breakdown of the patriarchal structuring of household production. In fact, both of the latter cases may be explained as reactions to the extraordinary success of that form in dominating the housing stock and the spatial array of urban residences. In the former case, older people frequently find it inefficient to continue to live in, maintain, and pay taxes on a large house, which previously operated as the workplace for rearing children. The homogeneity of single-family neighborhoods makes it nearly impossible to find alternative housing in the same area. The choice to migrate to another region entirely may be more a hallmark of the destruction of any meaningful or accessible social unit larger than the family than a product of the cessation of wage work or the search for a warmer, healthier climate. Similarly, the migration to small towns of nuclear families not tied to urban labor markets signals the inefficiency of household production imposed by the patriarchal urban form, particularly its destruction of access for women to collective help with household production, to jobs, and to urban amenities.

However, just as patriarchy will not disappear without organized resistance, the trends we have just discussed cannot be counted on anarchically to undermine the patriarchal structuring of households and urban space. The sobering reality of this form of urban spatial structure is its permanence. It is literally constructed in brick and concrete. Therefore, its existence continues to constrain the possibilities open to women and men seeking to form new types of households and to reorder the household division of labor. The fact that housing, the primary workplace for social reproduction, is also the major asset for many people tends to reinforce the single-family, patriarchal shape of housing and neighborhoods. Builders, and people buying from them, worry that innovative housing forms will not have a resale value. People interested in suprafamilial communal living situations have generally had to either migrate to rural areas where they could construct their own housing or to older central city areas where they could convert large old housing units (originally built for extended families or boarders). Efforts by lesbians and other organized groups (e.g., a church group in Detroit) to take over entire neighborhoods are severely hampered by the individualization of land-ownership patterns and the legal sanctity of property. Developers, through urban renewal, may use public domain powers to clear entire sections of urban land for private business development, but community and neighborhood groups have no such access to eminent domain for efforts to collectivize living space for the tasks of social reproduction. The dominance of the single-family, detached dwelling in a decentralized urban spatial structure reinforces people's ideas of what forms of household structure are possible and penalizes them materially if they wish to pursue other visions.

Since the resurgence of the women's movement in the late 1960s, women who wish to change their living arrangements and household responsibilities, or to increase their options for doing so, have found it necessary to attack the structural determinants of patriarchal household urban form directly.

[...]

In many urban areas, women have set up cooperative child care and other forms of cooperatives that help alleviate the burdens of housework. Women have found regulations of the welfare state that presume the patriarchal family is the normal household unit and that are frequently oppressive to men as well as women.

The widespread activism of women in urban struggles around housing, child care, and neighborhood preservation has generally been neglected in the literature on urban social movements. Feminist critics of this literature point out that because women's struggles have frequently been over issues bearing on conditions of social reproduction, they are invisible to the students of urban social movements, who identify urban problems as capitalist, not patriarchal, phenomena. Furthermore, the literature frequently overemphasizes the role of men in such movements and misrepresents the goals and strategy of the organizations. While leadership may be male, the main organizing in many urban struggles has been accomplished by women, who tend to know their neighborhoods better, who build collectivization of household labor into many community group organizing efforts, and who seem to opt for a form of organization that permits them to continue their household work.

[...]

TOWARD A FEMINIST NATIONAL URBAN POLICY

Since urban spatial structure and housing form is such an important constraint on women's ability to change household work roles, national urban policy should be a major target for feminists. The Department of Housing and Urban Development should rank high with departments concerned with labor, health, education, and welfare as a major agency charged with policy responsibilities central to women. HUD's programs should be scrutinized for their impacts on women and new initiatives should be proposed and demanded.

What would a national urban policy that addressed women's issues look like? Perhaps if we could map this out, we could better assess current national urban policy. The answer grows out of our criticisms of the current structuring of urban space for household production and from a more general critique of social structures that impede women's progress and that could be addressed at the urban level. First of all, the most pressing issue for both women's involvement in the labor force and for economizing on household production involves child care. Without some form of socialized child care, women with young children will remain tied primarily to household production locations. Three solutions are possible: sharing informally through extended-family arrangements and neighborhood co-ops; publicly producing child care; and privately producing, for profit, child care. Each currently exists on a limited basis. Each has serious consequences for both the quality of child-care services and the labor-force participation of women. No thorough study of the implications of each as a prototype system has been done. Clearly, the problem is urban in nature: Should child care be provided cooperatively in small neighborhood complexes or single-family homes, by the public sector in public buildings (like elementary education) on a larger neighborhood scale, or in private enterprises in neighborhood or regional shopping centers? The alternative location – at the plant, office, or shop – has received little attention, even though it has interesting possibilities for parent/worker involvement in child care and for efficiency in journey-to-work trips.

A second issue, also of significance, is the type of housing available in urban areas. Women living without men but with children (a rapidly growing group), and groups of single adults, have difficult times both in adapting existing housing to their needs and in obtaining access to it. Restricted credit opportunities are only one of several discriminating barriers. We have no way of knowing what types of households people would choose to live in, given a choice, but we can say that these choices have not existed in the past, nor do they exist in the present. Federal policy might (as it does in the energy business) invest in research and development and experimental projects with various forms of collective and nonpatriarchal housing.

A third set of policies would encourage and subsidize the reintegration of uses in urban space to enhance the efficiency of the household production sector. These policies might pioneer and provide incentives to small-scale commercial development; to the decentralization of jobs in small establishments; to efficiencies in the use of urban space, such as more park space in place of endless private front yards; fine-grained transportation systems; etc. While I leave the design of these to my imaginative sisters, I might suggest one criterion that could be used to judge the desirability of such projects (net of other costs): the elimination of unnecessary labor time expended in individual travel (excluding public transportation time which can be used to read newspapers or books or to socialize), in individual yard improvement and grooming, in individual meal preparations, child care, and so on. Other criteria must address both the quality of output and the quality of women's working conditions.

[...]

Finally, a feminist urban policy would establish a new research agenda. This would explore the theoretical understanding of the relationship between patriarchal household form, urban housing, and spatial structure. It would pursue extensive empirical work to document the cost in women's labor time and the working conditions within the household resulting from various aspects of urban structure. It would develop and introduce a new type of cost/benefit calculus for judging the efficiency and social welfare of such public projects as transportation systems, housing, and urban infrastructure (parks, streets, utilities). This calculus would evaluate the effects of such projects on the household division of labor and its productivity. It would design, propose, and experiment with policies and projects that would directly address household production concerns, increase women's choices, and alleviate the onerous division of household labor. It would investigate the proper levels of government or form of collective organization that should be charged with reshaping elements of urban structure. Such a research agenda, which would undoubtedly uncover many creative possibilities, would end the invisibility of women's concerns within both the academy and the agencies that shape urban policy.

" 'Race,' Space, and Power: The Survival Strategies of Working Poor Women"

from *Annals of the Association of American Geographers* (1998)

Melissa R. Gilbert

Editors' Introduction

The readings in Part Three addressed the spatial dimensions of post-World War II poverty in U.S. cities, pointing in particular to the serious decline in employment opportunities in the cores of older cities, where mostly poor minorities live. The rise of offshore manufacturing, suburbanization, misguided urban renewal policies, and racially-biased practices of the real estate industry and lending institutions contributed to what Loïc Wacquant calls the "impacted ghetto" – a space disconnected from social, economic, and political opportunities.

In this reading, Melissa Gilbert further examines the consequences of the "disconnectedness" of the contemporary ghetto from the point of view of working women. Gilbert takes issue with the "spatial entrapment" thesis, which claims poor urban women are "cut off" – in terms of distance and mobility – from the better-paying jobs to be found in the suburbs. Trapped within the confines of the inner city, poor women have access to jobs that pay low wages and offer little, if any, advancement. Poor women who live in urban ghettos and who have no reliable means of transportation (to the suburbs) are rendered powerless in the face of structural economic change.

Gilbert concurs that structural economic factors have produced severe disadvantages for poor women and others who call ghettos home. She disagrees, however, with the theory's portrayal of residents as trapped – as passive subjects of economic forces with little or no recourse to any situation but despair. Gilbert draws on feminist perspectives of intersectionality, which see race, class, gender, and sexuality as mutually interacting, as opposed to discrete, separate spheres of social identity. The lives of poor women are more complicated than the spatial entrapment and other, similar labor theories allow. Poor women might be economically "trapped" in ghettos, isolated from jobs in the suburbs and deprived the mobility to get to them. But not necessarily: They may be "rooted" in inner city neighborhoods as well and, as Gilbert claims, that rootedness can be a resource for women in their active efforts to secure economic, social, and personal well-being.

Women use personal networks of family members and friends to learn about jobs, places to live, and facilities for childcare. Because they have limited mobility and access outside the inner city, the networks for poor, mostly minority women tend to be more integrated and overlapping (work-based, church-based, etc.) than those for their suburban, mostly white counterparts. Drawing on a case study of African American and white women in Worcester, Massachusetts, Gilbert examines how poor women's rootedness in the urban core and their use of networks as a survival strategy are both enabling and constraining. Localized and intensive social networks are often crucial to finding jobs, homes, and childcare that fit the realities of limited mobility. Because these networks tend to be insular and recurring, however, the range of opportunities tends to be narrow and constricted.

Melissa Gilbert is Associate Professor in the Department of Geography and Urban Studies at Temple University in Philadelphia. She has written in the areas of labor market strategies, theoretical and policy debates concerning urban poverty, and labor and community organizing. Her publications include "Identity, Difference, and the Geographies of Working Poor Women's Survival Strategies," in K. Miranne and A. Young, eds, *Gendering the City: Women, Boundaries and Visions of Urban Life* (Lanham, MD: Rowman and Littlefield, 2000: 65–87); "Place, Politics, and the Production of Urban Spaces: A Feminist Critique of the Growth Machine Thesis," in A. Jonas and D. Wilson, eds, *Twenty-One Years After: Critical Perspectives on the Growth Machine* (Albany, NY: SUNY Press, 1999: 95–108); and "The Politics of Location: Doing Feminist Fieldwork at 'Home,'" *The Professional Geographer* 46, 1 (1994): 90–96.

The Personal Responsibility and Work Opportunity Reconciliation Act of 1996, more commonly referred to as "welfare reform," eliminated the federal guarantee of cash assistance to poor people and replaced it with a system that contains stringent work requirements and time-limited assistance. One of the underlying assumptions of welfare reform is that poor women with children will become economically self-sufficient through employment. Yet against this optimistic projection are the facts that most women are in sex-segregated occupations with the attendant low-wages and lack of opportunity for advancement, that there is still a significant gender wage gap, and that many women's wages are less than adequate to support a family. Indeed, the restructuring of the U.S. economy since the 1960s has led to a polarization of wages, an increase in low-wage service-sector jobs, and increases in part-time employment, all of which have further disadvantaged many women. The result of these trends is that many women who work for wages are little or no more financially secure than they would be if they had received payments from the now-defunct program, Aid to Families with Dependent Children (AFDC).

[...]

It is within this larger context that feminist geographers have made important contributions to the analysis of how labor-market inequalities are reproduced through space and in places by collecting overwhelming evidence that white women are more spatially constrained than white men. This is exemplified by women's shorter commutes to work, which negatively affect their employment opportunities. Kim England (1993) has termed this argument the "spatial-entrapment" thesis. Research examining the effects of "race" on women's levels of spatial containment, however, has demonstrated that many African-American women have longer commutes to work than white women and that this also negatively affects their employment opportunities. Clearly the fact that African-American women are more economically disadvantaged points to a limitation in the spatial-entrapment thesis. I suggest that this paradox requires us to reconceptualize the links between space and power underlying the thesis.

First, the longer commutes of many African-American women make clear that equating mobility with power and immobility with powerlessness is too simplistic to capture the spatiality of women's daily lives. Instead, I argue that no spatiality is inherently with or without power. Second, conceptualizing power as having a single source, as in, for example, gender relations, flattens out differences among women and minimizes the complexity of their lives. Rather, I argue, power should be conceptualized in terms of a multiplicity of interconnected, mutually transformative, and spatially constituted social relations. This reconceptualization of the links between space and power suggests that we move beyond the duality that equates mobility with power and immobility with powerlessness, exploring instead how mobility and immobility are related to multiple power relations. We can then begin to examine how rootedness, a potential outcome of spatial boundedness, may be a resource as well as a constraint, depending on the constellation of power relations. In doing so, we can provide a more nuanced analysis of the opportunities for and barriers to women's economic security than is now provided by the spatial-entrapment thesis.

To explore how the spatial boundedness of women's lives and its consequences vary, I examine the role of place-based personal networks in the survival strategies of working poor African-American and white women with children in Worcester, Massachusetts. The study illustrates how gender intersects with "race" to shape the spatiality of women's lives and the ways in which rootedness may be a resource as well as a constraint. Using the data from Worcester, I first evaluate the women's experiences in terms of the spatial-entrapment thesis, determining the spatial extent of African-American and white women's daily lives and its impact on their employment opportunities. Then I examine how women use the spatial boundedness of their lives to develop place-based personal networks – in the family, workplace, community, and neighborhood, and as one important aspect of their survival strategies – to ensure the health, safety, and security of themselves and their families. I show how women use personal networks to connect them to employment, housing, and childcare, and more broadly to ensure their economic and emotional well-being, while also demonstrating how these same networks can operate as constraints in terms of the kinds of jobs, childcare, and housing to which they are connected. I conclude by arguing that to understand the differences in the spatial boundedness of African-American and white women requires moving beyond the spatial-entrapment thesis to explore how the relationships among rootedness, networks, and survival strategies are shaped by racism.

[...]

RECONCEPTUALIZING THE LINKS BETWEEN SPACE AND POWER IN WOMEN'S DAILY LIVES

People are spatially bounded, and geographers and others conceptualize the degree of boundedness in terms of mobility and immobility. Associating mobility with power and immobility with powerlessness, however, is too simplistic to capture the spatiality of daily life.

Theorizing spatiality as a potential resource as well as a constraint highlights women's agency and adds complexity to our analyses of their lives. It also suggests that we need a conceptualization of power that will allow us to better explore differences among women. Feminists have paid increasing attention to the ways in which gender interacts with other power relations, such as sexuality, age, ethnicity, class, ableism, and "race," thereby conceptualizing power in terms of multiple axes of interconnected and mutually transformative relations.

[...]

Using this conceptualization, it becomes impossible to understand power as a dualism (all or nothing), and therefore to simplistically "read off" a given set of spatial outcomes for a particular set of power relations.

[...]

Building on this insight, I will examine how poor women – women who are often seen as either the victims of patriarchal structures or as passive and atomized individuals dependent on government subsidies – use rootedness in the construction of their survival strategies. I use the term "strategies" to refer to the everyday decisions and practices of poor women attempting to ensure the economic and emotional well-being of themselves and their families. By comparing the strategies of African-American and white women in Worcester, I will show the ways in which women's survival strategies are shaped by their location within a constellation of power relations.

[...]

GENDER, "RACE," AND THE GEOGRAPHY OF SOCIAL NETWORKS

There is an extensive literature documenting the significance of personal networks that people draw upon for emotional, social (including childcare and transportation), and financial support, as well as for housing and employment. Additionally, networks have been evaluated in terms of their existence and potential utility, or social capital, rather than their actual use. Women's networks tend to be focused on kin and neighbors, while men's networks tend to be focused on non-kin, particularly coworkers. These differences are due to the different structural locations of women and men in the family and labor market.

In addition, there is extensive literature examining the social networks of African-Americans, including comparisons of African-Americans' and whites' networks. Particular attention has been paid to the role of the extended family and church in the lives of African-Americans, because both have historically been important institutions in mitigating the effects of living in a racist society.

[...]

While some sociologists have conceptualized people's embeddedness in social networks, feminist geographers have explicitly analyzed the spatial dimensions of embeddedness, such as the impact of local context on spatially grounded networks and the role of space in shaping information flows.

[...]

An important aspect of spatial rootedness is the social networks that people develop in places. Research has shown that networks are a significant component of people's survival strategies; that the use of networks, as well as the type of networks used, can vary by gender, class and "race"; and that the spatial dimensions of networks have important consequences for women's economic opportunities and survival strategies.

This paper builds on these insights to more thoroughly examine the different contexts in which women create networks, the ways in which different networks are used for different purposes, and their place-based interrelatedness. In doing so, I explore the ways in which the spatiality of women's networks may be enabling as well as constraining, and the impact of racism on these processes.

THE STUDY AREA

Worcester, Massachusetts is a propitious context for this study because economic restructuring has had widespread impacts on its working-class people's economic opportunities. Worcester, like many cities in old industrial regions of the U.S., has experienced a decline in manufacturing jobs and an increase in service jobs. Research has also shown that Worcester has high levels of occupational sex segregation and wage disparities. These economic conditions have had severe impacts on African-American families, who are disproportionately likely to fall below the poverty line. Worcester has also disproportionately experienced a rise in female-headed households.

[...]

Worcester has always had a diversity of ethnic groups, with Irish, French Canadians, Scandinavians, Italians, and Poles being particularly prevalent. More recent arrivals to the city include Puerto Ricans, Asians, and African-Americans from the U.S. South. African-Americans were historically located in the neighborhood of Main South and along Belmont Street. Two historically African-American churches and a more recent congregation are located in or near these neighborhoods. When Interstate 290 was built through the city in the 1960s, it displaced many African-American families living in the Belmont Street area. Furthermore, a public-housing project, Plumly Village, was built in 1970 in the same area. Community leaders saw these actions as a deliberate attempt to break up the African-American community after the riots of the 1960s in many other U.S. cities.

DATA AND METHODS

Exploring how gender intersects with racism to shape the spatial boundedness of working poor women's lives and how rootedness may be a resource as well as a constraint in their survival strategies required collecting primary data. My goal was to interview African-American and white women living in Worcester who were working in low-waged jobs and who had or were raising children. I sought to interview enough women in each racialized group to allow a quantitative analysis of differences between racialized groups of women. . . .

[...]

I interviewed 26 African-American and 27 white women living primarily within the city of Worcester during the summer and fall of 1991. The average age of the women was 34, and their average years of education were 13.5. African-American women had significantly fewer years of education. With the exception of three who had lost their jobs between the time the interview was arranged and when it was conducted, all were employed. Most of the women worked full-time as clerical workers or in service occupations. In terms of income, there were no significant differences by racialized group, but African-American women had more children and larger households on average than white women. Most of the women had children requiring some kind of childcare. Seventy-eight percent of white women and 50 percent of African-American women were single parents. Only a few of these women received child support.

The nature of the research questions required that I obtain detailed qualitative information about the daily lives of the women in the study and their analyses of the meaning of, and reasoning behind, their activities and strategies. Therefore, I chose to conduct in-depth, interactive interviews. The semistructured interviews, averaging approximately two hours, contained both closed- and open-ended questions aimed at determining the spatial extent of daily life, the use of personal networks, and the importance of these networks in their survival strategies. Moreover, detailed employment, childcare, and residential histories were collected for each woman, thereby adding historical depth to a small sample. The interviews were tape recorded and transcribed.

[...]

THE SPATIAL EXTENT OF WOMEN'S DAILY LIVES

An analysis of the job histories of the women in this study revealed that the average travel time to work for all jobs held (N=186) was 14.9 minutes; women often traveled 10 minutes or less to their jobs and rarely traveled more than 30 minutes. Therefore, the experiences of the women in my study support the empirical claims of the spatial entrapment thesis; women with children generally have shorter commutes to work than do men.

Yet African-American women in this study appear to be more spatially entrapped than are white women. Interestingly, and contrary to previous research, African-American women traveled significantly less time to work than did white women. African-American women traveled an average of 11.9 minutes to their jobs (N=81), while white women traveled an average of 17.8 minutes to their jobs (N=105).

As further evidence of the spatial entrapment of women with children, we can examine women's childcare trips which, although neglected by the spatial-entrapment thesis, help to explain why many women with children have short commutes to work. Women's journey-to-work times increase substantially when the journey-to-childcare is included, suggesting that childcare trips limit women's ability to travel farther distances to work by increasing their overall commuting time.

By subtracting the direct travel-time-to-work from the travel time including childcare trips, we get a clear picture of the considerable time that the childcare trip adds to the journey-to-work. In an analysis of all jobs requiring a childcare trip, the average increase was 18.2 minutes. African-American women's childcare trips added less time to the journey-to-work than did white women's trips. White women spent significantly more time commuting (39.05 minutes), including childcare, than did African-American women (27.13 minutes), providing further evidence of the latter's greater spatial entrapment.

[...]

Susan Hanson and Geraldine Pratt (1995) have demonstrated that the length of the journey-to-work is strongly associated with a woman's occupation type. They found that women in female-dominated occupations are significantly more likely to work closer to home than women in other occupations. My findings support theirs and additionally show that women in female-dominated occupations added less time to their work trips because of childcare trips than did women in gender-integrated occupations. More than 58 percent of the women in my study said they would not spend more time traveling to work if they could find a job with higher wages. Women in female-dominated occupations were least likely to say that they would travel farther for higher wages, suggesting that they were more spatially entrapped. There were no significant differences, however, between African-American and white women.

[...]

THE INTERRELATEDNESS OF HOUSING, EMPLOYMENT, AND CHILDCARE DECISIONS

[...]

Many women, regardless of the presence or absence of a partner, must fulfill the multiple roles of employee, mother, and family provider. Consequently, many women's employment, childcare, and housing decisions are complex and often interrelated. Most of the women in my study made their housing decisions prior to their employment and childcare decisions, so these were determined by their residential location. Nearly all of the women I interviewed (97 percent) were making their employment decisions from a fixed residential location. Childcare factors, however, were sometimes part of women's housing decisions; nearly 13 percent of the responses as to why someone was sharing a residence were because they needed help with their children.

The employment decisions of women were strongly influenced by their responsibilities as mothers and family providers, including the hours worked, the type of work, and the specific job. The necessity of fulfilling multiple roles is evident in the hours women chose to work. While most of the women worked full-time, they overwhelmingly would have preferred working part-time due to their domestic and child-raising responsibilities. Nearly all of the women (94 percent) who worked full-time, but wanted to work part-time, could not do so because of financial reasons. Moreover, women's responsibilities as mothers and family providers affected the kinds of jobs that they do. Seventeen percent of the women chose their type of work because of their children. Some women selected certain fields because the hours allowed them to be home after school or because the job would not require overtime or travel. A number of women became family daycare providers in order to combine work and child raising or because they could not find jobs that would pay them enough to be able to afford childcare.

[...]

An additional 17 percent of the women chose the type of work because of the need to provide for their children. For example, some women would choose certain fields or employers they believed to be secure, or they would pick a growth field that did not require much training so that they could get off public assistance as quickly as possible. Most women wanted to do different jobs (83 percent), but could not because of their family responsibilities. When asked why they could not get their desired jobs, 58 percent of the responses were because of child responsibilities, while 30 percent said that they did not have the money to go back to school. Their stated reasons included that they could not afford the childcare necessary to attend school, were waiting for their children to be grown, or that the job did not pay enough or have the benefits that the women needed to support their families.

[...]

The intertwined nature of women's employment, housing, and childcare solutions sets spatial limits on their daily lives. Access to transportation is another factor in the spatial extent of women's daily-activity patterns and the nature of their employment, housing, and childcare decisions.

[...]

While 70 percent of the women employed at the time of the study drove themselves to work, only 12 percent had access to a car all of the time; 71 percent had access to a car sometimes, and 17 percent never had access to a car; here there were no significant differences between African-American and white women.

In sum, the women's experiences generally support the spatial-entrapment thesis. Interestingly, African-American women appeared to be more spatially entrapped than white women. This finding may be due to their more disadvantaged labor force position or to the fact that the African-American women in this study had on average more children than the white women.

[...]

ROOTEDNESS AS A RESOURCE AND/OR CONSTRAINT IN WOMEN'S SURVIVAL STRATEGIES

[...]

Women's networks

Women's networks are multifaceted; the community, workplace, and neighborhood are all places in which women create networks. I characterize women's networks as having four components: kin-based, work-based, community-based, and neighborhood-based. Women met their community-based contacts in a variety of contexts in the process of going about their daily lives. A common way for women to meet people in the community was through their children's friends or activities. Other places that women met people in their larger community included local bars, childbirth classes, and the welfare office. The kinds of community contacts mentioned most frequently were among people with whom the respondents had grown up or attended school, as well as church-based contacts. While approximately half of the women in both racialized groups attended church regularly, African-American women (26.9 percent) were more likely than white women (16.7 percent) to have said that they were actively involved in church activities. An African-American woman not originally from Worcester explained the importance of her church networks:

> I was always told, and anyplace I go, I have to find and establish a church family and that's what you do. Because people, I'm not going to say all people in the church are good, but you know once you establish yourself . . . I think that you get a lot of help.

[. . .]

The workplace is also an important context in which women build networks. More than two-thirds of the women had people at their workplace that they considered to be friends. Most of their friends were other women, although, not surprisingly, women in gender-integrated occupations had more male friends at work than did women in female-dominated occupations. Most of the friends were people that the women had met at the workplace. Interestingly, the workplace appears to be an important site of integration for African-American women. African-American women were most likely to meet women from other racialized groups at work, and most of their workplace friends were white. Few of the white women's workplace friends, however, were from other racialized groups. Some white women may be more open to interracial friendships than others, but occupational segregation by "race" can limit the opportunities for interaction.

[. . .]

While many women formed friendships that involved talking about personal problems with women from other racialized groups, they were more likely to form close friendships and socialize outside of work with women from the same racialized group.

As an extension of women's traditional gender roles, women traditionally have been seen as active participants within the neighborhood. Yet the neighborhood was less likely than the community or workplace to be a place in which the Worcester women developed networks, most likely because the women I interviewed were employed. As one African-American woman said, "I'm too busy. I'm not here often enough to make real friends." White women, however, were more likely than African-American women to develop neighborhood-based networks. African-American women (63 percent) were much more likely than white women (36.7 percent) to say that they did not feel part of a neighborhood. African-American women were more likely than white women to say that they knew no one in their current neighborhoods (40 percent versus 20 per cent), and were less likely to say that they had friends (44 percent versus 56 percent) or acquaintances (16 percent versus 24 percent) in the neighborhood. Nearly one-half of all of the women, however, had neighbors that they considered to be more than acquaintances. Most of their neighborhood friends were other women.

[. . .]

In sum, women's networks, which are multifaceted and often interrelated, are embedded in daily life. For both African-American and white women, the workplace was more important than the neighborhood as a site of social networks. African-American women, however, were more likely to develop church-based networks, while white women were more likely to develop neighborhood-based networks. It appears that the nature of residential segregation in Worcester contributes to this pattern. In addition, the nature of residential segregation and occupational segregation by "race," combined with the small racialized minority population in Worcester, contributes to the fact that African-American women's networks generally were more integrated than those of white women.

[. . .]

The impact of networks on women's survival strategies

While the spatial boundedness of women's lives helps to determine where and with whom women develop networks, it also affects women's survival strategies indirectly because their use of place-based networks can be enabling as well as constraining. Personal contacts, which connect women to jobs, childcare, and housing, and more generally are used for economic and emotional support, can be a resource as well as a constraint, often simultaneously.

The use and type of personal contacts played an important role in connecting women to employment. More than one-third of all jobs (N=213) were found through personal contacts. The use of personal contacts, however, was somewhat more likely to lead to female-dominated occupations. While family-based personal contacts were used most often, work-based contacts were also important. Such contacts affected women's employment outcomes. Work-based contacts led to jobs in male-dominated or gender-integrated occupations, while those obtained through kin contacts were more likely to lead to female-dominated jobs.

African-American women relied somewhat more on personal contacts than did white women, and they tended to rely more heavily on kin to connect them to jobs than did white women, who, in turn, relied on a variety of personal contacts, including community- and neighborhood-based contacts. African-American women would seem to be more disadvantaged through their use of networks. Thus, while personal networks are a resource women use to find jobs, the type of network used affects their employment opportunities.

[...]

There are important differences in the impersonal strategies African-American and white women used to attain employment. African-American women were most likely to change jobs with the same employer, while white women were most likely to respond to a newspaper advertisement.

[...]

Sixty percent of all childcare arrangements were found through personal contacts. While these were most likely to be kin-related, neighbors and community-based contacts also connected women to childcare, while work-based contacts rarely did. The use of contacts in general, and the type of contacts used, tended to connect women to different kinds of childcare. Impersonal contacts (e.g., newspapers, yellow pages, etc.) not surprisingly, led to formal childcare arrangements, while personal contacts overwhelmingly resulted in informal arrangements, particularly when kin-based. Such networks can operate as either a resource or a constraint. On the one hand, family members were most likely not to charge for their services, and they may have more flexible hours.

[...]

Informal childcare can also be a constraint, however, sometimes offering limited hours and no subsidies, as neither local nor federal governments subsidize unlicensed childcare.

[...]

Women used personal contacts to find housing to a greater extent than they did to find jobs and childcare. For all women, family members predominated and were likely to lead to the least expensive housing. African-American women were somewhat more likely than white to use personal contacts and to rely on kin, while white women, although most likely to rely on kin, were more likely than the other group to rely on neighbors.

[...]

In sum, an analysis of women's employment, childcare, and housing strategies highlights two points. First, personal networks created by women locally and intensively may be the key to a woman finding a job-housing-childcare combination that will allow her to be self supporting. At the same time, the local insularity of the network, relative to more widespread impersonal sources of information, may constrain the set of opportunities considered. Second, there are differences in the strategies African-American and white women developed to find jobs, childcare, and housing. African-American women relied more heavily on personal contacts, particularly kin, to connect them to jobs, childcare, and housing than did white women, who tended to use a variety of strategies, for example, newspapers or the yellow pages, in addition to personal contacts. These strategies can help African-American women deal with the effects of living in a racist and segregated society.

[...]

In addition to connecting women to jobs, childcare, and housing, the people in women's personal networks, primarily other women, are an important part of their economic and emotional survival more generally. Women were most likely to rely on family members for support in emergencies and for both financial and non-financial assistance. African-American women were less likely to rely on family members for financial support, suggesting, given their generally higher reliance on family members, that their families have less resources.

[...]

The African-American women who said that they participated heavily in church activities also said that they had or could rely on the church for different kinds of economic support (e.g., financial assistance or childcare in an emergency) as well as emotional or spiritual sustenance.

[...]

[M]any women use the spatial boundedness of their everyday lives to develop networks in place, which is an indication of rootedness. Women's use of place-based networks shows that rootedness can be both enabling and constraining, often simultaneously, suggesting that it is too simplistic to equate immobility with powerlessness. Furthermore, we can see that the manner in which rootedness can be a resource or constraint depends upon the constellations of power relations. African-American and white women's strategies differed in important ways because African-American women's strategies in Worcester reflect the constraints they experience from individual and institutionalized racism in the housing and labor market. African-American women's use of rootedness – their reliance on networks, particularly kin-based, and some African-American women's use of the church in their strategies – supports the previous research that has documented the importance of the extended family and church in many African-Americans' lives as strategies for living in a racist society. While other African-Americans play a crucial role in African-American women's networks, the fact that their networks are often segregated can constrain their opportunities.

EXPLAINING THE SPATIAL ENTRAPMENT OF AFRICAN-AMERICAN WOMEN

African-American women's daily activity patterns were more spatially bounded than were white women's due to a complex and interrelated set of factors, including residential segregation, the "racial" composition of women's networks, and the relationship between women's daily activity patterns and the nature and spatial extent of their networks.

[...]

The fact that African-American women were spatially segregated means that they were making their employment and childcare decisions from a more limited residential base than were white women.

[...]

The level of "racial" segregation of women's networks also contributes to differences in the spatial extent of women's daily activity patterns. Most of the personal contacts women used to find employment and childcare were of the same racialized group. Furthermore, 87 percent of both African-American and white women's childcare providers were of the same racialized group. Since most African-Americans in Worcester live in a few areas of the city, the spatial extent of African-American women's networks and their daily activity patterns, which are mutually constituted, are limited. This highlights the significance of place and context in shaping the relationship between space and multiple relations of power, in this case, racism and gender. Therefore it becomes important to ask how mobility and immobility are related to historically and geographically situated constellations of power relations. As my analysis of African-American and white women's spatial entrapment in Worcester has illustrated, we can better answer this question by moving beyond the spatial-entrapment thesis to examine how women use rootedness in their survival strategies.

CONCLUSIONS

I have presented an analysis of the role of personal networks in the survival strategies of working poor African-American and white women with children in Worcester, Massachusetts in order to demonstrate two limitations of the spatial-entrapment thesis. First, equating mobility with power and immobility with powerlessness is too simplistic to capture the spatiality of women's daily lives. Second, conceptualizing power as having a single source masks differences among women and impoverishes our analyses of the spatiality of women's lives. Instead, I have demonstrated that while patriarchal structures of inequality often result in the spatial entrapment of women, the spatial boundedness of women's lives can be both enabling and constraining, as women actively use rootedness in the construction of their survival strategies. Important aspects of rootedness are the networks that people develop in places. By examining women's use of personal networks in the creation of their survival strategies, it becomes apparent that the relationship between power and space is more complicated than suggested by the spatial-entrapment thesis.

I have also demonstrated that racism intersects with gender to shape the spatial boundedness of women's daily lives and the ways spatial rootedness is used in women's survival strategies. African-American and white women differed in terms of their use of personal networks, the kinds of personal networks used, and the spatial boundedness of their daily lives. Analyzing the differences in the spatiality of African-American and white women's survival strategies contributes to a better understanding of the ways that racism intersects with gender and class in shaping the spatiality of daily life in different places.

[. . .]

REFERENCES

England, K. V. L. 1993. "Suburban Pink Collar Ghettos: The Spatial Entrapment of Women?" *Annals of the Association of American Geographers* 83: 225–42.

Hanson, S., and Pratt, G. 1988. "Spatial Dimensions of the Gender Division of Labor in a Local Labor Market." *Urban Geography*, 9: 180–202.

"Gender and Space: Lesbians and Gay Men in the City"

from *International Journal of Urban and Regional Research* (1997)

Sy Adler and Johanna Brenner

Editors' Introduction

In 1983, urban sociologist Manuel Castells published *The City and the Grassroots: A Cross-Cultural Theory of Urban Social Movements*. The book is a series of case studies of collective action undertaken by Castells and his students to demonstrate how social identity and cultural lifestyles may form the basis of urban social movements. Scholars have long noted that urban conflicts, such as those between renters and landlords, are rooted in larger class-based struggles. Castells set out to show that conflicts around ethnic and national movements and sexuality and gender relationships are important sources of urban change as well. When, for example, particular territories, such as neighborhoods, become associated with specific social group identities, movements may organize to claim a stake in conventional district electoral politics. Such was the case of the Castro District as discussed in the book's chapter, "Cultural Identity, Sexual Liberation, and Urban Structure: The Gay Community in San Francisco." Castells shows that the Castro's space-based social and cultural activities – the clustering of gay bars and ensuing nightlife, the surfeit of gay-owned and gay-friendly businesses and gay clientele, and numerous festivals and celebrations – operated as the basis for successful political organizing. Openly gay Castro businessman Harvey Milk was elected to the city's Board of Supervisors in 1977, largely due to organized efforts of the gay community in the Castro area (Milk and Mayor George Moscone were shot and killed by conservative supervisor Dan White in 1978).

Castells' study of San Francisco became the benchmark study for later work that addressed the importance of cultural identity to the study of urban politics and movements. It also served as the standard for case studies of sexuality, community and urban social change that have since been examined in cities across the globe.

In this selection, Sy Adler and Johanna Brenner replicate Castells' methods in a study of a lesbian community in an intentionally unnamed U.S. city. Castells found lesbian communities less territorially visible and distinct than their gay male counterparts, due primarily to gender differences in the use and importance of urban space. Adler and Brenner depart from Castells' gender-difference explanation, noting that other variables must be taken into account in the study of lesbian communities. They see gender differences in the use of space as reflective of deeper, more significant causes: "the capacity to dominate space or claim territory depends on available wealth and restrictions placed by male violence on women's access to urban space."

Adler and Brenner suggest that the so-called "invisibility" of the lesbian community and, hence, its absence from territorial-based grassroots politics is a function of scholars looking in the wrong places (i.e. formal, organized, public expressions of community and representational/electoral politics). Women have long provided membership and leadership in social movements, the authors point out. The kinds of political issues that women address, however, tend to involve organization and mobilization through interpersonal networks (which may not be territorially clustered – think of the use of the Internet). Many of these issues – personal/household concerns about parenting and public policy, discrimination in the workplace, and gender-based violence – are not unique to the lesbian community; they are also the concerns of heterosexual feminists.

Sy Adler is Professor of Urban Studies and Planning and Johanna Brenner is Professor of Sociology and Coordinator of the Women's Studies Program at Portland State University. Adler conducts research on local, regional, and state institutions, and processes involving land use, transportation, health care, and natural resources, particularly in the Pacific Northwest. Johanna Brenner is author of *Women and the Politics of Class* (New York: Monthly Review Press, 2000). Her research interests are women and welfare and the impact of recent welfare reform on low-income women.

Recommendations on the issue of sexuality and the city include: David Bell and Gill Valentine, eds, *Mapping Desire: Geographies of Sexualities* (London: Routledge, 1995); Beatriz Colomina, ed., *Sexuality and Space* (Princeton: Princeton Architectural Press, 1992); Nancy Duncan, ed., *BodySpace: Destabilizing Geographies of Gender and Sexuality* (London: Routledge, 1996); Gordon Brent Ingram, Anne-Marie Bouthillette, and Yolanda Retter, eds, *Queers in Space: Communities, Public Places, Sites of Resistance* (Seattle: Bay Press, 1997); and Elizabeth Lapovsky Kennedy and Madeline D. Davis, *Boots of Leather, Slippers of Gold: The History of a Lesbian Community* (London: Routledge, 1993).

In the past two decades gay neighborhoods have become familiar parts of the urban landscape. Although these areas may include lesbians, gay men dominate their distinct subcultures, their businesses and their residences, their street life and their political activities. In his 1983 book, *The City and the Grassroots*, Manuel Castells argues that the predominance of gay men in the creation of distinctly homosexual urban neighborhoods reflects a profound gender difference. In relationship to space, gay men and lesbians, he says, behave first and foremost as men and women. Men seek to dominate space, while women attach more importance to networks and relationships, rarely having territorial aspirations: "Lesbians, unlike gay men, tend not to concentrate in a given territory, but establish social and interpersonal networks." Gay men require a physical space in order to conduct a liberation struggle, while lesbians are "placeless" and tend to create their own rich, inner world.

Lesbians are also politically different from gay men, according to Castells. They do not acquire a geographical basis for urban political objectives, because they create a political relationship with higher, societal levels. Lesbians "are far more radical in their struggle . . . [and] more concerned with the revolution of values than with the control of institutional power."

Castells' analysis makes several assumptions that we question. First, is it true that lesbians do not concentrate in a given territory? Second, does the absence of a publicly identifiable lesbian neighborhood reflect gender differences in interests, needs and values, or differences in resources available to gay men and lesbians? Third, do differences in the political orientation of politically active gay men and lesbians reflect gender differences in relationship to space or differences in political alliances, specifically the involvement of lesbians in feminist politics that include straight women?

The literature on differences between gay men and lesbians in relation to urban space is generally ambiguous about the existence of lesbian spatial concentrations. In support of his argument, Castells cites Deborah Wolf's 1980 study of the lesbian community in San Francisco, which, according to Wolf, "is not a traditional community in the sense that it has geographical boundaries." However, Wolf also noted that lesbians did tend to live in particular parts of town, and that "These areas bound each other and have in common a quality of neighborhood life." Wolf pointed out that since lesbians tend to be poor they live in older, ethnically mixed working-class areas, in low-rent apartment buildings or small, low-rent houses. "Since much of the socializing in the [lesbian] community consists of visiting friends, women without cars try to live near each other, so that gradually, within a small radius, many lesbian households may exist." Indeed, Castells himself noted that lower incomes mean less choice regarding the location of home and work, and that lesbian concentrations were emerging in the kinds of areas described by Wolf, particularly low-rent neighborhoods with counter-cultural communities.

[. . .]

Denyse Lockard (1985), who studied lesbians in a large south-western city, argues that lesbians create communities without a territorial base or geographical boundaries. However, she also argues that one of the four defining features of a lesbian community is

> its institutional base, the gay-defined places and organizations which characterize a community, and provide a number of functions for community members. It is this institutional base that provides the means by which a community can mobilize its members for action as a minority group in the larger society, and that is visible, at least to some degree, to the outside world.

Lockard does not explore whether these crucial institutional bases are themselves spatially concentrated, creating a semi-visible lesbian urban space, as is the case for the gay male community.

H.P.M. Winchester and Paul White (1988) did find discernible concentrations of lesbian facilities, with the most exclusively lesbian institutions located in a poor district of the inner city. Although these lesbian places are not public, they are known to those in contact with the lesbian community. Winchester and White assume, based only on anecdotal evidence, that there are no residential concentrations. Yet Wolf noted that in San Francisco most of the lesbian bars and community projects were located in the same neighborhoods as lesbian household concentrations. Thus, we would argue that there is at least some indication that there is a concentration of lesbian residence and activity space in urban areas.

In addition to questioning whether the differences in relationship to space between lesbians and gay men are as great as the literature assumes, we also question the explanations put forward for the differences that do exist. Mickey Lauria and Lawrence Knopp (1985) critique Castells' argument that the more visible concentration of gay men in cities expresses an innate male territorial imperative. However, they adopt what we consider to be an equally problematic approach. They argue that there has been a greater tolerance of relationships of depth, physicality and affection between women than between men, and even acceptance of lesbian sexuality in certain circumstances. Therefore, "gay males, whose sexual and emotional expressiveness has been repressed in a different fashion than lesbians, may perceive a greater need for territory."

While not denying this possibility, we would argue that before concluding that social-psychological needs and interests explain the absence of visible lesbian urban neighborhoods, we need to consider differences in capacity to dominate space, a variable reflecting available wealth as well as restrictions placed by male violence on women's access to urban space. The creation of visible, distinct neighborhoods requires more than residential concentration and the development of a network of voluntary and service organizations. To take over urban space requires also the control of residential and business property. Castells documents the role of gay businessmen, especially owners of taverns and retail stores and gay real estate entrepreneurs, in the development of the Castro district. It may be that gender differences in access to capital account, at least in part, for the fact that tendencies among lesbians in San Francisco towards concentration in residence and community institutions have not been followed by lesbian-owned businesses and real estate. In addition, as Gill Valentine (1989) argues, "women's fear of male violence . . . is tied up with the way public space is used, occupied and controlled by different groups at different times." These restrictions on the use of public space, which lesbians experience as women, would also account for the lack of visibility of their neighborhoods.

Finally, we question Castells' argument that lesbians relate to space primarily as women, that they are "placeless," uninterested in local politics, and therefore more radical than gay men. Localist politics and place-based organizing are hardly foreign to urban women, as Castells himself recognizes in his discussion of other urban communities. Martha Ackelsberg (1984) has noted that "Much recent research . . . has documented the prominent, if not predominant, role of women in urban struggles over what have been termed 'quality of life' issues." Women have historically provided the membership and leadership of neighborhood-based urban movements. Through community-based networks of friends and neighbors, many urban women become activists on issues at the interface of personal/household concerns and employer and public policy decisions, developing political consciousness in the process. Women are often brought into political activity as an extension of their family responsibilities. We would argue that at least a significant minority of lesbians share such concerns. They are certainly far more likely than gay men to carry such responsibilities, to have children, and thus be likely to confront the same kinds of private and collective consumption issues that bring heterosexual women into local politics. We disagree with Castells' argument that the more radical dimension sometimes present in lesbian political activism is an expression of fundamental differences between men and women in core values and preferences. Lesbian activists' interest in more than "representation" within the existing political system reflects the influence and organization of radical lesbian feminism within the community.

LOCATING THE LESBIAN NEIGHBORHOOD

Our main purpose in this paper is to clarify the ambiguities surrounding the existence and significance of spatial concentration among lesbians in urban communities. In order to do so, we attempted to replicate Castells' method of analysis as closely as possible in a study of a lesbian community in another United States city. In constructing our study, we sought help from lesbians knowledgeable about the lesbian community and from lesbian organizations who were asked for permission to use their membership lists. While wanting to cooperate, they were reluctant to give information, especially about residential locations. Their response is understandable, given the political environment. As in many other parts of the country, gay rights activism is countered by well-organized right-wing activists who are powerful enough in the state to overturn legislation protecting gays and lesbians from discrimination. In addition, attacks on gay men and lesbians by youth gangs are not infrequent in the city, and some of these gangs are connected to right-wing organizations. In this climate, while individuals were willing to identify themselves, they felt they could not risk identifying others who might or might not wish to be "out." Therefore, our informants, and the organizations who shared their mailing lists with us, did so only under the condition that we would do as much as we could to disguise the location of the community. For this reason, we have not identified the city, nor the street boundaries of the lesbian neighborhood.

When beginning our study, we recognized that the requirement of anonymity would create problems: our study would be more difficult to replicate and other scholars familiar with the area would not be able to suggest unique qualities of the city that we overlooked. We decided to proceed because it is generally so hard to obtain the kind of direct evidence of gay and lesbian residence which mailing lists provide. In our judgment, the value of the data outweighed the problems that disguising the location of the lesbian neighborhood would impose.

Castells sought to determine the areas in which gay men were concentrated, and to characterize these parts of San Francisco in socio-spatial terms. He then related this characterization to the social and urban characteristics of the rest of the city in order to understand the factors that influenced the settlement pattern of gay men. We sought first to establish the existence of concentrations of lesbians, and then to characterize the areas within which we found concentrations. We used the explanatory variables uncovered by Castells to see if these would also explain the lesbian settlement pattern that we found.

Castells used five different sources of information to locate areas of concentration of gay men: (1) key informants from the gay male community; (2) the presence of multiple male households on voter registration files; (3) gay bars and other social gathering places; (4) gay businesses, stores and professional offices; and (5) votes for gay candidate Harvey Milk in a local electoral contest. Census tract-based maps were prepared utilizing each of these sources, showing the distribution of concentrations of gay men across the city. Castells found that all five maps displayed a similar spatial pattern, permitting him to conclude that gay residential areas had been validly identified.

To identify areas of lesbian concentration, we used: (1) key informants from the lesbian community; (2) the location of lesbian bars and other social gathering places; (3) the location of lesbian businesses, professional services and social service agencies; and (4) two mailing lists of lesbian organizations.

Our 10 key informants unanimously agreed that spatially defined concentrations of lesbians did, in fact, exist. One neighborhood was identified by all informants as *the* lesbian community or "the ghetto," while some identified another neighborhood as a second lesbian community. This neighborhood was close to the first. Our informants identified street boundaries and also located these areas on a census tract map of the city.

Lesbian bars and social gathering places, businesses, professional and service agencies were identified by informants and from a variety of listings and advertisements in local women- and gay-oriented handbooks and resource guides, newspapers and magazines. The addresses from these sources were assigned to census tracts.

The addresses from the mailing lists of the two lesbian organizations, a total of 1150 names, were similarly assigned, making the assumption that the spatial distribution of people on these lists mirrors the distribution of the lesbian population. It might be argued that these lists are a biased sample of the lesbians in the city, since they include only those women who identify as lesbians. While this is true, such a bias does not undermine a test of our hypothesis, since we are arguing that lesbian-identified women do seek to establish spatially based communities. On the other hand, if Castells is correct in arguing that politically active lesbians tend to be more "placeless" than politically active gay men, then this bias would work against our hypothesis.

We chose not to use Castells' measure of household composition based on voter registration lists. We think that household composition is less valid an indicator of lesbian households than of gay male households, particularly when it is impossible to control for age. Mother–daughter households, for example, are not uncommon, particularly where the mother is elderly. A neighborhood with a high proportion of households in which two men live together can more reliably be assumed to be a gay male neighborhood than can a neighborhood with a high proportion of households in which two women live together.

We assigned the addresses on the mailing lists to census tracts, totaled addresses within census tracts, and then rank-ordered the tracts. The rank-order correlation for the two lists was (0.7), indicating a substantial agreement between them. In both cases 15 per cent or less of the tracts contained 50 per cent of the addresses. We selected for further analysis 12 tracts (constituting 10 per cent of the tracts in the city) that appeared among the top 13 on both lists, containing about 40 per cent of the addresses.

Our concerns about bias in the mailing lists were further allayed when we examined the spatial pattern that emerged from other sources, including the informant interviews and the locations of businesses, services and gathering-places. As in Castells' study of San Francisco, the various sources converged. Based on these diverse sources we were able to locate a major concentration of lesbian households in a set of 11 contiguous inner city census tracts east of the central business district, and another concentration in one inner city tract north-west of downtown. This tract is located in the city's "gay" neighborhood.

THE CHARACTERISTICS OF THE LESBIAN NEIGHBORHOOD

In order to further indicate the existence of lesbian concentration and to identify the features of this neighborhood that might account for this concentration, we compared the 12 tracts we found to contain concentrations of lesbian households with the rest of the city, using 1980 census data. Castells argued that gay settlement in San Francisco was opposed by "property, family, and high class: the old triumvirate of social conservatism." However, Castells did not address the extent to which life-cycle stage and related financial circumstances shaped the pattern of gay location. The gay men who were moving into the Castro area when Castells did his study were predominantly younger men, whose limited financial means would make it very difficult to purchase homes or pay high rents. We discuss below the ways in which financial limitations shape the pattern of lesbian concentration. Following Castells, we also hypothesized that areas with high proportions of home owning, of traditional family households and of high rents would be more difficult for lesbians to move into.

We found that the census tracts where lesbians concentrate have significantly lower levels of owner-occupied housing, lower rent levels and lower proportions of traditional families than the remainder of the city. Of the 12 tracts, only three have a higher proportion of home ownership than the city average, and only one is significantly higher: 75 per cent compared to 53 per cent for the city as a whole. This apparent anomaly is explained, in our view, by the fact that this tract is one-third black and has the highest median household income of all tracts with 400 or more blacks. An interracial, middle-income tract with relatively higher quality housing, this area would be more open to lesbians than one economically similar but socially more homogeneous. Lesbians with relatively higher incomes can purchase homes of higher quality than those available in the other tracts without having to locate outside the lesbian neighborhood.

While "property, family and wealth" are barriers to lesbian settlement, counter-cultural neighborhoods appear to be most open. Castells and others all found lesbians located in counter-cultural areas in the cities they studied. We found this to be true in our city also. In addition to the obvious cultural features of the area covered by our 12 census tracts – particularly the presence of counter-cultural institutions such as theatres, coffee shops, studios that present progressive/fringe entertainment, alternative businesses like food and bicycle repair coops, radical and feminist bookstores, etc. – census data indicate its counter-cultural character. Compared to the rest of the city, this area has a much higher proportion of people living in non-traditional households.

The other striking feature of this area is the high proportion of women living alone and of female-headed households. It is impossible to know, of course, whether or not these are lesbian households. And indeed it is impossible to know whether lesbians are more likely to live in this neighborhood because it is a "women's community," or whether it is a "women's community" because large numbers of lesbians live there, or both.

In our 12 census tracts, women from young to middle age (20–54) were significantly more likely to be living alone than women of the same age elsewhere in the city. "Female, non-family householders not living alone" may of course be living with a man, but we think it reasonable to assume that in the majority of cases, non-married heterosexual couples would identify the man as household head when forced to choose. Alternatively, these may be households of room-mates rather than lesbian households. Still, given the other evidence, the higher proportion of such households indicates that this is much more a women's community than the rest of the city.

Although there were fewer families with children living in the lesbian tracts than in the rest of the city, a significantly greater proportion of these families were female-headed. Given that these are also relatively low-rent areas, it might be argued that this figure only reflects the impact of single mothers' poverty.

[...]

However, the median income of female-headed households, no spouse present with children under 18 in our 12 tracts, is not significantly different from the income of similar female-headed households elsewhere in the city. This seems to us to indicate that single mothers are locating in these tracts for reasons other than (or at least in addition to) the relatively low rents.

Within the 12 tracts, three show a particularly high concentration of residences, businesses and services, accounting for 50 per cent of the total addresses located in the 12 tracts. These three tracts overlap the area which our key informants identified as the lesbian neighborhood. They appear to constitute a spatial "core" for the lesbian community. They rank very high on indirect measures of lesbian households.

To explain this configuration we hypothesized that these "core" tracts would be especially high in those features that encourage lesbian settlement: low proportion of owner-occupied housing, low rents, non-traditional households and families. Two of the tracts, as expected, are ranked low in proportion of home ownership and in their median rents, and are relatively low in their proportion of traditional family households. While they have more traditional family households than some tracts, they are also relatively high in the proportion of women-headed families. The third tract is somewhat anomalous, having low rents but a relatively high proportion of traditional families and owner-occupied housing compared to the other tracts. On the other hand, it is still more counter-cultural than the city as a whole and its owner-occupied houses are among the least expensive in the 12 tracts. Perhaps the low cost of homes has encouraged lesbians wishing to become home owners to risk moving into a more family-oriented area in order to be in the neighborhood core. Lesbians with children may be attracted to such an area as well.

THE HIDDEN NEIGHBORHOOD

It seems clear that there is a spatial concentration of lesbians in our city, a neighborhood that many people know about and move into to be with other lesbians. But the neighborhood has a quasi-underground character; it is enfolded in a broader counter-cultural milieu and does not have its own public subculture and territory.

To claim an urban territory takes more than residential concentration. It requires, in addition to this: (1) visibility – gay places, especially retail businesses and services run by and for gay people; (2) community activity – fairs, block parties, street celebrations etc., some kind of public, collective affirmation of the people who live in the neighborhood, even if it is only strolling out in the evening; (3) organization – of businesses and residents to defend the neighborhood's interests, relate to city government, financially support the community activities which create and maintain the urban subculture, giving the neighborhood its distinct character. Businesses and churches are often key providers of the financial and organizational infrastructure for community activity. Castells' history of the Castro district demonstrated the importance of gay businesses in developing the street life of San Francisco.

Castells also noted that the presence of gay men in the real estate and other professions in San Francisco was a major factor facilitating the development of a gay neighborhood. Lauria and Knopp (1985) similarly point to the importance of more affluent gay men in providing the capital necessary for gentrification. As gay men became landlords, the amount of rental housing available to gay men increased, drawing more residents to the area, increasing the clientele for gay retail businesses and the participants in neighborhood street life. Knopp (1990) also analyses the leading role played by gay real estate speculators and developers in the emergence and consolidation of New Orleans's gay community. If lesbians have not yet taken the next steps toward creating a distinct neighborhood, perhaps this reflects their lack of capital more than their lack of interest.

Castells and Lauria and Knopp argue that gay men use their spatial concentration as a base for political mobilization. Castells argues that lesbians, whose political values and interests are both more global and more radical than those of gay men, have less need for this kind of place-based organizing. It is true that lesbians are well represented in radical feminist political activities and organizations. But lesbians are also active in mainstream politics such as the Gay/Lesbian Task Force of the American Civil Liberties Union, the Right to Privacy Political Action Committee, the Women's Political Caucus and so forth. In the November 1988 election a major lesbian community organization contributed fully 50 out of 200 volunteers who worked to elect a pro-lesbian and gay (although not openly lesbian) candidate to the state assembly from their district. That lesbians have not yet fielded a lesbian candidate may reflect more their lack of visibility than their lack of interest in affecting local politics. Without visibility it is difficult to assess the extent of their voting base and thus their potential for successfully running an openly lesbian candidate. On the other hand, they are able to use their spatial concentration to mobilize around issues of concern to them. In the November election an anti-gay ballot measure barely lost in the city as a whole (52.7 per cent "no"), but in the lesbian neighborhood (our 12 tracts) it was defeated by a 46 per cent margin (73 per cent "no").

CONCLUSION

While we reject Castells' characterization of the gender differences between gay men and lesbians, we do think there are important dimensions along which gay men and lesbians may relate differently to urban space and urban politics. These differences do reflect the fact that lesbians are women. First, as we have already argued, lesbians, like other women, tend not to have access to capital. Second, lesbians are more likely than gay men to be primary caretakers of children. Their choices about where to live will have to take their children's needs into account. When they do locate in a lesbian neighborhood, they bring a set of interests and styles of sociability very different from those of people without children. This diversity, while a strength for the community as a whole, might also militate against the development of a distinctive urban subculture. Third, lesbians share with other women a vulnerability to male physical and sexual violence. Of course, it is true that gay men have had to defend themselves from attacks on their places and persons. And lesbians have ably resisted assaults, for example on lesbian bars. Still, it seems to us that women are more at risk in public places than are men, and that this in turn may limit lesbians' interest in or ability to create the kind of vibrant street culture that makes an urban community.

Finally, we agree with Castells that, compared to the politics of the gay male community, lesbian politics has tended to be more global and less restricted to representation in the existing political system, therefore less "place-based." But in explaining this difference, we would emphasize, in addition to restrictions on lesbians' access to space, the impact of feminism on lesbian politics and culture (and the impact of lesbians on the women's movement and its culture of resistance). Lesbians organize as lesbians, just as gay men organize as gays. But lesbians also organize as oppressed women. Their politics and their culture reflect this double vision. Lesbians contributed a significant core of activists to the women's movement from the beginning. Lesbian writers, poets, artists and musicians have played a major role in the development of a women's culture which draws audiences of both straight women and lesbians. Pressure from lesbians in feminist organizations has made the women's movement less homophobic, while issues of lesbian rights have become more integrated into feminist politics. Lesbians who created lesbian culture and urban communities in the early 1970s did so in connection with a movement against male domination whose critique went far beyond simple demands for incorporation into the existing society. While the relationship between straight women feminists and lesbian feminists has sometimes been more conflictual than collaborative, participation in the feminist movement has supported the more radical political tendencies within the lesbian community. (And, correlatively, lesbian feminist separatism constituted one of the more radical political tendencies within the women's movement.) In contrast, gay communities and culture preceded the political organization of gay men. The gay men's community is not homogeneous and does include political tendencies that favor more radical politics than the "interest group" strategies that are dominant. And lesbian communities are more heterogeneous, less counter-cultural and political than in the 1970s. Still, in so far as lesbian urban communities continue to express commitments and world views defined by feminist politics, which draw on but do not simply reflect gendered experience, differences in the urban culture and politics of lesbians and gay men will remain.

REFERENCES

Ackelsberg, M. 1984. "Women's Collaborative Activities and City Life: Politics and Policy." In J. Flammang, Ed., *Political Women: Current Roles in State and Local Government*. Beverly Hills, CA: SAGE.

Castells, M. 1983. *The City and the Grassroots: A Cross-Cultural Theory of Urban Social Movements*. Berkeley, CA: University of California Press.

Knopp, L. 1990. "Exploiting the Rent Gap: The Theoretical Significance of Using Illegal Appraisal Schemes to Encourage Gentrification in New Orleans." *Urban Geography*, 11: 48–64.

Lauria, M. and L. Knopp. 1985. "Toward an Analysis of the Role of Gay Communities in the Urban Renaissance." *Urban Geography*, 6: 152–69.

Lockard, D. 1985. "The Lesbian Community: An Anthropological Approach." *Journal of Homosexuality*, 11: 83–95.

Valentine, G. 1989. "The Geography of Women's Fear." *Area*, 21: 385–90.

Winchester, H. P. M. and P. E. White. 1988. "The Location of Marginalized Groups in the Inner City." *Environment and Planning D: Society and Space*, 37–54.

Wolf, D. 1980. *The Lesbian Community*. Berkeley, CA: University of California Press.

"Freeing South Africa: The 'Modernization' of Male–Male Sexuality in Soweto"

from *Cultural Anthropology* (1998)

Donald L. Donham

Editors' Introduction

As we have learned from the readings in Part 4, the interplay between gender and sexual identities and urban space is multifaceted. Expressions of identity – be they performative, political, cultural, or some combination of the three – structure the use of urban space, its representation and direction of change. Likewise, urban space – its location, built environment, zoning and other legal statutes – influence the formation and expressions of social identity. This latter point is made clear in the following selection, in which anthropologist Donald Donham examines how apartheid-era segregation of settlement space in South Africa shaped both collective and individual understandings of homosexuality and what effect apartheid's demise had on such understandings.

Racial segregation defined the spatial arrangements of apartheid. Yet black workers often needed to be located within commuting distance in order to provide labor to the white-dominated economy. In places like Soweto rural black men, who otherwise had no right to live there, were permitted to reside without their wives in all-male hostels. Donham explains how apartheid-era spatial segregation and the separation of sexes coalesced over time to strongly influence the localized construction of homosexuality. A same-sex sexual culture developed around the urban hostels, in which black male workers took on *skesanas* – effeminate, sexually submissive males – as "wives." While the sexual or gendered identity of the sexually dominant male workers remained uncontested, the *skesanas* were collectively thought of as ostensibly girls (gender) or some biologically mixed "third" sex. Ceremonies were conducted to induct new *skesanas* as wives and an elaborate sexual-social hierarchy existed within the hostels. This same-sex culture continued for as long as hostels housed black workers laboring in white-dominated industries, businesses, and households.

The end of apartheid in the early 1990s swept away the institutional practices that sustained the sexual culture that developed around the hostels, thereby challenging localized notions of homosexuality. Wives joined their husbands. Hostels became sites of political infighting. New meanings and ways of looking at sexuality were required, including the notion that both the sexually dominant and submissive partners were to be considered gay. In addition, new opportunities, such as the annual gay pride march in Johannesburg, allowed for the public, right-based expression of sexual identity in post-apartheid South Africa. Gay South Africans made connections with counterparts in Europe and the US, where political mobilization matured earlier. Donham concludes on a positive, yet cautious, note. The undoing of apartheid, especially the spatial segregation of "races," brought forth a cultural redefinition (a modernization, if you will) of sexuality. Yet the facts remain that blacks are overwhelmingly poor and global gay culture is represented as white and middle class. Women, especially lesbians, are culturally restricted, prohibiting a fully westernized notion of sexual emancipation.

Donald Donham is a cultural anthropologist who has written extensively on the effects relationships between culture and economy have on changing forms of power. He is Professor of Anthropology at the University of California, Davis. He was a Fellow at the Woodrow Wilson International Center for Scholars in Washington, DC in 1999–2000.

In February 1993, a black man in his mid-thirties named Linda (an ordinary male name in Zulu) died of AIDS in Soweto, South Africa. Something of an activist, Linda was a founding member of GLOW, the Gay and Lesbian Organization of the Witwatersrand. Composed of both blacks and whites, GLOW was and is the principal gay and lesbian organization in the Johannesburg area. Because Linda had many friends in the group, GLOW organized a memorial service at a member's home in Soweto a few days before the funeral.

Linda's father, who belonged to an independent Zionist church, attended and spoke. He recalled Linda's life and what a good person he had been, how hard he had worked in the household. But then he went on, in the way that elders sometimes do, to advise the young men present: "There was just one thing about my son's life that bothered me," he said. "So let me tell you, if you're a man, wear men's clothes. If you're colored, act colored. Above all, if you're black, don't wear Indian clothes. If you do this, how will our ancestors recognize [and protect] you?" Linda had been something of a drag queen, with a particular penchant for Indian saris.

To Linda's father and to his church, dress had ritual significance. One might even say that there was an indigenous theory of "drag" among many black Zionist South Africans, albeit one different from that in North America. To assume church dress not only indicated a certain state of personhood, it in some real sense effected that state. Writing on Tswana Zionists, who like the Zulu have been drawn into townships around Johannesburg, Jean Comaroff asserted, "The power of uniforms in Tshidi perception was both expressive and pragmatic, for the uniform instantiated the ritual practice it represented."

If dress had one set of associations within Zionist symbolism, it had others for a small group of young black South African activists who saw themselves as "gay." To the members of GLOW present, most of whom were black, Linda's father's comments were insulting. Most particularly, they were seen as homophobic. As the week wore on, GLOW began to organize to make their point and to take over the funeral.

As Saturday neared – nearly all Soweto funerals are held on the weekend – tensions rose. There were rumors that there might be an open confrontation between the family and GLOW. Along with Paul, a member of GLOW from Soweto, I attended, and the following is a description of what transpired, taken from a letter that I wrote home a few days later to my American lover:

> The funeral was held in a community center that looked something like a run-down school auditorium. There was a wide stage on which all of the men of the church, dressed in suits and ties, were seated behind the podium. In front of the podium was the coffin. And facing the stage, the women of the church were seated as an audience – dressed completely in white. (Independent churches have distinctive ways of dressing especially for women, but also sometimes for men).
>
> To the right (from the point of view of the seated women) was a choir of young girls – again all in white: white dresses and white hats of various kinds (most were the kind of berets that you have seen South African women wear). I stood at the very back of the hall, behind the seated women, along with most of the members of GLOW and various other men and women, most dressed up. This last group was apparently made up of friends and relatives who were not members of the church.
>
> I had arrived late, about 9:30 in the morning (the service had begun at about 9:00). I was surprised to see, behind the coffin, in front of the podium, a GLOW banner being held by two members. There were flowers on the coffin and around it. Throughout the service, including the

sermon, the two GLOW members holding the banner changed periodically. From the back, two new people marched up through the ladies in white to take the place of the two at the banner. Then those who had been relieved came back through the congregation to the back of the church.

One GLOW member videotaped the funeral from the back. About six or seven of the members who had come were white. It was hard to tell exactly, but there were probably 10 or 15 black members. Quite a few, both white and black, wore GLOW T-shirts (the back of which said, "We can speak for ourselves"). Finally, two or three of the black members wore various stages of drag. One, Jabu, was especially notable in complete, full regalia – a West African-style woman's dress in a very colorful print with a matching and elaborately tied bonnet, one edge of which read, "Java print." Wearing a heavy gold necklace, she walked up and down the aisle to hold the banner at least twice – in the most haughty, queenly walk. It was almost as if she dared anybody to say anything. She made quite a contrast with the stolid, all-in-white ladies seated in the audience (one of whom was heard to comment to a neighbor, "She's very pretty, isn't she [referring to Jabu]? But look at those legs!").

When we arrived, Simon Nkoli, one of the first black gay activists in South Africa, was speaking. Simon was dressed in an immaculately white and flowing West African (male) outfit with gold embroidery. He spoke in English, and someone translated simultaneously (into Zulu). His speech was about gay activism in South Africa and the contributions that Linda, his dead friend, had made. At points in his speech, Simon sang out the beginning lines of hymns, at which point the congregation immediately joined in, in the [style] of black South African singing, without instruments and in part-harmony.

After Simon, there were other speeches by the ministers of the church. They emphasized that Linda was a child of the church, that his sins had been forgiven, and that he was in heaven. Diffusing any trace of tension between the church and GLOW, one of the ministers rose and apologized on behalf of Linda's father for offending the group earlier in the week.

[. . .]

Because there are so many funerals in Soweto on the weekend (probably 200 at Avalon cemetery alone) and because the cemetery had only one entrance (the better to control people), the roads were clogged and it took us an hour to go a few miles. The members of GLOW got out of the bus and *toi toied*, the distinctive, punctuated jogging-dancing that South African blacks have developed in antiapartheid demonstrations.

[. . .]

There was something about the routinized way that so many people had to bury their dead and leave (others were waiting) that brought home to me, in a way that I had not anticipated, what apartheid still means in many black people's lives. . . . The South African police stood in the background. Continually, another and another group arrived, and as each rushed to its gravesite, red dust began to cover us all. The sun got hotter.

[. . .]

After the graveside service, GLOW members gathered at one of the member's houses in Soweto and proceeded to get drunk. I had had enough. Driving back to Johannesburg (it's a little over 30 minutes), I almost had an accident. Tired and with my reflexes not working for left-hand-of-the-road driving, I turned into oncoming traffic. By the end of the day, I felt overwhelmed. Another gay man dead – yet another. And his burial had brought together, for me, a mind-numbing juxtaposition of peoples and projects, desires and fears – Zionist Christians and gay activists, the first, moreover, accommodating themselves to and even apologizing to the second. Could anything comparable have happened in the United States? A gay hijacking of a funeral in a church in, say, Atlanta?

APARTHEID AND MALE SEXUALITY

Although engaged in another research project, in my free time with friends like Paul, I thus stumbled onto a series of questions that began to perplex me: Who was Linda? In the letter quoted above, I had unproblematically identified Linda as "gay." But in *his* context, was he? And if so, how did he come to see himself as so? And I quickly confronted questions of gender as well. Did Linda consider himself as male? And if so, had he always done so?

As issues like these began to pose themselves, I soon realized that for black men in townships around Johannesburg, identifying as gay was both recent and tied up, in unexpectedly complex ways, with a much larger historical transformation: the end of apartheid and the creation of a modern nation; in a phrase, the "freeing" of South Africa.

This story, more than any other, constitutes for most South Africans (certainly black South Africans) what Stuart Hall referred to as a "narrative of history." It structures identity, legitimates the present, and organizes the past. There are indeed few places on earth in which modernist narratives of progress and freedom currently appear so compelling. This undoubtedly results, at least in part, because apartheid itself was an antimodernist project that explicitly set itself against most of the rest of the "developed" world.

[. . .]

So how did Linda become gay? I never met or interviewed Linda, but fortunately, for the purposes of this article, before he died Linda wrote an extraordinarily self-revealing article with Hugh McLean entitled "*Abangibhamayo Bathi Ngimnandi* (Those Who Fuck Me Say I'm Tasty): Gay Sexuality in Reef Townships." The collaboration between Hugh and Linda – both members of GLOW – was itself a part of the transformations I seek to understand: the creation of a black gay identity, Linda's "coming out," and the "freeing" of South Africa.

To begin with, Linda did not always consider himself – to adopt the gender category appropriate at the end of his life – to be "gay." If anything, it was female gender, not sexuality as such, that fit most easily with local disciplinary regimes and that made the most sense to Linda during his teenage years. Indeed, in apartheid-era urban black culture, gender apparently overrode biological sex to such a degree that it is difficult, and perhaps inappropriate, to maintain the distinction between these two analytical concepts below.

Let me quote the comments of Neil Miller, a visiting North American gay journalist who interviewed Linda:

> Township gay male culture, as Linda described it, revolved around cross-dressing and sexual role-playing and the general idea that if gay men weren't exactly women, they were some variation thereof, a third sex. No one, including gay men, seemed to be quite sure what gay meant – were gay men really women? men? or something in between? . . . When Linda was in high school word went out among his schoolmates that he had both male and female sex organs. Everyone wanted to have sex with him, he claimed, if only to see if the rumors were true. When he didn't turn out to be the anatomical freak they had been promised, his sexual partners were disappointed. Then, there was the male lover who wanted to marry Linda when they were teenagers. "Can you have children?" the boy's mother asked Linda. The mother went to several doctors to ask if a gay man could bear a child. The doctors said no, but the mother didn't believe them. She urged the two boys to have sex as frequently as possible so Linda could become pregnant. Linda went along with the idea. On the mother's orders, the boys would stay in bed most of the weekend. "We'd get up on a Saturday morning, she'd give us a glass of milk, and she'd send us back to bed," Linda told me. After three months of this experiment, the mother grew impatient. She went to yet another doctor who managed to convince her that it was quite impossible for a man, even a gay man, to bear a child. Linda's relationship with his friend continued for a time until finally the young man acceded to his mother's wishes and married a woman, who eventually bore the child Linda could never give him.

The description above uses the word gay anachronistically. In black township slang, the actual designation for the effeminate partner in a male same-sex coupling was *stabane* – literally, a hermaphrodite. Instead of sexuality in the Western sense, it was local notions of sexed bodies and gendered identities – what I shall call sex/gender in the black South African sense – that divided and categorized. But these two analytical dimensions, gender and sex, interrelated in complex ways. While she was growing up, Linda thought of herself as a girl, as did Jabu, the drag queen at Linda's funeral about whom I shall have more to say below. Even though they had male genitalia, both were raised by their parents as girls and both understood themselves in this way.

If it was gender that made sense to Linda and Jabu themselves (as well as to some others close to them, such as parents and "mothers-in-law"), strangers in the township typically used sex as a classificatory grid. That is, both Linda and Jabu were taken by others as a biologically-mixed third sex. Significantly, as far as I can tell, neither ever saw themselves in such terms.

[. . .]

If an urban black South African boy during the 1960s and 1970s showed signs of effeminacy, then there was only one possibility: she was "really" a woman, or at least some mixed form of woman. Conversely, in any sexual relationship with such a person, the other partner remained, according to most participants, simply a man (and certainly not a "homosexual").

This gendered system of categories was imposed on Linda as she grew up:

> I used to wear girls' clothes at home. My mother dressed me up. In fact, I grew up wearing girls' clothes. And when I first went to school they didn't know how to register me.

Miller recorded the following impressions:

> Linda didn't strike one as particularly effeminate. He was lanky and graceful, with the body of a dancer. The day we met he was wearing white pants and a white cotton sweater with big, clear-framed glasses and a string of red African beads around his neck. But even as an adult, he was treated like a girl at home by his parents. They expected him to do women's jobs – to be in the kitchen, do the washing and ironing and baking. "You can get me at home almost any morning," he told me. "I'll be cleaning the house." There were girls' shopping days when he, his mother, and his sister would go off to buy underclothes and nighties. Each day, he would plan his mother's and father's wardrobes. As a teenager, Linda began undergoing female-hormone treatments, on the recommendation of a doctor. When he finally decided to halt treatments, his father, a minister of the Twelfth Apostle Church, was disappointed. It seemed he would rather have a son who grew breasts and outwardly appeared to be a girl than a son who was gay. Even today, Linda sings in the choir at his father's church-in a girl's uniform.
>
> "What part do you sing?" I asked him.
>
> "Soprano, of course," he replied. "What did you think?"

The fact that Linda wore a girl's uniform in church into the early 1990s offers some insight into his father's remarks that caused such a stir in GLOW. His father was not, it seems, particularly concerned with "cross-dressing." Phrasing the matter this way implies, after all, a naturally given bodily sex that one dresses "across." To Linda's family, he was apparently really a female. What the father was most upset about was dressing "across" race, and the implications that had for ancestral blessings.

In sum, black townships during the apartheid era found it easier to understand gender-deviant boys as girls or as a biologically mixed third sex. By the early 1970s, a network of boys who dressed as girls existed in Soweto, many of whom came to refer to themselves in their own slang as skesana.

[. . .]

Skesanas dressed as women and adopted only the receptive role in sexual intercourse. Here is Linda speaking:

> In the township they used to think I was a hermaphrodite. They think I was cursed in life to have two organs. Sometimes you can get a nice *pantsula* [tough, macho guy] and you will find him looking for two organs. You don't give him the freedom to touch you. He might discover that your dick is bigger than his. Then he might be embarrassed, or even worse, he might be attracted to your dick. This is not what a *skesana* needs or wants. So we keep up the mystery. We won't let them touch and we won't disillusion them . . . I think it makes you more acceptable if you are a hermaphrodite, and they think your dick is very small. The problem is, the *skesanas* always have the biggest dicks. And I should know . . .

It would be a mistake to view this system of sex/gender categories as *only* being imposed upon skesanas. In adopting their highly visible role, skesanas sometimes used the traditional subordinate role of the woman to play with and ultimately to mock male power. According to Linda,

> On a weekend I went to a shebeen [informal drinking establishment] with a lady friend of mine. I was in drag. I often used to do this on the weekends – many *skesanas* do it. We were inside. It seemed as if four boys wanted to rape us, they were *pantsulas* and they were very rough. One of them proposed to my friend and she accepted. The others approached me one by one. The first two I didn't like so I said no! I was attracted by the third one, so I said yes to him. As we left the shebeen, my one said to me, "If you don't have it, I'm going to cut your throat." I could see that he was serious and I knew I must have it or I'm dead. So I asked my friend to say that she was hungry and we stopped at some shops. I went inside and bought a can of pilchards [inexpensive fish]. I knew that the only thing the *pantsula* was interested in was the hole and the smell. *Pantsulas* don't explore much, they just lift up your dress and go for it. We all went to bed in the one room. There were two beds. The one *pantsula* and my friend were in one and I was in the other bed with this pantsula. . . . Sardines is one of the tricks the *skesanas* use. We know that some *pantsulas* like dirty pussy, so for them you must use pilchards, but not Glenrick [a brand] because they smell too bad. Other *pantsulas* like clean pussy, so for them you can use sardines. For my *pantsula* I bought pilchards because I could see what kind he was. So before I went to bed I just smeared some pilchards around my anus and my thighs. When he smelled the smell and found the hole he was quite happy. We became lovers for some months after that. He never knew that I was a man, and he never needed the smell again because he was satisfied the first time.

Although the connection would have been anathema to the Puritan planners of apartheid, skesana identity was finally tied up with the structure of apartheid power – particularly with the all-male hostels that dotted Soweto. In these hostels, rural men without the right to reside permanently in Soweto and without their wives lived, supposedly temporarily, in order to provide labor to the white-dominated economy. From the 19th century onward, there is evidence that at least some black men in these all-male environments saw little wrong with taking other, younger workers as "wives." In these relationships, it was age and wealth, not sex, that organized and defined male-male sexual relationships; as boys matured and gained their own resources, they in turn would take "wives." This pattern has been described among gangs of thieves on the Rand in the early 20th century and among gold mine workers into the 1980s.

Certainly, in Soweto in the 1960s, hostels populated by rural men had become notorious sites for same-sex sexual relations. Township parents warned young sons not to go anywhere nearby, that they would be swept inside and smeared with Vaseline and raped. To urban raised skesanas like Linda, however, these stories apparently only aroused fantasy and desire. Linda described a "marriage ceremony" in which she took part in one of the hostels, as follows:

> At these marriage ceremonies, called *mkehlo*, all the young *skesanas* . . . sit on one side and the older ones on the other. Then your mother would be chosen. My mother was MaButhelezi. These things would happen in the hostels those days. They were famous. The older gays [sic] would choose you a mother from one of them. Then your mother's affair [partner] would be your father. Then your father is the one who would teach you how to screw. All of them, they would teach you all the positions and how to ride him up and down and sideways . . .

MODERNITY AND SEXUALITY IN THE "NEW" SOUTH AFRICA

By the early 1990s, a great deal had changed in South Africa and in Linda's life. Nelson Mandela had been released from prison. It was clear to everyone in South Africa that a new society was in process of being born. This clarity had come, however, only after more than a decade and a half of protracted, agonizing, and often violent struggle – a contest for power that upended routines all the way from the structures of the state down to the dynamics of black families in Soweto. As a result, the cultural definitions and social institutions that supported the sex/gender system in which Linda had been raised had been shaken to its roots.

By the 1990s, Linda and his friends no longer felt safe going to the hostels; many rural men's compounds in the Johannesburg region had become sites of violent opposition to the surrounding black townships, the conflict often being phrased in terms of the split between the Inkatha Freedom Party and the ANC. Also, as the end of apartheid neared, rural women began to join their men in the hostels, and the old days of male–male marriages were left behind. Looking back from the 1990s, Linda commented,

> This [male–male marriage] doesn't happen now. You don't have to be taught these things. Now is the free South Africa and the roles are not so strong, they are breaking down.

I will make explicit what Linda suggested: with the birth of a "free" South Africa, the notion of sexuality was created for some black men, or more precisely, an identity based on sexuality was created. The classificatory grid in the making was different from the old one. Now, both partners in a same-sex relationship were potentially classified as the same (male) gender – and as "gay."

Obviously, this new way of looking at the sexual world was not taken up consistently, evenly, or completely. The simultaneous presence of different models of same-sex sexuality in present-day South Africa will be evident by the end of this article. Whatever the overlapping ambiguities, it is interesting to note who took the lead in "modernizing" male–male sexuality in black South Africa: it was precisely formerly female-identified men like Linda and Jabu. But if female-identified men seem to have initiated the shift, a turning point will be reached when their male partners also uniformly identify as gay. It is perhaps altogether too easy to overstate the degree to which such a transformation has occurred in the United States itself, particularly outside urban areas and outside the white middle and upper classes.

If one sexual paradigm did not fully replace another in black townships, there were nonetheless significant changes by the early 1990s. Three events, perhaps more than others, serve to summarize these changes. First was the founding of a genuinely multiracial gay rights organization in the Johannesburg area in the late 1980s – namely, GLOW. Linda was a founding member. Second, around the same time, the ANC, still in exile, added sexuality to its policy of nondiscrimination. As I shall explain below, the ANC's peculiar international context – its dependence on foreign support in the fight against apartheid – was probably one of the factors that inclined it to support gay rights. According to Gevisser,

> ANC members in exile were being exposed to what the PAC's [Pan-Africanist Congress] Alexander calls "the European Leftist position on the matter." Liberal European notions of gender rights and the political legitimacy of gay rights had immense impact on senior ANC lawyers like Albie Sachs and Kader Asmal, who have hence become gay issues' strongest lobbyists within the ANC.

Finally, a third event that heralded change was the first gay pride march in Johannesburg in 1990, modeled on those held in places like New York and San Francisco that celebrated the Stonewall riots of 1969. Linda and his friends participated, along with approximately one thousand others. This annual ritual began to do much, through a set of such internationally recognized gay symbols as rainbow flags and pink triangles, to create a sense of transnational connections for gay South Africans.

How was Linda's life affected by these changes? Exactly how did sexuality replace local definitions of sex/gender in her forms of self-identification? According to Linda himself, the black youth uprising against apartheid was the beginning:

> Gays are a lot more confident now in the townships. I think this happened from about 1976. Before that everything was very quiet. 1976 gave people a lot of confidence . . . I remember when the time came to go and march and they wanted all the boys and girls to join in. The gays said: We're not accepted by you, so why should we march? But then they said they didn't mind and we should go to march in drag. Even the straight boys would wear drag. You could wear what you like.

As black youth took up the cause of national liberation and townships became virtual war zones, traditional black generational hierarchies were shaken to the core. Black youth came to occupy a new political space, one relatively more independent of the power of parents. But as such resistance movements have developed in other times and places, gender hierarchies have sometimes been strengthened. In resisting one form of domination, another is reinforced. In the black power movement in the United States during the 1960s, for example, masculinist and heterosexist ideals were sometimes celebrated.

Why did this reaction, with respect to gender, not take place in South Africa? One respect in which the South African case differs, certainly compared to the United States in the 1960s, is the extent to which the transnational was involved in the national struggle. Until Mandela was released, the ANC was legally banned in South Africa. Leaders not in prison were based outside the country, and there can be no doubt that the ANC could not have accomplished the political transition that it did without international support. In this context, the international left-liberal consensus on human rights – one to which gay people also appealed – probably dampened any tendency to contest local racial domination by strengthening local gender and sexuality hierarchies. Any such move would certainly have alienated antiapartheid groups from Britain to Holland to Canada to the United States.

But the significance of the transnational in the South African struggle was not only material. The imaginations of black South Africans were finally affected – particularly, in the ways in which people located themselves in the world. And it was precisely in the context of transnational antiapartheid connections that some skesanas like Linda, particularly after they were in closer contact with white gay people in Johannesburg, became aware for the first time of a global gay community – an imagined community, to adapt Benedict Anderson's phrase, imaginatively united by "deep horizontal bonds of comradeship."

How did this occur? Perhaps the incident, more than any other, that catalyzed such associations, that served as a node for exchange, was the arrest of Simon Nkoli. Nkoli, by the 1980s a gay-identified black man, was arrested for treason along with others and tried in one of the most publicized trials of the apartheid era – the so-called Delmas treason trials. After Nkoli's situation became known internationally, he became a symbol for gay people in the antiapartheid movement across the globe. For example, in December of 1986, while he was in prison, Nkoli was startled to receive more than 150 Christmas cards from gay people and organizations around the world.

According to Gevisser,

> In Nkoli, gay anti-apartheid activists found a ready-made hero. In Canada, the Simon Nkoli Anti-Apartheid Committee became a critical player in both the gay and anti-apartheid movements. Through Nkoli's imprisonment, too, progressive members of the international anti-apartheid movement were able to begin introducing the issue of gay rights to the African National Congress. The highly respectable Anti-Apartheid Movements of both Britain and Holland, for example, took up Nkoli's cause, and this was to exert a major impact on the ANC's later decision to include gay rights on its agenda.

These cultural connections and others eventually helped to produce changes in the most intimate details of skesanas' lives. To return to Linda, gay identity meant literally a new gender and a new way of relating to his body. In Linda's words,

> Before, all skesanas wanted to have a small cock. Now we can relax, it does not matter too much and people don't discuss cocks as much. . . . Before, I thought I was a woman. Now I think I'm a man, but it doesn't worry me anyway. Although it used to cause problems earlier.

In addition to how he viewed his body, Linda began to dress differently:

> I wear girl's clothes now sometimes, but not so much. But I sleep in a nightie, and I wear slippers and a gown – no skirts. I like the way a nightie feels in bed.

Consider the underneath-of-the-iceberg for the intimacies that Linda described: it is difficult to reconstruct the hundreds of micro-encounters, the thousands of messages that must have come from as far away as Amsterdam and New York. Gevisser outlines some of the social underpinnings of this reordering:

> The current township gay scene has its roots in a generalized youth rebellion that found expression first in 1976 and then in the mid-1980s. And, once a white gay organization took root in the 1980s and a collapse of rigid racial boundaries allowed greater interaction between township and city gay people, ideas of gay community filtered into the already-existent township gay networks. A few gay men and lesbians, like Nkoli, moved into Hillbrow. As the neighborhood started deracializing, they began patronizing the gay bars and thus hooking into the urban gay subculture – despite this subculture's patent racism. GLOW'S kwaThema chapter was founded, for example, when a group of residents returned from the Skyline Bar with a copy of Exit [the local gay publication]: "When we saw the publicity about this new non-racial group," explains Manku Madux, a woman who, with Sgxabai, founded the chapter, "we decided to get in touch with them to join."

The ways in which an imagined gay community became real to black South Africans were, of course, various. In Jabu's case, he had already come to see himself differently after he began work in a downtown hotel in Johannesburg in the late 1970s:

> Well, I joined the hotel industry. I started at the Carlton Hotel . . . There was no position actually that they could start me in. I won't say that being a porter was not a good job; it was. But they had to start me there. But I had some problems with guests. Most of them actually picked up that I'm gay. How, I don't know. Actually . . . how am I going to word this? People from foreign countries, they would demand my service in a different way . . . than being a porter . . . We had Pan Am, British Airways, American Airways coming. Probably the whole world assumes that any male who works for an airline is gay. I used to make friends with them. But the management wasn't happy about that, and they transferred me to the switchboard.

The assumption of a gay identity for Jabu affected not only his view of the present but also of the past. Like virtually all forms of identification that essentialize and project themselves backward in history, uniting the past with the present (and future) in an unchanging unity, South African black gay identity does the same. According to Jabu, being gay is "natural"; gay people have always been present in South African black cultures. But in his great-grandmother's time, African traditional cultures dealt with such things differently:

> I asked my grandmother and great-grandmother (she died at the age of 102). Within the family, the moment they realized that you were gay, in order to keep outside people from knowing, they organized someone who was gay to go out with you, and they arranged with another family to whom they explained the whole situation: "Okay, fine, you've got a daughter, we have three sons, this one is gay, and then there are the other two. Your daughter is not married. What if, in public, your daughter marries our gay son, but they are not going to have sex. She will have sex with the younger brother or the elder brother, and by so doing, the family will expand, you know." And at the end of the day, even if the next person realizes that I am gay, they wouldn't say anything because I am married. That is the secret that used to be kept in the black community.

[. . .]

It goes without saying that the category of gay people in South Africa is hardly homogeneous, nor is it the same as in Western countries. Being black and gay and poor in South Africa is hardly the same as being black and gay and middle-class, which is again hardly the same as being white and gay and middle-class, whether in South Africa or in North America. Despite these differences, there is still in the background a wider imagined community of gay people with which all of these persons are familiar and, at least in certain contexts, with which they identify. How this imagined community becomes "available" for persons variously situated across the globe is a major analytical question.

In Paul Gilroy's analysis of the black diaspora, he writes suggestively of the role of sailors, of ships, and of recordings of black music in making a transnational black community imaginatively real. As black identity has been formed and reformed in the context of transnational connections, black families have typically played some role – complex to be sure – in reproducing black identity. Gay identity is different to the degree that it does not rely upon the family for its anchoring; indeed, if anything, it has continually to liberate itself from the effects of family socialization.

This means, ipso facto, that identifying as gay is peculiarly dependent upon and bound up with modern media, with ways of communicatively linking people across space and time. In North America, how many "coming out" stories tell about trips to the public library, furtive searches through dictionaries, or secret readings of novels that explore lesbian and gay topics? A certain communicative density is probably a prerequisite for people to identify as gay at all, and it is not improbable that as media density increases, so will the number of gay people.

In less-developed societies of the world today, then, transnational flows become particularly relevant in understanding the formation of sexual communities. Sustained analysis of these connections has hardly begun, but I would suggest that we start not with ships but with airplanes, not with sailors (although they undoubtedly played their role here as well, particularly in port cities) but, in the South African case, with tourists, exiled antiapartheid activists, and visiting anthropologists; and finally not with music, but with images, typically erotic images – first drawings, then photographs, and now videos, most especially of the male body.

Given the composition of the global gay community, most of these images are of the *white* male body. For black men, then, identifying as gay must carry with it a certain complexity absent for most white South Africans. Also, the fact that international gay images are overwhelmingly male probably also affects the way that lesbian identity is imagined and appropriated by South African women, black and white. In any case, it could be argued that these kinds of contradictory identifications are not exceptional under late capitalism; they are the stuff of most people's lives. And lately the flow of images has been greatly accelerated; South African gays with access to a modem and a computer – admittedly, a tiny minority so far – can now download material from San Francisco, New York, or Amsterdam.

[...]

The overlapping of the national and gay questions means that gay identity in South Africa reverberates – in a way that it cannot in the United States – with a proud, new national identity. Let me quote the reaction of one of the white gays present at Linda's funeral:

> As I stood in Phiri Hall behind the black gay mourners behind the hymn-singing congregants, I felt a proud commonality with Linda's black friends around, despite our differences; *we were all gay, all South African.*

[...]

PART FIVE

Globalization and Urban Change

Plate 10 Chinatown, New York City, 1998 by Chien-Chi Chang. A newly arrived immigrant eats noodles on a fire escape (reproduced by permission of Magnum Photos).

INTRODUCTION TO PART FIVE

This is the global age, as is often said. But what does globalization mean for cities? In this section, we include readings that address the mostly structural economic processes tied to globalization and their consequences for cities. This approach has been the mainstay of research on globalization. We also include readings that exemplify more recent efforts to open up the field of globalization studies to include factors other than economic and structural ones. The emphasis on cultural aspects of globalization addresses a different set of questions, such as, what does globalization mean to everyday life in the city?

The most prominent feature of structural analyses of globalization – with an impact on the ranking of cities and the economic, social and political transformations within cities – is the movement or flow of international finance, foreign direct investments, and other forms of capital. The accelerated flow of capital is the defining attribute of globalization. These flows do not occur on their own, technological automation notwithstanding. There are key actors and institutions with vested interests in the mobility of capital that profit from making the flow of money and goods proceed as smoothly as possible. Transnational corporations are perhaps the most well known of these actors because their profits are directly tied to investment flows and favorable production and trade arrangements across the globe. Financial markets, such as the New York Stock Exchange, function to move financial resources to their potentially most profitable position (and, hence, location). Supranational organizations, such as the World Bank and the International Monetary Fund (IMF), serve to insure the implementation of national economic policies that favor investment (and, conversely, disinvestment such as the ease of moving factories offshore). Finally, transnational trading blocks, such as the North American Free Trade Agreement (NAFTA) and the European Community (EC), work to lessen political obstacles that impede flows of capital across borders.

Within this framework of capital flows, the cities that fare best are those that "capture" capital, if only temporarily, and provide the crucial services that facilitate profit accumulation and continued investment. Global cities are "infrastructural nodes." John Friedmann and Saskia Sassen concur that a city's ranking in the global hierarchy is dependent upon its level of organizations and firms who command and control key financial decisions. Also important are telecommunications infrastructure and the availability of financial markets. Given the hypermobility of capital, there is a real incentive for cities to lower the economic and political barriers to doing business (e.g., real estate, payroll, revenue and other taxes). Neoliberal political ideology and policy trumpets deregulation, fiscal conservancy with regards to social services, and the unfettered marketplace (see Part Seven for resistance to neoliberalism and its effects).

In addition to the flow of capital, globalization is marked by the flow of people both within nations and regions and between them. In lesser-developed countries, globalization has fueled the growth of industrial parks and Enterprise Zones (manufacturing spaces where limited regulation encourages relocation of factories from other countries). The resulting employment opportunities (among other factors, such as agribusiness) have encouraged an influx of migrants from the countryside, overburdening many large cities. Global cities, too, have experienced a growth in migrants from the southern hemisphere (e.g., Latin America, the Caribbean, North Africa) as opportunities, both real and imagined, draw individuals and families from impoverished areas.

The flows of both capital and people have altered the social geography between cities and within cities. The economic imbalance between the core regions of the world and the periphery is heightened by capital flows in search of profits and cheap labor markets. The class structure within global cities is increasingly polarized as well. At the top of the class hierarchy sit the corporate elites and wealthy professional class whose fortunes are connected to investment capital, real estate transactions, and high-end service provisions. An army of service workers is found at the bottom of the hierarchy. These individuals, predominantly minorities, increasingly foreign-born, and more often than not female, hold low-skilled jobs with little chance of advancement.

Globalization has produced structural changes in the economy, politics, and social life of the city – movements of people, a shift in manufacturing from the core to the periphery, and increased polarization globally and locally. But the analysis of structural changes is not the only means of addressing or understanding globalization, as the readings by Michael Peter Smith, Paul Stoller and Jasmin-Tahmaseb McConatha, and Jan Lin demonstrate.

How is globalization – often characterized as macro level and abstract – actually experienced by those who live in cities? The structural approaches to the study of globalization tend to paint a grim picture of hidden forces that compel changes in the way people live and work and meet little or no local resistance. As individuals living and socializing in different places, we rarely think of forces or processes occurring at different scales. Globalization, as the remaining authors inform us, is experienced locally in real circumstance and, therefore, exists only in its articulation locally. Glocal is a term better suited than the global-local dichotomy to fully capture this dynamic. Michael Peter Smith, a key proponent of this approach, calls for an empirical research agenda, which examines "the global" from the ground-up. We see this approach at work in the selection by Stoller and McConatha, who study West African immigrant traders in New York City. These migrants capitalize on transnational networks that connect New York and Africa to improve their collective and individual lot while selling their wares on the streets of Manhattan. Here we are privy to global processes at work in ways both innovative and unexpected.

Globalization can also affect community redevelopment in unexpected (both positive and negative) ways. Jan Lin examines the growing importance of Asian and Latino enclaves in constituting and conveying the global processes in U.S. immigration gateway cities. Globalization and ethnic commerce are leading edges of postindustrial urban growth, counterbalancing some of the decline brought by deindustrialization and runaway factories. Ethnic places were in a previous era generally shunned and subject to slum clearance, but are now promoted by urban managers pursuing growth machine strategies in the new cultural economy. Intra-group conflicts can occur, however, between overseas and local ethnic interests. Immigrant mom-and-pop businesses and poor residents are also subject to displacement by the process of urban redevelopment in central cities.

There are numerous books on globalization and global cities. For an additional preliminary overview of the topics, the following may be useful: Mark Abrahamson, *Global Cities* (New York: Oxford University Press, 2004); Peter Marcuse and Ronald van Kempen, eds, *Globalizing Cities: A New Spatial Order?* (Oxford: Blackwell 2000); H. V. Savitch and Paul Kantor, *Cities in the International Marketplace* (Princeton University Press, 2002; Janet L. Abu-Lughod, *New York, Chicago and Los Angeles: America's Global Cities* (Minneapolis: University of Minnesota Press, 1999); Roger Keil, Gerda R. Wekerle, and David V. J. Bell, eds, *Local Places in the Age of the Global City* (Montreal: Black Rose Books, 1996); and John Eade, ed., *Living in the Global City: Globalization as Local Process* (London: Routledge, 1996); Paul L. Knox and Peter J. Taylor, eds, *World Cities in a World-System* (Cambridge: Cambridge University Press, 1995); and Michael Peter Smith and Joe R. Feagin, eds, *The Capitalist City: Global Restructuring and Community Politics* (Oxford: Blackwell, 1987).

"The World City Hypothesis"

from *Development and Change* (1986)

John Friedmann

Editors' Introduction

In what he calls a "world city hypothesis," John Friedmann lays out the architecture of the global system of cities, presenting a set of theses that explain how city form is connected to global economic forces. Cities are ranked – from core, global to peripheral, regional – according to their position within a "new international division of labor." The most important cities within this hierarchy are those that carry out advanced economic functions, such as serving as a global financial center and headquarters for multinational corporations. A city's standing within the global economic system, in turn, strongly influences a host of economic, social, and political changes that takes place within it, from types of employment to funding for parks or museums.

Friedmann notes that the organization of the world city system is far from haphazard; key patterns may be detected. Global capital tends to use certain cities throughout the world as "basing points" in which investments and the development of markets predominate. There is a linear character to the distribution of global cities along an East–West axis. The Asian sub-system of global cities stretches from Tokyo to Singapore. The core cities of the American subsystem are New York, Chicago and Los Angeles, with linkages north to Toronto and south to Mexico City and Caracas. Finally, the European sub-system includes London, Paris and cities in the Rhine Valley, with linkages to a few cities in the southern hemisphere.

The economic sector fueling the global economy and, hence, the organization of the global city hierarchy, is investment flows and the support services behind them, such as advertising, accounting, insurance and legal services. Within global cities, a profound division of labor tends to exist: a workforce of highly educated professionals who specialize in making complex business decisions (what are often referred to as "command and control functions" in the global cities literature) co-exists with a larger labor force of low-skilled workers employed in the manufacturing, personal services, hospitality and entertainment industries.

An additional feature of the global system of cities is the flow of people – migration. In secondary or semi-peripheral global cities, migrants tend to flock to inner city neighborhoods or suburban fringe settlements from the countryside out of economic necessity. As a result, these cities have experienced exponential growth rates which have deepened urban poverty, pollution, and environmental degradation. Because global cities are "growth poles" for a variety of industries at multiple levels, they too function as magnets for immigrants from developing countries in search of employment. In response to massive flows of immigrants or to prevent them, global cities have enacted a variety of legal restrictions to limit the numbers of potential arrivals.

The geography of the contemporary world system of cities is marked by socioeconomic polarization that may be mapped at three scales. The first scale is global in which a small number of core, very wealthy countries (and especially notable cities within them) exist in contradistinction to a larger number of peripheral, extremely poor countries. The second scale is regional in which polarization is evident where second-level global cities with high levels of wealth and urban amenities are surrounded by impoverished rural regions often lacking in basic infrastructure. The third scale of polarization exists within cities themselves where enormous income gaps between corporate professionals and low-skilled workers predominate.

John Friedmann is a leading figure in research on urban planning. He founded the Program on Urban Planning in the Graduate School of Architecture and Planning at UCLA and served as its head for a total of 14 years between 1969 and 1996. He is a recipient of a Guggenheim Award (1976) and an American Collegiate Schools of Planning Distinguished Planning Educator Award (1988). He is currently Professor Emeritus of Urban Planning at UCLA. Friedmann is author of several books and articles, including *Urbanization, Planning, and National Development* (Beverly Hills, CA: Sage Publications 1973); *Life Space and Economic Space: Essays in Third World Planning* (Oxford: Transaction Press 1988); *Empowerment: The Politics of Alternative Development* (Cambridge, MA: Blackwell 1992); "World City Formation: An Agenda for Research and Action" (with Goetz Wolff), *International Journal of Urban and Regional Research* (1982); and, "Where We Stand: A Decade of World City Research," in Paul L. Knox and Peter J. Taylor, eds, *World Cities in a World-System* (Cambridge: Cambridge University Press 1995).

Some fifteen years ago, Manuel Castells and David Harvey revolutionized the study of urbanization and initiated a period of exciting and fruitful scholarship. Their special achievement was to link city forming processes to the larger historical movement of industrial capitalism. Henceforth, the city was no longer to be interpreted as a social ecology, subject to natural forces inherent in the dynamics of population and space; it came to be viewed instead as a product of specifically social forces set in motion by capitalist relations of production. Class conflict became central to the new view of how cities evolved.

Only during the last four or five years, however, has the study of cities been directly connected to the world economy. This new approach sharpened insights into processes of urban change; it also offered a needed spatial perspective on an economy which seems increasingly oblivious to national boundaries. My purpose in this introduction is to state, as succinctly as I can, the main theses that link urbanization processes to global economic forces. The world city hypothesis, as I shall call these loosely joined statements, is primarily intended as a framework for research. It is neither a theory nor a universal generalization about cities, but a starting-point for political enquiry. We would, in fact, expect to find significant differences among those cities that have become the "basing points" for global capital. We would expect cities to differ among themselves according to not only the mode of their integration with the global economy, but also their own historical past, national policies, and cultural influences. The economic variable, however, is likely to be decisive for all attempts at explanation.

The world city hypothesis is about the spatial organization of the new international division of labour. As such, it concerns the contradictory relations between production in the era of global management and the political determination of territorial interests. It helps us to understand what happens in the major global cities of the world economy and what much political conflict in these cities is about. Although it cannot predict the outcomes of these struggles, it does suggest their common origins in the global system of market relations. There are seven interrelated theses in all. As they are stated, I shall follow with a comment in which they are explained, or examples are given, or further questions are posed.

1. *The form and extent of a city's integration with the world economy, and the functions assigned to the city in the new spatial division of labour, will be decisive for any structural changes occurring within it.*

Let us examine each of the key terms in this thesis.

(a) City. Reference is to an economic definition. A city in these terms is a spatially integrated economic and social system at a given location, or metropolitan region. For administrative or political purposes, the region may be divided into smaller units, which underlie, as a political or administrative space, the economic space of the region.

(b) Integration with the world capitalist system. Reference is to the specific forms, intensity, and duration of the relations that link the urban economy into the global system of markets for capital, labour and commodities.

(c) Functions assigned to it in the new spatial division of labour. The standard definition of the world capitalist system is that it corresponds to a single (spatial) division of labour. Within this division, different localities – national, regional, and urban subsystems – perform specialized roles. Focusing only on metropolitan economies, some carry out headquarter functions, others serve primarily as a financial centre, and still others have as their main function the articulation of regional and/or national economies with the global system. The most important cities, however, such as New York, may carry out *all* of these functions simultaneously.

(d) Structural changes occurring within it. Contemporary urban change is for the most part a process of adaptation to changes that are externally induced. More specifically, changes in metropolitan function, the structure of metropolitan labour markets, and the physical form of cities can be explained with reference to a worldwide process that affects the direction and volume of transnational capital flows, the spatial division of the functions of finance, management and production or, more generally, between production and control, and the employment structure of economic base activities.

These economic influences are, in turn, modified by certain endogenous conditions. Among these the most important are: first, the *spatial patterns of historical accumulation*; second, *national policies*, whose aim is to protect the national economic subsystem from outside competition through partial closure to immigration, commodity imports and the operation of international capital; and third, certain *social conditions*, such as *apartheid* in South Africa, which exert a major influence on urban process and structure.

2. *Key cities throughout the world are used by global capital as "basing points" in the spatial organization and articulation of production and markets. The resulting linkages make it possible to arrange world cities into a complex spatial hierarchy.*

Several taxonomies of world cities have been attempted. In Table 1, a different approach to world city distribution is attempted. Because the data to verify it are still lacking, the present effort is meant chiefly as a means to visualize a possible rank ordering of major cities, based on the presumed nature of their integration with the world economy.

When we look at Table 5.1, certain features of the classification spring immediately into view.

(a) All but two primary world cities are located in core countries. The two exceptions are São Paulo (which articulates the Brazilian economy) and the city-state of Singapore which performs the same role for a multi-country region in South-East Asia.
(b) European world cities are difficult to categorize because of their relatively small size and often specialized functions. London and Paris are world cities of the first rank but beyond that, classification gets more difficult. By way of illustration, I have included as world cities of the first rank the series of closely linked urban areas in the Netherlands focused on the Europort of Rotterdam, the West German economy centred on Frankfurt, and Zurich as a leading world money market.
(c) The list of secondary cities in both core and semi-periphery is meant to be only suggestive. Within core countries, secondary cities tend on the whole to be somewhat smaller than cities of the first rank, and some are more specialized as well (Vienna, Brussels and Milan). In semi-peripheral countries, the majority of secondary world cities are capital cities. Their relative importance for international capital depends very much on the strength and vitality of the national economy which these cities articulate.

The complete spatial distribution suggests a distinctively linear character of the world city system which connects, along an East–West axis, three distinct sub-systems: an Asian sub-system centred on the Tokyo–Singapore axis, with Singapore playing a subsidiary role as regional metropolis in Southeast Asia; an American sub-system based on the three primary core cities of New York, Chicago and Los Angeles, linked to Toronto in the North and to Mexico City and Caracas in the South, thus bringing Canada, Central America and the small Caribbean nations into the American orbit; and a West European sub-system focused on London, Paris and the Rhine valley axis from Randstad and Holland to Zurich. The southern hemisphere is linked into this sub-system via Johannesburg and São Paulo (see Figure 1).

Table 1 World city hierarchy[a]

Core Countries[b]		*Semi-Peripheral Countries*[b]	
Primary	*Secondary*	*Primary*	*Secondary*
London* I	Brussels* III		
Paris* I	Milan III		
Rotterdam III	Vienna* III		
Frankfurt III	Madrid* III		
Zurich III			Johannesburg III
New York I	Toronto III	São Paulo I	Buenos Aires* I
Chicago II	Miami III		Rio de Janeiro I
Los Angeles I	Houston III		Caracas* III
	San Francisco III		Mexico City* I
Tokyo* I	Sydney III	Singapore* III	Hong Kong II
			Taipei* III
			Manila* II
			Bangkok* II
			Seoul* II

* National capital. Population size categories (referring to metro-region): I 10–20 million; II 5–10 million; III 1–5 million.

[a] Selection criteria include: major financial centre; headquarters for TNCs (including regional headquarters); international institutions; rapid growth of business services sector; important manufacturing centre; major transportation node; population size. Not all criteria were used in every case, but several criteria had to be satisfied before a city would be identified as a world city of a particular rank. In principle, it would have been possible to add third- and even fourth-order cities to our global hierarchy. This was not done, however, since our primary interest is in the identification of only the most important centres of capitalist accumulation.

[b] Core countries are here identified according to World Bank criteria. They include nineteen so-called industrial market economies. Semi-peripheral countries include for the most part upper middle income countries having a significant measure of industrialization and an economic system based on market exchange.

3. *The global control functions of world cities are directly reflected in the structure and dynamics of their production sectors and employment.*

The driving force of world city growth is found in a small number of rapidly expanding sectors. Major importance attaches to corporate headquarters, international finance, global transport and communications; and high level business services, such as advertising, accounting, insurance and legal. An important ancillary function of world cities is ideological penetration and control. New York and Los Angeles, London and Paris, and to a lesser degree Tokyo, are centres for the production and dissemination of information, news, entertainment and other cultural artifacts.

In terms of occupations, world cities are characterized by a dichotomized labour force: on the one hand, a high percentage of professionals specialized in control functions and, on the other, a vast army of low-skilled workers engaged in manufacturing, personal services and the hotel, tourist and entertainment industries that cater to the privileged classes for whose sake the world city primarily exists.

In the semi-periphery, with its rapidly multiplying rural population, large numbers of unskilled workers migrate to world city locations in their respective countries in search of livelihood. Because the "modern" sector is incapable of absorbing more than a small fraction of this human mass, a large "informal" sector of microscopic survival activities has evolved.

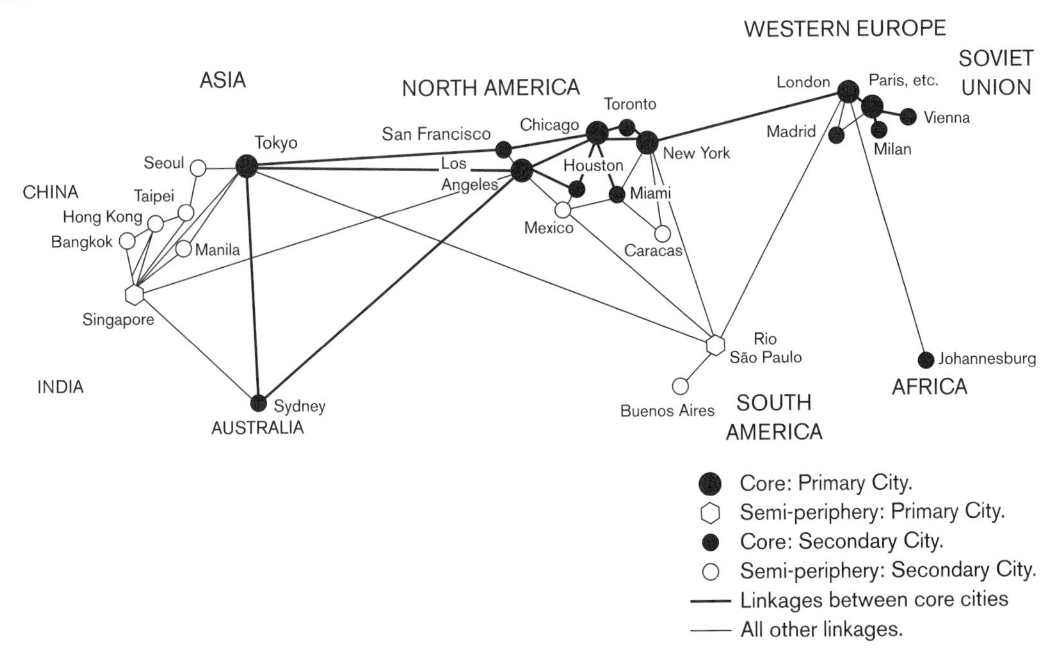

Figure 1 The hierarchy of world cities

4. World cities are major sites for the concentration and accumulation of international capital.

Although this statement would seem to be axiomatic, there are significant exceptions. In core countries, the major atypical case is Tokyo. Although a major control centre for Japanese multi-national capital, Japanese business practices and government policy have so far been successful in preventing foreign capital from making major investments in the city.

In the semi-periphery, the economic crisis since 1973 has led to massive international indebtedness, originally incurred in the hope of staving off economic disaster in the teeth of a world-wide recession of unprecedented depth and duration. A combination of declining per capita incomes, slow growth in the core of the world economy, IMF-imposed policies, the high cost of capital, capital repatriation, capital flight and obligatory loan repayments in some cases amounting to more than 35 per cent of export earnings have contributed in a number of Latin American countries to *a net export* of capital. If this trend, extraordinary for the postwar period, should persist, the semi-periphery is bound to backslide into peasant-peripheral status. Although strenuous attempts are being made to reverse this tidal drift into economic insolvency, declining living standards for the middle classes, immiseration for the poor and the collapse of the world economic system as it presently exists, the outcome is not at all certain.

5. World cities are points of destination for large numbers of both domestic and/or international migrants.

Two kinds of migrants can be distinguished: international and interregional. Both contribute to the growth of primary core cities, but in the semi-periphery world cities grow chiefly from interregional migration.

In one form or another, all countries of the capitalist core have attempted to curb immigration from abroad. Japan and Singapore have the most restrictive legislation and, for all practical purposes, prohibit permanent immigration. Western European countries have experimented with tightly controlled "guest worker" programmes. They, too, are jealous of their boundaries. And traditional immigrant countries, such as Canada and Australia, are attempting to limit the influx of migrants to workers possessing professional and other skills in high demand. Few if any countries have been as open to immigration from abroad as the United States, where both legal and illegal immigrants abound.

In semi-peripheral countries, periodic attempts to slow down the flow of rural migrants to large cities have been notably unsuccessful. Typically, therefore, urban growth has been from 1.5 to 2.5 times greater than the overall rate of population increase, and principal (world) cities have grown to very large sizes. Of the thirty cities in Table 1, eight have a population of 15 ± 5 million and another six a population of 7.5 ± 2.5 million. Absolute size, however, is not a criterion of world city status, and there are many large cities even in the peasant periphery whose size clearly does not entitle them to world city status.

6. World city formation brings into focus the major contradictions of industrial capitalism – among them spatial and class polarization.

Spatial polarization occurs at three scales. The first is global and is expressed by the widening gulf in wealth, income and power between peripheral economies as a whole and a handful of rich countries at the heart of the capitalist world. The second scale is regional and is especially pertinent in the semi-periphery. In core countries, regional income gradients are relatively smooth, and the difference between high and low income regions is rarely greater than 1:3. The corresponding ratio in the semi-periphery, however, is more likely to be 1:10. Meanwhile, the income gradient between peripheral world cities and the rest of the national economies which they articulate remains very steep. The third scale of spatial polarization is metropolitan. It is the familiar story of spatially segregating poor inner-city ghettos, suburban squatter housing and ethnic working-class enclaves. Spatial polarization arises from class polarization. And in world cities, class polarization has three principal facets: huge income gaps between transnational elites and low-skilled workers, large-scale immigration from rural areas or from abroad and structural trends in the evolution of jobs.

In the income distribution of semi-peripheral countries, the bottom 40 per cent of households typically receive less than 15 per cent of all income and control virtually none of the wealth. These data refer to countries that, overall, have low incomes when measured on the scale of a Western Europe or the US. In many of the primary cities of the core, however, the situation is not significantly better. In Los Angeles and New York, for example, huge immigrant populations are seriously disadvantaged.

In the semi-periphery, the massive poverty of world cities is underscored by the relative absence of middle-income sectors. The failure of semi-peripheral world cities to develop a substantial "middle class" has often been noted. Although there are important salaried sectors in cities such as Buenos Aires, their economic situation is subject to erosion by an inflationary process that is almost always double-digit and in some years rises to more than 200 per cent! Middle sectors have also become increasingly vulnerable to unemployment.

The basic structural reason for social polarization in world cities must be looked for in the evolution of jobs, which is itself a result of the increasing capital intensity of production. In the semi-periphery, most rural immigrants find accommodation in low-level service jobs, small industry and the "informal" sector. In core countries, the process is more complex. Given the downward pressure on wages resulting from large scale immigration of foreign (including undocumented) workers, the number of low-paid, chiefly non-unionized jobs rises rapidly in three sectors: personal and consumer services (domestics, boutiques, restaurants and entertainment), low-wage manufacturing (electronics, garments and prepared foods) and the dynamic sectors of finance and business services which comprise from one-quarter to one-third of all world city jobs and also give employment in many low-wage categories.

The whole comprises an *ecology* of jobs. As shown in Figure 2, the restructuring process in cities such as New York and Los Angeles involves the *destruction* of jobs in the high-wage, unionized sectors (EXODUS) and job *creation* in what Saskia Sassen-Koob calls the production of global control capability. Linked to these dynamic sectors are certain personal and consumer services (employing primarily female and/or foreign workers), while the slack in manufacturing is taken up by sweat shops and small industries employing non-union labour at near the minimum wage. It is this structural shift which accounts for the rapid decline of the middle-income sectors during the 1970s.

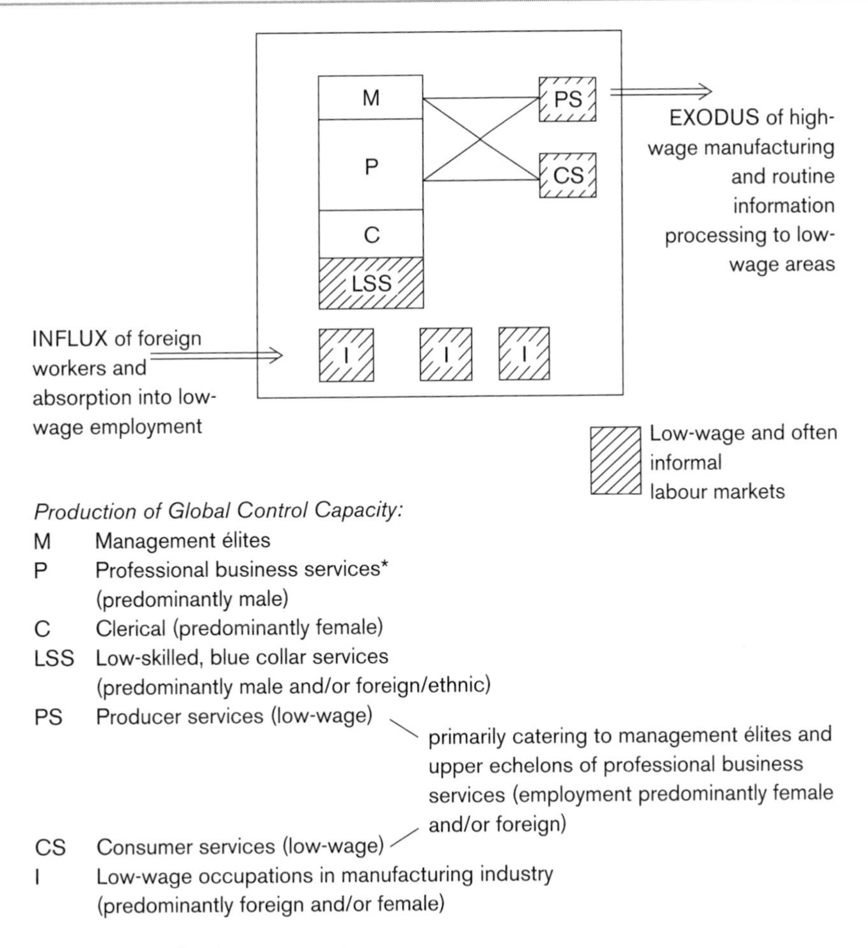

Figure 2 World city restructuring in core countries
* Many professional business services engage increasingly in international trade serving their clients, the transnational corporations, both at home and abroad. They include accounting, advertising, banking, communications, computer services, health services, insurance, leasing, legal services, shipping and air transport, and tourism. In 1981, US service exports equalled 50 per cent of merchandise exports and were still rising.

7. World city growth generates social costs at rates that tend to exceed the fiscal capacity of the state.

The rapid influx of poor workers into world cities – be it from abroad or from within the country – generates massive needs for social reproduction, among them housing, education, health, transportation and welfare. These needs are increasingly arrayed against other needs that arise from transnational capital for economic infrastructure and from the dominant Mites for their own social reproduction.

In this competitive struggle, the poor and especially the new immigrant populations tend to lose out. State budgets reflect the general balance of political power. Not only are corporations exempt from taxes; they are generously subsidized in a variety of other ways as well. At the same time, the social classes that feed at the trough of the transnational economy insist, and usually insist successfully, on the priority of their own substantial claims for urban amenities and services. The overall result is a steady state of fiscal and social crisis in which the burden of capitalist accumulation is systematically shifted to the politically weakest, most disorganized sectors of the population. Their capacity for pressing their rightful claims against the corporations and the state is further contained by the ubiquitous forces of police repression.

"The Urban Impact of Economic Globalization"

from *Cities in a World Economy*, second edition (2000)

Saskia Sassen

Editors' Introduction

What factors account for the determination of select cities as "global cities" and others not? For Saskia Sassen, a leading scholar in the study of cities in the world economy, the answer lies in a city's capacity to capture and maintain a strategic advantage in a world system of fast-moving international finance. The socioeconomic fate of cities (and their position in the global cities hierarchy described by Friedmann) is dependent in large part upon their role in international financial transactions or capital flows. Sassen directs us to the forces behind the emergence of global cities: the transformation and rising importance of foreign direct investment (FDI) since the mid-1970s. High levels of foreign investment in both developed and developing countries have contributed to double-digit rates of economic growth.

But globalization does not proceed through economic processes alone. As Sassen states, "We cannot take the world economy for granted and assume that it exists simply because international transactions do." Transnational corporations are the major source of financial flows and their decisions where to invest and where not to invest are driven by profit concerns. Their collective power to shape the world economy, relative to national governments, has raised considerable concern (see the selection by Kohler and Wissen). The rapidity of capital flows and the large sums of investment involved have similarly elevated the status of financial markets as "strategic organizers" of the global economy. In an effort to compete globally, governments have increasingly turned to national and international political arrangements to improve their standing in the economy of investment. Transnational trading blocks, such as the North American Free Trade Agreement (NAFTA) and the European Community (EC), have lowered the political barriers and economic costs of capital flows among participating nations.

Global cities and offshore banking centers are the strategic places within the world economy. Unlike the trading of goods and commodities, investment exchanges tend to be concentrated among a select set of nodes that can provide the services buyers and sellers of capital require. Flows of international capital rest on complex transactions – legal, financial, etc. – which, in turn, require a supportive infrastructure to provide them. The emergence of global markets for finance has resulted in the demand for specialized services for firms engaged in investment transactions. Global cities are those that operate as critical infrastructural nodes in the international flow of capital. They provide the necessary specialized services and sophisticated telecommunications that support the web of transactions that keep FDI moving. As such, they are "postindustrial production sites for the leading industries of this period – finance and specialized services." They also function as marketplaces where multinational corporations and governments can purchase a variety of financial, legal, and insurance services. Within a world economy of money flows, tax-free offshore banking centers are appealing. The Cayman Islands, Hong Kong, Bahrain, and Monaco, among others, offer minimal restrictions and regulations on banking. They have become specialized but nonetheless important nodes in the circuit of finance capital.

Saskia Sassen is the Ralph Lewis Professor of Sociology at the University of Chicago and has been Centennial Visiting Professor at the London School of Economics. She is the author of many books on cities and globalization, including *Denationalization: Economy and Polity in a Global Digital Age* (Princeton, NJ: Princeton University Press, 2003), *Global Networks/Linked Cities* (New York and London: Routledge, 2002), *The Global City*, 2nd Edition (Princeton, NJ: Princeton University Press, 2001), *Guests and Aliens* (New York: New Press, 1999), *Globalization and Its Discontents* (New York: New Press, 1998), co-authored with Anthony Appiah, *Losing Control? Sovereignty in an Age of Globalization* (New York: Columbia University Press, 1996), *Cities in a Global Economy* (Thousand Oaks, CA: Pine Forge Press, 1994), and *The Mobility of Capital: A Study in International Investment and Labor Flows* (New York: Cambridge University Press, 1988).

Neil Brenner and Roger Keil have edited a comprehensive reader on global cities entitled *The Global Cities Reader* (London and New York: Routledge 2003).

Profound changes in the composition, geography, and institutional framework of the global economy have had major implications for cities. In the 1800s, when the world economy consisted largely of trade, the crucial sites were harbors, plantations, factories, and mines. Cities were already servicing centers at that time: The major cities of the time typically developed alongside harbors, and trading companies depended on multiple industrial, banking, and other commercial services located in cities. Cities, however, were not the key production sites for the leading industries in the 1800s; the production of wealth was centered elsewhere. Today, international trade continues to be an important fact in the global economy, but it has been overshadowed both in value and in power by international financial flows, whether loans and equities or foreign currency transactions. In the 1980s, finance and specialized services emerged as the major components of international transactions. The crucial sites for these transactions are financial markets, advanced corporate service firms, banks, and the headquarters of transnational corporations (TNCs). These sites lie at the heart of the process for the creation of wealth, and they are located in cities.

Thus, one of the factors influencing the role of cities in the new global economy is the change in the composition of international transactions, a factor often not recognized in standard analyses of the world economy. The current composition of international transactions shows this transformation very clearly. For instance, foreign direct investment (FDI) grew three times faster in the 1980s than the growth of the export trade. Furthermore, by the mid-1980s, investment in services had become the main component in FDI flows, whereas before it had been in manufacturing or raw materials extraction. These trends became even sharper in the 1990s. By 1999, the monetary value of international financial flows was vastly larger than the value of international trade and FDI. The sharp growth of international financial flows has raised the level of complexity of transactions. This new circumstance demands a highly advanced infrastructure of specialized services and top-level concentrations of telecommunications facilities. Cities are central locations for both.

The first half of this chapter will present a somewhat detailed account of the geography, composition, and institutional framework of the global economy today. The second half will focus on two types of strategic places for international financial and service transactions: global cities and offshore banking centers. Finally, we will consider the impact of the collapse of the Pax Americana on the world economy and the subsequent shift in the geographical axis of international transactions.

THE GLOBAL ECONOMY TODAY

Here we emphasize new investment patterns and dominant features of the current period. The purpose is not to present an exhaustive account of all that constitutes the world economy today. It is rather to discuss what distinguishes the current period from the immediate past.

Geography

A key feature of the global economy today is the geography of the new types of international transactions. When international flows consist of raw materials, agricultural products, or mining goods, the geography of transactions is in part determined by the location of natural resources. Historically, this has meant that a large number of countries in Africa, Latin America, and the Caribbean were key sites in this geography. When finance and specialized services became the dominant component of international transactions in the early 1980s, the role of cities was strengthened. At the same time, the sharp concentration in these industries means that now only a limited number of cities play a strategic role.

The fact of a new geography of international transactions becomes evident in FDI flows – that is, investors acquiring a firm, wholly or in part, or building and setting up new firms in a foreign country. FDI flows are highly differentiated in their destination and can be constituted through many different processes. During the last two decades, the growth in FDI has been embedded in the internationalization of production of goods and services. The internationalization of production in manufacturing is particularly important in establishing FDI flows into developing countries.

Compared with the 1950s, the 1980s saw a narrowing of the geography of the global economy and a far stronger East–West axis. This is evident in the sharp growth of investment and trade within what is often referred to as the triad: the United States, Western Europe, and Japan. FDI flows to developed countries grew at an average annual rate of 24 percent from 1986 to 1990, reaching a value of US$129.6 billion in 1991, out of a total worldwide FDI inflow of US$159.3 billion. By the mid-1980s, 75 percent of all FDI stock and 84 percent of FDI stock in services was in developed countries. There was a sharp concentration even among developed countries in these patterns: The top four recipient countries (United States, United Kingdom, France, and Germany) accounted for half of world inflows in the 1980s; the five major exporters of capital (United States, United Kingdom, Japan, France, and Germany) accounted for 70 percent of total outflows. In the early 1990s, there were declines in most of these figures due to the financial crisis, but by the late 1990s, levels of investment had grown sharply, reaching US$233.1 billion in developed countries and US$148.9 in developing countries. Financial concentration is evident in a ranking of the top banks in the world, with only eight countries represented (see Exhibit 1).

Although investment flows in developing countries in the 1990s were lower than in developed countries, they were high in historic terms – a fact that reflects the growing internationalization of economic activity. International investment in developing countries lost share in the 1980s, although it increased in absolute value and regained share by the early 1990s. Yet the share of worldwide flows going to developing countries as a whole fell from 26 percent to 17 percent between the early 1980s and the late 1980s, pointing to the strength of flows within the triad (United States, Western Europe, and Japan); it grew in the 1990s, reaching 37.2 percent by 1997 before the financial crisis of the late 1990s. Most of the flow to developing countries has gone into East, South, and Southeast Asia, where the annual rate of growth rose on the average by over 37 percent a year in the 1980s and 1990s.

[. . .]

The other two major components of the global economy are trade and financial flows other than FDI. By its very nature, the geography of trade is less concentrated than that of direct foreign investment. Wherever there are buyers, sellers are likely to go. Finance, on the other hand, is enormously concentrated.

Composition

In the 1950s, the major international flow was world trade, concentrated in raw materials, other primary products, and resource-based manufacturing. In the 1980s, the gap between the growth rate of exports and that of financial flows widened sharply. Although there are severe problems with measurement, the increase in financial and service transactions, especially the former, is so sharp as to leave little doubt. For instance, worldwide

Exhibit 1 Top Banks in the World Ranked by Assets, 1998 (US$ millions)

Asset rank	*Name*	*City*	*Assets*	*Net income*
1	Bank of Tokyo (22) Mitsubishi (6)	Tokyo	752,318	352
2	Deutsche Bank (11)	Frankfurt	575,693	1,441
3	Sumitomo Bank (2)	Osaka	513,781	294
4	Dai-Ichi Kangyo Bank (1)	Tokyo	476,696	−1,531
5	Fuji Bank (4)	Tokyo	474,371	942
6	Sanwa Bank (5)	Tokyo	470,336	223
7	ABN Amro Holdings (16)	Amsterdam	444,410	1,790
8	Sakura Bank (3)	Tokyo	470,336	153
9	Industrial & Commercial Bank	Beijing	435,723	622
10	HSBC Holdings	London	405,037	5,330
11	Norinchukin Bank (9)	Tokyo	400,031	267
12	Industrial Bank of Japan (10)	Tokyo	399,509	110
13	Dresdner Bank (24)	Frankfurt	389,626	999
14	Banque Nationale de Paris (12)	Paris	358,187	743
15	Societe Generale (20)	Paris	342,760	876
16	Chase Manhattan	New York	336,099	2,461
17	Union Bank of Switzerland (26)	Zurich	325,082	−261
18	Commerzbank	Frankfurt	320,419	773
19	Barclays Bank (13)	London	318,551	2,807
20	National Westminster Bank (18)	London	317,411	752
21	Credit Lyonnais (9)	Paris	312,926	39
22	Mitsubishi Trust & Banking (15)	Tokyo	312,223	97
23	Westdeutsche Landesbank	Duesseldorf	305,879	441
24	Tokal Bank (14)	Nagoya	296,895	145
25	Cle. Financie de Paribas	Paris	293,437	838
26	Bank of China	Beijing	292,554	1,067
27	Citicorp (29)	New York	277,653	3,788
28	Swiss Bank	Geneva	268,161	−1,461
29	Sumitomo Trust & Banking (17)	Osaka	266,035	63
30	Bayerische Vereinsbank	Munich	256,371	527

Note: Numbers in parentheses denote standing in 1991. Five banks have dropped out of the top 50 from 1991: Mitsui Trust and Banking Co. Ltd. (19), Kyowa Saltama Bank Ltd. (23), Daiwa Bank (25), Yasuda Trust and Banking Co. Ltd. (27), Instituto Bancario San Paulo di Torino (28). Three banks are still in the top 50 but not reported in the table above: Long-Term Credit Bank (previous rank, 21/35), Credit Agricole (8/35), and Toyo Trust and Banking (30/48).
Source: Based on *Hoover's Handbook of World Business* (1998).

outflows of FDI nearly tripled between 1984 and 1987, grew another 20 percent in 1988, and grew yet another 20 percent in 1989. By 1990, total worldwide stock of FDI stood at US$1.5 trillion and at US$2 trillion by 1992. Furthermore, the shares of the tertiary sector grew consistently over the 1980s and 1990s while that of the primary sector fell. Between 1992 and 1997, worldwide FDI inflows grew by 56 percent. FDI worldwide outward stock stood at US$3.5 trillion in 1997.

Many factors have fed the growth of FDI: (1) Several developed countries became major capital exporters, most notably Japan; (2) the number of cross-border mergers and acquisitions grew sharply; and (3) the flow of services and transnational service corporations have emerged as major components in the world economy. Services, which accounted for about 24 percent of worldwide stock in FDI in the early 1970s, had grown to 50 percent of stock and 60 percent of annual flows by the end of the 1980s. The single largest recipient of FDI in services in the 1980s – the decade of high growth of these flows – was the European Community, yet another indication of a very distinct geography in world transactions. But it should be noted that these flows have also increased in absolute terms in the case of less developed countries.

Another major transformation has been the sharp growth in the numbers and economic weight of TNCs – firms that operate in more than one country through affiliates, subsidiaries, or other arrangements. The central role played by TNCs can be seen in the fact that U.S. and foreign TNCs accounted for 80 percent of international trade in the United States in the late 1980s. More than 143 countries have adopted special FDI regimes to attract FDI, up from 20 in 1982.

Institutional framework

How does the "world economy" cohere as a system? We cannot take the world economy for granted and assume that it exists simply because international transactions do. One question raised by the developments described above is whether the global economic activities occurring today represent a mere quantitative change or actually entail a change in the international regime governing the world economy. Elsewhere, I have argued that the ascendance of international finance and services produces a new regime with distinct consequences for other industries, especially manufacturing, and for regional development, insofar as regions tend to be dominated by particular industries. One consequence of this new regime is that TNCs have become even more central to the organization of the world economy, and the new, or vastly expanded older, global markets are now an important element in the institutional framework.

In addition to financing huge government deficits, the financial credit markets that exploded into growth in the 1980s served the needs of TNCs to a disproportionate extent. TNCs also emerged as a source for financial flows to developing countries, both through direct inflows of FDI and indirectly, insofar as FDI stimulates other forms of financial flows. In some respects, TNCs replaced banks. For better or for worse, the TNC is now a strategic organizer of what we call the world economy.

Global financial markets have emerged as yet another crucial institution organizing the world economy. The central role of markets in international finance, a key component of the world economy today, was in part brought about by the so-called Third World bank crisis formally declared in 1982. This was a crisis for the major transnational banks in the United States, which had made massive loans to Third World countries and firms incapable of repayment. The crisis created a space into which small, highly competitive financial firms moved, launching a whole new era in the 1980s in speculation, innovation, and levels of profitability. The result was a highly unstable period but one with almost inconceivably high levels of profits that fed a massive expansion in the volume of international financial transactions. Deregulation was another key mechanism facilitating this type of growth, centered in internationalization and in speculation. Markets provided an institutional framework that organized these massive financial flows. Notwithstanding two financial crises, one in 1990–91 and the second in 1997–98, the end of the 1990s saw a sharp growth in the value of financial transactions.

The formation of transnational trading blocs is yet another development that contributes to the new institutional framework. The two major blocs are the North American Free Trade Agreement (NAFTA) and the European Economic Community (EEC). According to the World Trade Organization (WTO), there were over 70 regional trade agreements by the late 1990s. The specifics of each of the two major trading blocs currently being implemented vary considerably, but both strongly feature the enhanced capability for capital to move across borders. Crucial to the design of these blocs is the free movement of financial services. Trade, although it has received far more attention, is less significant; there already is a lot of trade among the countries in each bloc, and tariffs are already low for many goods. The NAFTA and EEC blocs represent a further formalization of capital as a transnational category, one that operates on another level from that represented by TNCs and global financial markets. Finally, in 1993, the WTO was set up to oversee cross-border trade. It has the power to adjudicate in cross-border disputes between countries and represents potentially a key institutional framework for the governance of the global economy.

Considerable effort and resources have gone into the development of a framework for governing global finance. This includes the development of new institutional accounting and financial reporting standards, minimum capital requirements for banks, and efforts to institute greater transparency in corporate governance.

These realignments have had pronounced consequences. One consequence of the extremely high level of profitability in the financial industry, for example, was the devaluing of manufacturing as a sector – although not necessarily in all subbranches. Much of the policy around deregulation had the effect of making finance so profitable that it took investment away from manufacturing. Finance also contains the possibility for superprofits by maximizing the circulation of and speculation in money – that is, buying and selling – in a way that manufacturing does not (e.g., securitization, multiple transactions over a short period of time, selling debts). Securitization, which played a crucial role, refers to the transformation of various types of financial assets and debts into marketable instruments. The 1980s saw the invention of numerous ways to securitize debts, a trend that has continued in the 1990s with the invention of ever more complex and speculative instruments. A simple illustration is the bundling of a large number of mortgages that can be sold many times, even though the number of houses involved stays the same. This option is basically not available in manufacturing. The good is made and sold; once it enters the realm of circulation, it enters another set of industries, or sector of the economy, and the profits from subsequent sales accrue to these sectors.

These changes in the geography and in the composition of international transactions, and the framework through which these transactions are implemented, have contributed to the formation of new strategic sites in the world economy. This is the subject of the next section.

STRATEGIC PLACES

Four types of places above all others symbolize the new forms of economic globalization: export processing zones, offshore banking centers, high-tech districts, and global cities. There are also many other locations where international transactions materialize. Certainly, harbors continue to be strategic in a world of growing international trade and in the formation of regional blocs for trade and investment. And massive industrial districts in major manufacturing export countries, such as the United States, Japan, and Germany, are in many ways strategic sites for international activity and specifically for production for export. None of these locations, however, captures the prototypical image of today's global economy the way the first four do. Some geographers now speak of global city regions to capture this development.

Because much has been published about export processing zones and high-tech districts, and because they entail types of activity less likely to be located in cities than finance and services, we will not examine them in detail. As they are less known, let me define the first as zones in low-wage countries where firms from developed countries can locate factories to process and/or assemble components brought in from and re-exported to the developed countries. Special legislation was passed in several developed countries to make this possible. The central rationale for these zones is access to cheap labor for the labor-intensive stages of a firm's production process. Tax breaks and lenient workplace standards in the zones are additional incentives. These zones are a key mechanism in the internationalization of production.

Here we will focus briefly on global cities and offshore banking centers.

Global cities

Global cities are strategic sites for the management of the global economy and the production of the most advanced services and financial operations. They are key sites for the advanced services and telecommunications facilities necessary for the implementation and management of global economic operations. They also tend to concentrate the headquarters of firms, especially firms that operate globally. The growth of international investment and trade and the need to finance and service such activities have fed the growth of these functions in major cities. The erosion of the role of the government in the world economy, which was much larger when trade was the dominant form of international transaction, has displaced some of the organizing and servicing work from governments and major headquarters to specialized service firms and global markets in services and finance. Here we briefly examine these developments, first by presenting the concept of the global city and then by empirically describing the concentration of major international markets and firms in various cities.

The specific forms assumed by globalization over the last decade have created particular organizational requirements. The emergence of global markets for finance and specialized services, along with the growth of investment as a major type of international transaction, has contributed to the expansion in command functions and in the demand for specialized services for firms. Much of this activity is not encompassed by the organizational form of the TNC or bank, even though these types of firms account for a disproportionate share of international flows. Nor is much of this activity encompassed by the power of transnationals, a power often invoked to explain the fact of economic globalization. It involves work and workers. Here some of the hypotheses developed in our recent work are of interest, especially those that examine the spatial and organizational forms of economic globalization and the actual work of running transnational economic operations. This way of framing the inquiry has the effect of recovering the centrality of place and work in processes of economic globalization.

A central proposition in the global city model is that the combination of geographic dispersal of economic activities and system integration that lies at the heart of the current economic era has contributed to a strategic role for major cities. Rather than becoming obsolete because of the dispersal made possible by information technologies, cities instead concentrate command functions. To this role, I have added two additional functions: (1) Cities are postindustrial production sites for the leading industries of this period – finance and specialized services – and (2) cities are transnational marketplaces where firms and governments from all over the world can buy financial instruments and specialized services.

The territorial dispersal of economic activity at the national and world scale implied by globalization has created new forms of concentration. This territorial dispersal and ongoing concentration in ownership can be inferred from some of the figures on the growth of transnational enterprises and their affiliates. Exhibit 2 shows how vast the numbers of affiliates of TNCs are. This raises the complexity of management functions, accounting, and legal services, and hence the growth of these activities in global cities.

In the case of the financial industry, we see a similar dynamic of global integration: a growth in the number of cities integrated in the global financial network and a simultaneous increase of concentration of value managed at the top of the hierarchy of centers. We can identify two distinct phases. Up to the end of the 1982 Third World debt crisis, the large transnational banks dominated the financial markets in terms of both the volume and the nature of financial transactions. After 1982, this dominance was increasingly challenged by other financial institutions and the major innovations they produced. These challenges led to a transformation in the leading components of the financial industry, a proliferation of financial institutions, and the rapid internationalization of financial markets. The marketplace and the advantages of agglomeration – and hence, cities – assumed new significance beginning in the mid-1980s. These developments led simultaneously to (1) the incorporation of a multiplicity of markets all over the world into a global system that fed the growth of the industry after the 1982 debt crisis and (2) new forms of concentration, specifically the centralization of the industry in a few leading financial centers. Hence, in the case of the financial industry, to focus only on the large transnational banks would exclude precisely those sectors of the industry where much of the new growth and production of innovations has occurred. Also, it would again leave out an examination of the wide range of activities, firms, and markets that compose the financial industry beginning in the 1980s.

The geographic dispersal of plants, offices, and service outlets and the integration of a growing number of stock markets around the world could have been accompanied by a corresponding decentralization in control and central functions. But this has not happened.

Exhibit 2 Number of Parent Transnational Corporations and Foreign Affiliates, by Region and Country, Selected Years (1990–1997)

	Year	*Parent corporations based in country*	*Foreign affiliates located in country*
All developed countries	1990	33,500	81,800
	1996	43,442	96,620
Select countries			
Australia	1992	1,306	695
	1997	485	2,371
Canada	1991	1,308	5,874
	1996	1,695	4,541
Federal Republic of Germany	1990	6,984	11,821
	1996	7,569	11,445
France	1990	2,056	6,870
	1996	2,078	9,351
Japan	1992	3,529	3,150
	1996	4,231	3,014
Sweden	1991	3,529	2,400
	1997	4,148	5,551
Switzerland	1985	3,000	2,900
	1995	4,506	5,774
United Kingdom	1991	1,500	2,900
	1996	1,059	2,609
United States	1990	3,000	14,900
	1995	3,379	18,901
All developing countries	1990	2,700	71,300
	1996	9,323	230,696
Select countries			
Brazil	1992	566	7,110
	1995	797	6,322
China	1989	379	15,966
	1997	379	145,000
Colombia	1987	–	1,041
	1995	302	2,220
Hong Kong, China	1991	500	2,828
	1997	500	5,067
Indonesia	1988	–	1,064
	1995	313	3,472
Philippines	1987	–	1,952
	1995	–	14,802
Republic of Korea	1991	1,049	3,671
	1996	4,806	3,878
Singapore	1986	–	10,709
	1995	–	18,154
Central and Eastern Europe	1990	400	21,800
	1996	842	121,601
World Total	1990	36,600	174,900
	1996	53,607	448,917

Source: Based on UNCTAD, World Investment Report (1998:3,4).

If we organize some of the evidence on financial flows according to the places where the markets and firms are located, we can see distinct patterns of concentration. The evidence on the locational patterns of banks and securities houses points to sharp concentration. For example, the worldwide distribution of the 100 largest banks and 25 largest securities houses in 1991 shows that Japan, the United States, and the United Kingdom accounted for 39 and 23 of each, respectively. This pattern persists in the late 1990s, notwithstanding multiple financial crises in the world and in Japan particularly.

The stock market illustrates this pattern well. From Bangkok to Buenos Aires, governments deregulated their stock markets to allow their participation in a global market system. And they have seen an enormous increase in the value of transactions. Yet there is immense concentration in leading stock markets in terms of worldwide capitalization – that is, the value of publicly listed firms. The market value of equities in domestic firms confirms the leading position of a few cities. In September 1987, before the stock market crisis, this value stood at US$2.8 trillion in the United States and at US$2.89 trillion in Japan. Third ranked was the United Kingdom, with US$728 billion. The extent to which these values represent extremely high levels is indicated by the fact that the next largest value was for West Germany, a major economy where capitalization stood at US$255 billion, a long distance from the top three. What these levels of stock market capitalization represent in the top countries is indicated by a comparison with gross national product (GNP) figures: in Japan, stock market capitalization was the equivalent of 64 percent; in the United States, the equivalent of 119 percent; in the United Kingdom, the equivalent of 118 percent of GNP; and in Germany, 23 percent of GNP.

[. . .]

The concentration in the operational side of the financial industry is made evident in the fact that most of the stock market transactions in the leading countries are concentrated in a few stock markets. The Tokyo exchange accounts for 90 percent of equities trading in Japan. New York accounts for about two-thirds of equities trading in the United States; and London accounts for most trading in the United Kingdom. There is, then, a disproportionate concentration of worldwide capitalization in a few cities.

[. . .]

Offshore banking centers

Offshore financial centers are another important spatial point in the worldwide circuits of financial flows. Such centers are above all else tax shelters, a response by private sector actors to government regulation. They began to be implemented in the 1970s, although international tax shelters have existed in various incipient forms for a long time. The 1970s marked a juncture between growing economic internationalization and continuing government control over the economy in developed countries, partly a legacy of the major postwar reconstruction efforts in Europe and Japan. Offshore banking centers are, to a large extent, paper operations. The Cayman Islands, for example, illustrate some of these issues. By 1997, they were ranked as the seventh largest international banking operation in the world and the fifth largest financial center after London, Tokyo, New York, and Hong Kong, according to International Monetary Fund (IMF) data. They also were still the world's second largest insurance location with gross capital of US$8 billion in 1997. But even though that tiny country supposedly has well over 500 banks from all around the world, only 69 banks have offices there, and only 6 are "real" banks for cashing and depositing money and other transactions. Many of the others exist only as folders in a cabinet.

These offshore centers are located in many parts of the world. The majority of Asian offshore centers are located in Singapore and Hong Kong; Manila and Taipei are also significant centers. In the Middle East, Bahrain took over from Beirut in 1975 as the main offshore banking center, with Dubai following as a close second. In 1999, Abu Dhabi made a bid to create its own offshore financial center on Saadiyat Island. In the South Pacific, we find major centers in Australia and New Zealand and smaller offshore clusters in Vanuatu, the Cook Islands, and Nauru; Tonga and Western Samoa are seeking to become such centers. In the Indian Ocean, centers cluster in the Seychelles and in Mauritius. In Europe, Switzerland tops the list, and Luxembourg is a major center; others are

Cyprus, Madeira, Malta, the Isle of Man, and the Channel Islands. Several small places are struggling to compete with established centers: Gibraltar, Monaco, Liechtenstein, Andorra, and Campione. The Caribbean has Bermuda, the Cayman Islands, Bahamas, Turks and Caicos, and the British Virgin Islands.

Why do offshore banking centers exist? This question is especially pertinent given the massive deregulation of major financial markets in the 1980s and the establishment of "free international financial zones" in several major cities in highly developed countries. The best example of such free international zones for financial activity is the Euromarket, beginning in the 1960s and much expanded today, with London at the center of the hiromarket system. Other examples, as of 1981, were international banking facilities in the United States, mostly in New York City, that allowed U.S. banks to establish special adjunct facilities to accept deposits from foreign entities free of reserve requirements and interest rate limitations. Tokyo, finally, saw the development of a facility in 1986 that allowed transactions in the Asian dollar market to be carried out in that city; this meant that Tokyo got some of the capital being transacted in Hong Kong, Singapore, and Bahrain, all Asian dollar centers.

Compared with the major international centers, offshore banking centers offer certain types of additional flexibility: secrecy, openness to "hot" money and to certain "legitimate" options not quite allowed in the deregulated markets of major financial centers, and tax minimization strategies for international corporations. Thus, offshore centers are used not only for Euromarket transactions but also for various accounting operations aimed at tax avoidance or minimization.

In principle, the Euromarkets of London are part of the offshore markets. They were set up to avoid the system for regulating exchange rates and balance-of-payments imbalances contained in the Bretton Woods agreement of 1945. The Bretton Woods agreement set up a legal framework for the regulation of international transactions, such as foreign currency operations, for countries or banks wanting to operate internationally. Euromarkets were initially a Eurodollar market, where banks from the United States and other countries could do dollar transactions and avoid U.S. regulations. Over the last decade, other currencies have joined.

In finance, offshore does not always mean overseas or foreign; basically the term means that less regulation takes place than "onshore" – the latter describing firms and markets not covered by this special legislation. The onshore and offshore markets compete with each other. Deregulation in the 1980s brought a lot of offshore capital back into onshore markets, especially in New York and London – a not insignificant factor in convincing governments in these countries to proceed with deregulation of the financial markets in the 1980s.

[. . .]

In brief, offshore banking centers represent a highly specialized location for certain types of international financial transactions. They are also a buffer zone in case the governments of the leading financial centers in the world should decide to reregulate the financial markets. On the broader scale of operations, however, they represent a fraction of the financial capital markets now concentrated in global cities.

CONCLUSION: AFTER PAX AMERICANA

The world economy has never been a planetary event; it has always had more or less clearly defined boundaries. Moreover, although most major industries were involved throughout, the cluster of industries that dominated any given period changed over time, contributing to distinct structurings of the world economy. Finally, the institutional framework through which the world economy coheres has also varied sharply, from the earlier empires through the quasi-empire of the Pax Americana – the period of U.S. political, economic, and military dominance – and its collapse in the 1970s.

It is this collapse of the Pax Americana, when the rebuilt economies of Western Europe and Japan reentered the international markets, that we see emerging a new phase of the world economy. There is considerable agreement among specialists that in the mid-1970s new patterns in the world economy became evident. First, the geographical axis of international transactions changed from North–South to East–West. In this process, significant parts of Africa and Latin American became unhinged from their hitherto strong ties with world markets in commodities and raw materials. Second was a sharp increase in the weight of FDI in services and in the role played by international financial markets. Third was the breakdown of the Bretton Woods agreement, which had established the institutional framework under which the world economy had operated since the end of World War II.

These realignments are the background for understanding the position of different types of cities in the current organization of the world economy. A limited but growing number of cities are the sites for the major financial markets and leading specialized services firms. And a large number of other major cities have lost their role as leading export centers for manufacturing, precisely because of the decentralization of production.

"Power in Place: Retheorizing the Local and the Global"

from John Eade and Christopher Mele (eds), *Understanding the City* (2002)

Michael Peter Smith

Editors' Introduction

Our readings in this section thus far by two key global theorists, John Friedmann and Saskia Sassen, have presented us both the architecture of the world system of cities and the underlying global political and economic forces that have given its shape. As both contend, the fate of individual cities – from their economies to the quality of life in everyday social interaction – is increasingly tied to global finance and the producer services that facilitate the rapid exchange of capital. In many ways, then, global forces exist "out there," forcing localities – and their citizens – to respond.

Michael Peter Smith is the leading figure behind the call for a less deterministic approach to globalization. His work takes issue with the "top down" view many global theorists take and contends that changes in the social fabric of neighborhoods and the realm of everyday life cannot be easily understood through a view from outside or above. In this selection, Smith outlines an alternative approach to studying globalization – one that treats "the local" as a site where global processes are experienced and lived in varied and often unpredictable ways. First, he argues against the many theories that commonly use the term, "global–local," and tend to separate (global, political–economic) causal processes from (local, cultural and social) effects. Arguments that rely on the global–local dichotomy result in a rash determinism: local communities are seen as dominated by global capitalist economics. Localities, too, are poorly theorized – most seem like "sitting ducks" subject to the mostly negative effects of global capitalism. If certain localities are able to resist globalization, it is often said that such communities exist "outside the logic of globalization." Either way, there is, as Smith writes, "a systemic disjunction between local and global social processes."

Is globalization really experienced as something abstract or "out there"? Do global flows of finance capital now require social groups who reside within communities to act in specific and known ways? Smith believes not and asks that urbanists and global theorists alike adopt a social constructionist view in which the global–local interplay is a site of inquiry and investigation rather than a predictable or determined given. Smith sees the global–local as a criss-crossing of local and extralocal cultural, economic, and social processes that momentarily become "fixed" in particular locations. He calls for a "new way of conceptualizing the locality as the place where localized struggles, and alternative discourses on the meanings of 'global conditions' are played out." In doing so, we will avoid seeing globalization as "an inevitable force operating behind people's backs independently of their actual material and discursive practices."

Smith's approach would require researchers to examine up close and in detail the conditions by which globalization is "made meaningful in particular places at particular times." Localities have indeed been transformed by flows of capital and people and new and quicker means of communication. Many aspects of everyday life – from shopping for clothes, listening to and thinking about the day's news, and buying a car to purchasing a meal or choosing a CD – are no longer local phenomena. Smith sees these everyday practices as sites where the global and local intersect; hence globalization is experienced "on the ground" in different ways by different social groups and individuals. The job of urban researchers is to show how.

In one example, Smith points to a British study of Muslim neighborhoods in cities throughout the United Kingdom. Religious fundamentalism among the Muslim residents might be read as a cultural reaction or defensive position to the conditions of living in an increasingly global capitalist society. The authors find differently and Smith concurs. The routine practices for the Muslim residents instead reflect a criss-crossing of global and local knowledge and meanings produced by transnational linkages, including religious ceremonies, telephone conversations, television and radio programs, newspaper accounts, videos, and music. The practice of Islam among Muslim residents within British cities, then, is "viewed not as a purely local or 'tribal' reaction to the globalization of capitalism but as a transnational cultural formation, as much involved in the transnationalization of information, cultural exchange, and network formation, as are the networks of financial transactions that comprise what we now regard as the globalization of capital."

The next reading in this section, "City Life: West African Communities in New York," by Paul Stoller and Jasmin Tahmaseb McConatha is another example of recent research on the criss-crossing of the local and the global.

Michael Peter Smith is Professor and Associate Chair of the Department of Human and Community Development at the University of California, Davis. He is author of numerous articles on urban social theory, globalizations and transnationalism, international migration and urban racial and ethnic politics. His books include *Transnational Urbanism: Locating Globalization* (Malden, MA: Blackwell 2001); *Transnationalism from Below* (New Brunswick; NJ: Transaction Press 1998); *The Bubbling Cauldron: Race, Ethnicity, and the Urban Crisis* (with Joe R. Feagin), (Minneapolis: University of Minnesota Press, 1995); *The Capitalist City: Global Restructuring and Community Politics* (Oxford: Blackwell 1987); and, *The City and Social Theory* (Oxford: Blackwell 1980).

Readers may also wish to consult Alejandro Portes, "Globalization from Below: The Rise of Transnational Communities," in Don Kalb, Marco van der Land, and Richard Staring, *The Ends of Globalization: Bringing Society Back In* (Lanham, MD: Rowman and Littlefield, 2000).

Two dominant themes have informed the construction of the "local" in the discourse on the global–local interplay in urban studies. The local has been frequently represented as the cultural space of embedded communities and, inversely, as an inexorable space of collective resistance to disruptive processes of globalization. In writings ranging from classical urban sociology to contemporary discourses on globalization and place, the "locality" has been used to signify an embedded community. "Community" in turn is represented as a static, bounded, cultural space of being where personal meanings are produced, cohesive cultural values are articulated, and traditional ways of life are enunciated and lived. In classical urban sociological thought the "urban" served as a surrogate for the rational instrumentalism of the capitalist market and the bureaucratization of the life-world – the transformation of *gemeinschaft*-like social relations into the mediated impersonal ties of a *gesellschaft*-like urban society. In the contemporary period the "urban" has been replaced by the "global" as a metaphor for the central outside threat to the primary social ties binding local communities. "Globalization" has been represented as a new form of capitalist (post)modernity, a process inherently antagonistic to the sustainability of local forms of social organization and meaning-making.

I begin this essay by arguing that the schemata used by urban structuralists such as David Harvey and Manuel Castells to conceptualize the global–local connection reifies the theoretical terms in this dialectic, privileging (while marginalizing) the local as the place of culture or "community," while marking the global as the dynamic economic space of capital and information flows. I go on to argue that, ironically, key conceptualizations informing ethnographic practice in postmodern cultural studies and even the conceptualization of "everyday resistance" informing the politics of everyday life problematic in urban studies also rely upon a binary dichotomization of the global vs. the local.

In the remainder of this chapter I further disrupt this binary by showing the myriad ways in which social networks and practices that are transnational in scope and scale are constituted by their interrelations with, and thus their groundedness inside, the local. I then ask what constraints have been placed on urban theory and research by the persistence of this dichotomous way of thinking and what can be done, both epistemologically and methodologically, to overcome these constraints. The aim is to frame a more dynamic conception of locality, one more likely to capture the connections linking people and places to the complex and spatially dispersed transnational communication circuits now intimately affecting the ways in which everyday urban life is experienced and lived. Writing from a social constructionist standpoint, I recommend that the sociological imagination be enriched by an engagement with a multi-sited mode of ethnographic research that is historically contextualized and recognizes the importance of everyday practices without romanticizing the local or losing sight of the structures of power/knowledge created by human practice.

THE CONFINES OF THE LOCAL IN URBAN STRUCTURALISM

The representation of the local as a once firmly situated cultural space of community-based social organization now rendered unstable by the global dynamism of capitalist (post)modernity is well captured in the following excerpt from David Harvey's *The Condition of Postmodernity* (1989):

> Movements of opposition to the disruptions of home, community, territory, and nation by the restless flow of capital are legion . . . Yet all such movements, no matter how well articulated their aims, run up against a seemingly immovable paradox. . . . [T]he movements have to confront the question of value and its expression as well as the necessary organization of space and time appropriate to their own reproduction. In so doing, they necessarily open themselves to the dissolving power of money as well as to the shifting definitions of space and time arrived at through the dynamics of capital circulation. Capital, in short, continues to dominate, and it does so in part through superior command over space and time. The "othernesses" and "regional resistances" that postmodernist politics emphasize can flourish in a particular place. But they are all too often subject to the power of capital over the co-ordination of universal fragmented space, and the march of capitalism's global historical time that lies outside the purview of any particular one of them.

[. . .]

In this narration of the waning power of local cultural formations in the face of capitalist globalization the author of social change is never in doubt. Capital's superior global command over resources to reorganize time and space is opposed to the disorientation of defensive "local" social movements representing the interests of home, community, place, region, and even nation. The latter are represented as static forms of social organization, efforts to organize social life around "being" rather than "becoming." Defensive place-based movements are represented as cultural totalities expressing entirely place-bound identities in a world in which the dynamic flows of globalization exist entirely outside their purview. Oppositional movements representing "locality" may win some battles in what Harvey terms "postmodern politics." But they confront a restless adversary, whose processes of accumulation thrive on constantly disrupting the spatial and temporal arrangements upon which stable forms of local social organization might be constructed. Thus, in this grand narrative, in the final analysis, "capital" is the only agent of social change. Capitalist economic dynamics continue to dominate localities whose specific histories are relegated to the dustbin, rendering them fit only for periodic bouts of reactionary politics. As for the role of people in this grand narrative, we never know who lives, works, acts, and dies in Harvey's urban spaces since people are seldom represented as anything other than nostalgic romantics or cultural dupes.

Manuel Castells is another urban theorist who has represented the local as a political space of social movements defending threatened cultural and political meanings placed under siege by global economic and technological restructuring. At first glance his view of locality appears to be different from Harvey's theorization. In Castells' work, late-modernity is represented as an informational mode of development, a "space of flows" which accelerates global financial and informational linkages, converts places into spaces, and threatens to dominate local processes of cultural meaning. While the space of flows is a global space of economic and technological power, the space of cultural meaning and experience remains local. The global networks of wealth and power accumulate and exchange information instantaneously as a central source of institutional power. This boundary-penetrating process disrupts the sovereignty of the nation-state and threatens to marginalize the life-worlds of local cultural "tribes." As Castells (1984) expresses this argument:

> On the one hand the space of power is being transformed into flows. On the other hand, the space of meaning is being reduced to micro-territories of new tribal communities. In between cities and societies disappear. Information tends to be dissociated from communication. Power is being separated from political representation. And production is increasingly separated from consumption, with both processes being piece-mealed in a series of spatially distinct operations whose unity is only recomposed by a hidden abstract logic. The horizon of such a historical tendency is the destruction of human experience, therefore of communication, and therefore of society.

Following from this logic is a structuralist dialectic of domination and resistance. Global domination produces local resistance. Resistance to globalization is tied not to the agency of specific actors confronting unique historical conjunctures but to the very structural dynamic of the technological revolution which threatens to render the local "tribes" irrelevant to the new informational world that has come into being. In his view the nation-state is disintegrating as a space of internalized identity formation. Rather, two modes of identity formation are said to give rise to different types of communal resistance to globalization – "project identities" and "resistance identities." The former are viewed as encompassing such bases of social identity as religious fundamentalism and ethnic nationalism. It is the structural connection of vastly different cultural formations as "bypassed" cultural spaces, forged in the context of disintegrating national civil societies, that allows Castells to lump together social movements as diverse as the Zapatistas in Mexico, the militia and patriot movements in the USA, and the *Aum Shinrikyo* cult in Japan, treating them as functionally equivalent "social movements against the new global order," despite their historical differences in goals, ideologies, national and local contexts, and specific histories.

[. . .]

In one key respect Harvey and Castells converge – namely, both represent the local as a cultural space of communal understandings, a space where meaning is produced entirely outside the global flows of money, power, and information. People in these narrow social worlds make sense of their world and form their political identities in a culturally bounded microterritory, the locality. These local cultural meanings, in turn, are represented as generating identities inherently oppositional to the global restructuring of society and space. For both, then, "place" is understood as the site of cohesive community formations existing outside the logic of globalization. While Harvey and Castells differ in their assessments of whether globalization will annihilate or defensively energize these community formations, they both maintain a systemic disjunction between local and global social processes.

THE LIMITS OF THE POSTMODERN TURN TO THE LOCAL

In seeking to displace the reliance of social theory on grand narratives of macro-social development, such as those of Harvey and Castells, some social theorists embracing the label "postmodern" have turned to what Foucault (1977) has termed the "essentially local character of criticism." In my view, despite their rhetorical gestures in the direction of the desirability of grounded field work, some proponents of this turn to the local have posited an equally grand theory of local knowledge which privileges the ethnographic conversation as the only reliable route to personal knowledge, and personal knowledge as the only reliable measure of the "partial truths" about the workings of the world. This is the case for instance when postmodern anthropologist Stephen Tyler (1987) asserts, with totalistic rhetorical flourish, that "discourse is the maker of this world, not its mirror."

The image which first comes to mind in this move, which some might regard as "postcolonial" is the "travel writing" found in colonial anthropology which relied upon the discourse of thick ethnographic "description" to fix, construct, and hence master colonized objects. It vividly detailed the habits, customs, speech acts, and bodily practices of conquered peoples in order to comprehensively "know," and hence implicitly control, the colonial subject. Ironically, however, the simple inversion of this hierarchy has often been the case in the postmodern cultural turn in urban ethnography in which the "thick description" of ethnographic narrative is relied upon to romanticize "the postcolonial subject" as embodied in the everyday practices of such socially constructed "communities" as urban taggers, street gangs, or various insurgent social movements.

[. . .]

RETHINKING THE ETHNOGRAPHIC TURN

Must the ethnographic imagination be reduced to the pursuit of idiosyncratic *petit* narratives which then become ventriloquized voices of postcolonial resistance? This is not necessarily the case.

[. . .]

On the ground, there has been over a decade of richly textured transnational ethnographic story-telling. Throughout the 1990s, numerous multi-locational practitioners of a transnational form of ethnographic field work produced a body of research uniquely sensitive to the social construction of contextuality as well as identity. These new modes of ethnographic inquiry successfully asked quite big questions to little people and obtained very intriguing results. For example, by asking transnational migrants involved in circular migration networks to first construct their understanding of the opportunities and constraints they faced in the world(s) in which they lived, and then to talk about the ways in which they appropriated, accommodated to, or resisted the forms of power and domination, opportunity and constraint that they experienced as they traversed political and cultural borders, this mode of ethnographic practice opened up a discursive space for contextually situated ethnographic narratives that captured the emergent character of transnational social practices.

[. . .]

But is the alternative of simply listening closely to and "recording" ordinary peoples' characterizations of globalization enough to explain the links they may have to a changing structural context? In my view there is also a clear danger inherent in the postmodern ethnographic effort to assume a subject position of tentativeness, if not authorial passivity in "recording" the pure, authentic voice of the "marginal" and the "different." Cultural voices are historically contingent, not timelessly pure. The temptation to capture the essence of a "local voice", inscribing it as a heroic individual or collective challenge to the oppressive forces of global modernity precisely mirrors the problem of idealizing the "other culture" as a radically strange entity which characterized colonial discourse.

My version of social constructionist urban analysis is at home with the view of cultural practice as adaptive, inventive, and multi-valent. It thus rejects the view that "cultures" are closed local systems or singular wholes that can be penetrated by the simple ethnographic act of "being there" as much as it questions the view of "the global" as a single modernist/capitalist social space that can be captured only by a pure act of political–economic theoreticism. By my reading it is necessary to question all abstractions that tend to ignore the historically specific conditions of cultural production as these become localized, interrelated, and mutually constitutive in particular places at particular times.

TRANSNATIONAL NETWORKS AND THE LOCALIZATION OF POWER/KNOWLEDGE

[. . .]

Since human agency operates at multiple spatial scales, and is not restricted to "local" territorial or socio-cultural formations, the very concept of the "urban" thus requires re-conceptualization as a social space that is a pregnant meeting ground for the interplay of diverse localizing practices of regional, national, transnational, and even global scale actors, as these wider networks of meaning, power, and social practice come into contact with more purely locally configured networks, practices, meanings, and identities.

This way of envisioning the process of localization thus locates "globalization" in the realm of social practice and situates the global–local interplay in historically specific milieus. It extends the meaning of the global–local nexus to encompass not just the social actions of "global capitalists" interacting with "local communities" but of the far more complex interplay of cultural, economic, political, and religious networks that operate at local, trans-local, and transnational social scales but which intersect in particular places at particular times. Closer study of this interplay will, in my view, enable urban researchers to explain the formation of new "subject positions" and give due attention the multiple patterns of accommodation and resistance to dominant power relations and discursive spaces, particularly the patterns of dominance entailed in the current discourse on "globalization" itself.

My view of "culture" as a fluid and dynamic set of understandings produced by discursive and material practices is not restricted to the understandings developed in closed and self-contained local communities. The circuits of communication and everyday practice in which people are implicated are resources as well as limits. People's everyday lives are sites of criss-crossing communication circuits, many of which transcend the boundaries of local social and political life. They constitute sometimes separate, sometimes overlapping, and sometimes competing terrains for the contestation as well as the reproduction of cultural meanings, for resistance as well as accommodation to dominant modes of power and ideology. My emphasis upon the social construction of social relations through acts of accommodation and resistance to prevalent modes of domination has significant implications for urban research. We need a new way of conceptualizing the locality as the place where localized struggles, and alternative discourses on the meanings of "global conditions" are played out. This effort brings to the forefront of urban research the fact that "globalization" is a historical construct, not an inevitable economic force operating behind people's backs independently of their actual material and discursive practices. It requires examining at close range but in rich contextual detail the specific historical and contemporary conditions by which "global conditions" are made meaningful in particular places at particular times. Cities thus may be usefully conceptualized as local sites of cultural appropriation, accommodation, and resistance to "global conditions" as experienced, interpreted, and understood in the everyday lives of ordinary people and mediated by the social networks in which they are implicated.

In considerable measure, as a result of the growing transnational interconnectivity among localities, the "local" itself has become transnationalized as transnational modes of communication, streams of migration, and forms of economic and social intercourse continuously displace and relocate the spaces of cultural production. The rapid socio-cultural transformation of localities by the accelerated flows of transnational migration has thus significantly modified the social relations of "community" that had grounded traditional ethnographies in more or less convergent social spaces of cultural production and place. These growing disjunctures require us to develop new modes of urban research and refine old methods of social inquiry to better fit the new spaces of cultural production and reproduction in these times of growing transnational interconnectivity.

[. . .]

REIMAGINING THE POLITICS OF EVERYDAY LIFE

[. . .]

In my view, at the current transnational moment "the politics of everyday life" needs to be opened up more widely as a social and political imaginary. The "everyday" needs to be freed from its association with purely local phenomena. In transnational cities people's everyday urban experiences are affected by a wide variety of phenomena, practices, and criss-crossing networks which defy easy boundary setting. Multiple levels of analysis and social practice now inform the urban politics of everyday life throughout the world.

[. . .]

Consider the following example of emergent transnational urban practices and identities. Sociologists Martin Albrow, John Eade, Jorg Dürrschmidt, and Neil Washbourne (1997) have written insightfully on the social construction of the boundaries of imagined community in British Muslim neighborhoods in cities throughout the United Kingdom. Rather than viewing religious fundamentalism as a local expression of belonging and identity framed against economic globalization, as in the work of Barber (1995) and Castells (1997), these scholars connect the rhetorics of belonging found in various local British Muslim enclaves to a wider social construction of Islamic community (*umma*) transmitted by transnational religious and cultural networks. Everyday life in the Muslim neighborhoods is infused with knowledge and meanings produced in these transnational networks and encountered in the local neighborhoods on a daily basis. The social construction of belonging to a transnational Islamic community is produced and transmitted through a transnational network of social and technological linkages including religious ceremonies, telephone conversations, television and radio programs, newspaper accounts, videos, and music. As these scholars conclude, in everyday visits to relatives and friends, in interactions at work, and in other neighborhood forms of community involvement, local Muslims employ this transnational network, which is physically absent but hardly spiritually distant, to socially construct their "locality." In this construction Islam is viewed not as a purely local or "tribal" reaction to the globalization of capitalism but as a transnational cultural formation, as much involved in the transnationalization of information, cultural exchange, and network formation, as are the networks of financial transactions that comprise what we now regard as the globalization of capital.

In short, everyday life is neither a fixed spatial scale nor a guaranteed site of local resistance to more global modes of domination, whether capitalist or otherwise. Rather, in today's transnational times our everyday life-world is one in which "competing discourses and interpretations of reality are already folded into the reality we are seeking to grasp." Grasping this sort of reality now requires us to develop a transnational imaginary and to fashion perceptual tools capable of making sense of the new identities emerging from this politics of representation and boundary-setting.

[. . .]

TRANSNATIONALISM AND THE POLITICS OF PLACE-MAKING

In an effort to unbind the conceptualization of "place" from the conflation of locality and community, critical urban geographer Doreen Massey has advanced an imaginative response to the question of the interplay of the global and the local. Massey's view of place is decisively fluid. On the one hand, her critique of David Harvey's conception of "time-space compression" warns against the tendency to view the implosion of time and space that Harvey terms the condition of postmodernity as equally accessible to all. In her view different individuals and social groups are differently positioned vis-a-vis the flows and interconnections that constitute the "globalization" of capital and culture (1993). On the other hand, these flows and interconnections intersect in particular places at particular times, giving each place its own unique dynamism and making it possible for us to envision a "global" or "progressive" sense of place.

Theoretically, Massey depicts localities as acquiring their particularity not from some long internalized history or sedimented character but from the specific interactions and articulations of contemporary "social relations, social processes, experiences, and understandings" that come together in situations of co-presence, "but where a large portion of those relations . . . are constructed on a far larger scale than what we happen to define for that moment as the place itself . . .". When understood as articulated moments among criss-crossing networks of social relations and understandings, places do not possess singular but multiple and contested identities. Place-making is shaped by conflict, difference, and social negotiation among differently situated, and at times antagonistically related social actors, some of whose networks are locally-bound, others whose social relations and understandings span entire regions and transcend national boundaries. Massey, in short, provides key theoretical ingredients for conceptualizing the transnational interconnectivity I seek to locate in this essay.

Massey's approach would trace the trajectories of both residents' and non-residents' routes through a place as well as identifying "their favourite haunts within it, the connections they make (physically, or by phone or post, or in memory and imagination) between here and the rest of the world . . .". This is a good way to grasp the fluidity, diversity, and multiplicity of any place and the ways in which social relations affecting that place are stretched out over space and memory (i.e., time). It is also a good way to avoid an essentialist construction of localities as closed communities, as ontological "insides," constructed against a societal or global "outside" by tracing connections between the locality and what Arjun Appadurai (1991) has called the "global ethnoscape."

Gupta and Ferguson (1997a and 1997b) have offered another clear-headed critique of the scholarly conflation of place and culture that is germane to my effort to contextualize emergent transnational social relations and to situate them in the field of urban studies. Gupta and Ferguson point out that representations of localities as cohesive community formations fail to recognize and deal with a variety of boundary penetrating social actors and process now very much a part of the transnational world in which we live. Left out of such localized communitarian narratives are the border dwellers who live along border zones separating localities, regions, and nation-states. These social actors engage with actors and networks based on the other side of juridical borders in processes of intercultural borrowing and lending which anthropologists now call "transculturation." The "locality as community" problematic is equally inattentive to the socio-cultural and political implications of the growing number of border crossers – i.e., migrants, exiles, refugees, and diasporas – who now orchestrate their lives by creating situations of co-presence that link social networks across vast geographical distances across the globe. Such border penetrating processes go a long way toward helping explain, though they by no means exhaust, the difference-generating relations of power that constitute cultural and political identity and difference within localities defined as both political jurisdictions and as socio-cultural spaces.

[. . .]

Viewing locality (like ethnicity and nationality) as complex, contingent, and contested outcomes of political and historical processes, rather than as timeless essences also challenges the theoretical framing of "locality" as an inexorable space of resistance to globalization. Instead of opposing autonomous local cultures, be they tribes, militias, urban formations, or regions to the economic domination of global capital, the homogenizing movement of cultural globalization, or the hegemonizing seductions of global consumerism, Gupta and Ferguson recommend paying close attention to the ways in which dominant global cultural forms may be appropriated and used or even significantly transformed "in the midst of the field of power relations that links localities to a wider world."

Germane to this larger question, are several more particular ones:

- How are perceptions of locality and community discursively constructed in different time-space configurations?
- How are the understandings springing from these perceptions internalized and lived?
- What role in producing politically salient differences within localities is played by the cultural, political, and economic connections localities have with worlds "outside" their borders that configure their interdependence?
- What roles do the global and local mass media play in framing the understandings and practices within socially constructed communities and their constitutive fields of otherness?

Having raised these questions I would further suggest that we leave open to sociological and historical investigation the character of the contextualizing socio-spatial interdependence in which particular localities are enmeshed at particular times. Specifically, I agree with Gupta and Ferguson that it is possible for local interventions to "significantly transform" dominant cultural forms. I would therefore leave open the question of whether or not the criss-crossing relations of power that come together on the ground in localities must necessarily be understood hierarchically. When Gupta and Ferguson turn from the question of the patterning of power relations across space to the issue of the social construction of space as place, they acknowledge the contingency of power relations.

The open-ended questions they raise in response to this issue are, in my view, as germane to questions of the spatial distribution of power as they are to issues of identity and place-making. These questions are: "With meaning-making understood as a practice, how are spatial meanings established? Who has the power to make places of spaces? Who contests this? What is at stake?"

It is precisely questions such as these that can move urban researchers interested in the social construction of locality beyond essentialist assumptions about the equivalence of locality and culture. For example, research on the politics of urban heritage has produced a spate of studies on "the making of place" by a wide variety of political actors including local neighborhood groups, government officials, and business interests as well as wider networks of social practice such as architectural activists, historic preservationists, and global developers. In particular places these actors collude and collide in contests over the cultural meanings of place. Sociologically, historically, and ethnographically grounded case studies bring into focus the issue of the politics of representation thereby modifying a discourse on globalization and community that has been dominated by agency-less narratives of urban and regional change that tend to exclude non-capitalist actors and their representations of space and place from consideration.

[...]

BRINGING THE SOCIAL BACK IN

In this essay I have sought to move urban studies beyond naturalistic constructions of "locality" which view the local as an inherently defensive community formation. I have tried to show that even the most material elements of any locality are subject to diverse readings and given different symbolic significance by differently situated social groups and their corresponding discursive networks. The result is a highly politicized social space where representations of place are constructed and contested. I have argued that the prevailing structuralist schemata used to conceptualize the global–local connection by leading urban theorists such as Harvey and Castells has tended to reify the terms in this dialectic. In so doing they have reproduced a totalizing binary framework in which the global is equated with the abstract, universal, and dynamic (i.e., "capital"), while the local is invested with concreteness, particularity, and threatened stability (i.e., "community"). Such a discourse of capital versus community treats the global a priori as an oppressive social force while constructing localities in more positive, albeit more static if not anemic terms.

Ironically, key conceptualizations informing ethnographic practice in postmodern cultural studies and even the conceptualization of "everyday resistance" informing the politics of everyday life problematic in urban studies also rely upon binary dichotomizations of global power versus local culture which overlook the ways in which transnational social networks and practices are constituted by their interrelations with, and thus their groundedness inside the local. The theoretical frames of reference critiqued in this essay tend to ignore the considerable interplay of spatial scales and discursive practices to be found in any locality. They elide or underestimate the intricacy involved in sorting out the social interactions and processes at multiple spatial scales that constitute the complex politics of place-making under contemporary conditions of transnational interconnectivity.

To overcome the conceptual confinement of the global–local dichotomy and restimulate the sociological imagination, I recommend the further refinement of a multi-sited, trans-local mode of transnational ethnographic practice, inflected with the domination-accommodation-resistance motif, and combined with a historicized approach to political-economic and social relations. This recombinant method of social inquiry can be used to make sense of the power relations and meaning-making practices I have elsewhere termed "transnational urbanism." The social spaces through which social actors move and within which they operate to give meaning to "place" are crucial and increasingly trans-local and multi-scaled, thus making trans-local ethnography and carefully historicized political-economic and social research both necessary and possible. A fusion of these heretofore distinct, if not opposed, approaches to urban studies is a good way to bring "the social" back into urban theory and research, where it belongs.

REFERENCES

Albrow, Martin, John Eade, Jorg Dürrschmidt, and Neil Washbourne. 1997. "The Impact of Globalization on Sociological Concepts: Community, Culture and Milieu." In John Eade, ed., *Living the Global City: Globalization as Local Process.* London: Routledge: 20–36.

Appadurai, Arjun. 1991. "Global Ethnoscapes: Notes and Queries for a Transnational Anthropology." In R. G. Fox, ed., *Recapturing Anthropology: Working in the Present.* Santa Fe: School of American Research Press: 191–210.

Barber, Benjamin R. 1995. *Jihad vs. McWorld.* New York: Times Books.

Castells, Manuel. 1984. "Space and Society: Managing the New Historical Relationships." In Michael Peter Smith, ed., *Cities in Transformation.* London: SAGE: 235–60.

——. 1997. *The Power of Identity.* Oxford: Blackwell.

Foucault, Michel. 1977. *Power/Knowledge.* C. Gordon, ed. New York: Pantheon Books.

Gupta, Akhil and James Ferguson. 1997a. "Beyond 'Culture': Space, Identity and the Politics of Difference." In Akhil Gupta and James Ferguson, eds, *Culture, Power and Place.* Durham, NC: Duke University Press, 33–51.

——. 1997b. "Culture, Power and Place: Ethnography at the End of an Era." In Akhil Gupta and James Ferguson, eds, *Culture, Power and Place.* Durham, NC: Duke University Press, 3–29.

Harvey, David. 1989. *The Condition of Postmodernity.* Oxford: Blackwell.

Massey, Doreen. 1993. "Power-Geometry and a Progressive Sense of Place." In J. Bird, et al., eds, *Mapping the Futures: Local Cultures, Global Change.* London: Routledge: 59–69.

Tyler, Stephen. 1987. *The Unspeakable.* Madison: University of Wisconsin Press.

"City Life: West African Communities in New York"

from *Journal of Contemporary Ethnography* (2001)

Paul Stoller and Jasmin Tahmaseb McConatha

Editors' Introduction

A key facet of globalization is international migration – the movement of large numbers of people across borders. As in the past, contemporary immigrants to the United States build communities in their new cities and towns. Yet these communities differ from the ethnic enclaves of the late nineteenth and earlier twentieth century migrations. Due to relatively inexpensive airfares, modern communications, and the availability of goods and services on a global basis, many of today's immigrants readily construct "transnational communities" – spaces, both imagined and real, in which the worlds of homeland and host country intermingle. According to Peggy Levitt, who has written extensively on the topic, "transnational social expectations, cultural values, and patterns of human interaction . . . are shaped by more than one social, economic, and political system" (2001: 197).

In this selection, Paul Stoller and Jasmin Tahmaseb McConatha present a community study of West African street vendors in New York City. The traders' community is truly transnational. West African traders have constructed what Stoller and McConatha label, an "invented community," linked by shared economic pursuits and Africanity. Traders form networks based on common interests of religion and ethnicity. They are also linked to their families and to economic networks back home in West Africa. While living in New York City, the traders participate in an imagined community which draws on their collective memory of home and is reiterated through Islamic religious practices.

Stoller and McConatha organize their chapter to clearly show how West African traders "live" transnationalism. They address the social, cultural, and psychological dimensions of West African city life, showing how Muslim traders have built communities from their personal and economic networks that span nations.

West African traders live in different neighborhoods throughout New York City but for most their initial days and weeks in the U.S. are spent in "vertical villages" – a handful of low-rent single room occupancy (SRO) residences in Manhattan populated by immigrants from West Africa. There, new arrivals are immediately linked to an intact fellowship of individuals with knowledge of the City's economic, social, and cultural terrain. Yet, for all its benefits fellowship cannot replace family. There is a palpable sense of loss of family connection among traders and their primary activity is to maintain contact with wives, parents, children, and siblings back home. Such contact includes letters, telephone calls, as well as remittances (money earned by traders and sent to family members in West Africa).

In addition to family ties, the practice of Islam ranks high among priorities for traders. Indeed, Islam bridges the social and cultural distance between West Africa and New York City – religious practices and cooperation and trust among fellow Muslims have empowered traders to cope with social isolation in America. Stoller and McConatha point out that Islam is the foundation of the traders' transnational communities: "personal networks in which kinship, ethnicity, and nationality affect the density of contact and degree of trust and co-operation are baseline community forms." Formal clubs and associations based on national origin also help sustain communities but, as the authors point out, these organizations tend to reflect class and status divisions from back home.

The existence of a transnational community clearly helps traders adjust to life in New York, to maintain ties with home, and to manage their businesses. It provides a necessary social structure for individual traders to tap into. How successfully or unsuccessfully individuals make use of the community is the final topic Stoller and McConatha address. They use the concept "cultural competence" to discuss the individual trader's ability "to adapt to an environment by making effective choices and plans and by controlling – as much as possible – the events and outcomes of daily life." Some traders have succeeded better than others due to unequal levels of cultural competence. Two forms of competence are key to managing life in New York City. Linguistic competence in English enables traders to construct transnational business, expand their trading operations, and "charm" customers. Cultural competence – knowledge of the diverse economic and social customs and value systems that operate in New York City – is as important, as some traders have learned the hard way.

Paul Stoller is Professor of Anthropology and Sociology at West Chester University. He has conducted research among Songhay-speaking peoples in the Republic of Niger and in New York City for more than two decades. He is the author of *Money Has No Smell: The Africanization of New York City* (Chicago: University of Chicago Press 2002).

Jasmin Tahmaseb McConatha is Professor of Psychology at West Chester University. Her research interests include life span development, cross-cultural psychology, and cultural influences on communication patterns.

See also Peggy Levitt, "Transnational Migration: Taking Stock and Future Directions," *Global Networks*, 1, 3 (2001), 195–216 and *The Transnational Villagers* (Berkeley, CA: University of California Press, 2001).

The notion of community has long been central to the ethnographic enterprise. In sociology, the landmark study of community was Robert and Helen Lynd's *Middletown* (1929). In their highly descriptive book, however, the Lynds did not attempt to construct a theory of community. Robert Redfield took up that task in the 1940s and 1950s when he published *The Folk Culture of Yucatan* (1941) and *The Little Community* (1955). Redfield depicted the community of Teplotzlan as bounded, harmonious, and homogenous. His work inspired the criticism of Oscar Lewis who had studied the same community but found a very different kind of society. Where Redfield uncovered harmony, homogeneity, and social adjustment, Lewis found violence, sociopolitical schisms, and social maladjustment (see Lewis 1951). As Sherry Ortner (1997) has recently pointed out, Lewis's criticism still rings true to contemporary scholars who have, for the past fifteen years, underscored the fragmentation and hybridity of sociocultural processes and organizations. Although the concept of community has been plagued with epistemological and conceptual problems, Ortner thinks it is a notion well worth preserving in the era of globalized, transnational settings. For her,

> the importance of community studies is this: such studies have the virtue of treating people as contextualized social beings. They portray the thickness of people's lives, the fact that people live in a world of relationships as well as a world of abstract forces and disembodied images.

In this article, we attempt to extend Ortner's (1997) insights to dispersed communities of West African street vendors in New York City. Several of the emergent "postcommunities" that Ortner identified conform to community forms found among West African traders in New York City. Congregating at the African market on 125th Street or now at the Malcolm Shabazz Harlem market, West African traders have constructed an "invented community" of "African brothers" linked by shared economic pursuits and Africanity. The traders also participate in what Ortner called *translocal communities*. Despite being dispersed over great distances, they are linked through common interests like religion or ethnicity. As mobile merchants who follow the circuit of African American and Third World cultural festivals, West African traders from New York City form periodic links of support and friendship with fellow Africans as they travel across the United States. They are also linked to their families and to economic networks back home in West Africa. Another form of community in which the traders participate is a *community of mind* – of memory – a community engendered primarily through Islam. No matter where they are, West African traders, most of whom are Muslims, try to pray five times a day and obey the various dietary and behavioral dictates of Islam. One of the central themes of Islam, in fact, is the "community of faithful," the *Ummah*. The themes of this community of the mind are reinforced through daily prayer, daily behavior toward others, and *Jumma* services during which localized communities of believers are asked to assemble at the Friday mosque. During these services, the imam delivers readings from the *Qu'ran* and sometimes speaks to the faithful about the values of their religion.

Membership in these various and fluid forms of community can enable West African traders to cope with the deep cultural alienation they experience in New York City. Consider briefly the case of Moussa Boureima, a West African trader from Niger, who has sold caps, gloves, and scarves at the Malcolm Shabazz Harlem market since 1994. He suffers from rheumatism and back pain but is hesitant to go to a hospital or see a physician. As an undocumented immigrant, he distrusts public agencies, which will ask him for identification and perhaps compromise his precarious immigration status. He puts off seeing a physician out of shame and embarrassment; his English is poor, and the doctor may not be able to understand him. In New York City, his life is socially, culturally, and linguistically alien to him. How does Moussa Boureima cope? In this article, we suggest that the degree of his adaptive success – his well-being – in New York City results from his cultural competence as well as from his ability to make use of support-system resources provided by various forms of West African communities.

Cultural competence, for us, refers to a person's ability to adapt to an environment by making effective choices and plans and by controlling – as much as possible – the events and outcomes of daily life. Although community support systems have been shown to diminish feelings of loneliness, help resolve social problems, and buffer cultural dislocation, our data indicate that they afford only the potentiality for these outcomes. It is up to the individual to tap community resources – however they might be structured – to foster his or her well-being. Boureima, for example, can find help for his medical bills from informal associations constructed by his compatriots but only if he chooses to seek treatment.

[...]

HOUSED FRUSTRATIONS

Since coming to New York City in September 1990, Boube Mounkaila, whose robust health obviates the need for modern or traditional medical treatment, has lived in three apartments. For six weeks, he lived in a single-room occupancy (SRO) hotel in Chelsea, which he disliked. "Too many roaches and bandits," he said, "and the place smelled bad, too." Since then, he has lived in Harlem, first in a building on 126th Street and then in an apartment on Lenox between 126th and 127th Streets. He lived alone in this rundown apartment building, the hallways of which stank of stale onions and cabbage. His neighbors included two other Nigerian immigrants who lived on his floor, a Senegalese immigrant married to a Puerto Rican woman who was a U.S. citizen, two Senegalese merchants, and a large Puerto Rican family. He lived in a one-room, second story walk-up that had no toilet or bath or shower facilities. Each floor featured one toilet and one bath/shower situated at the end of a dark hallway.

In 1994, Boube said he paid his landlord four hundred dollars per month for the room, a sum that did not include utilities. At that time, he sent a monthly money order of five hundred dollars to support his wife, daughter, and his mother. This sum also helped to support his two sisters, their husbands, and their children. His father had died some years earlier. These expenses created a considerable financial strain. Like that of most peddlers, Boube's concern about money has seasonal ups and downs. At the 125th Street market, he might take in seven hundred to eight hundred dollars on the weekends in the spring, summer, and fall. During the week, however, he might gross two hundred on a good day. In the winter, these sums decline precipitously. He also expressed concern over his living conditions.

> "They take advantage of us," he said. "Look at this place. A small room without a toilet. And I pay $400 per month plus utilities. They know that many of us don't have papers, and they expect us to pay in cash and not cause problems. What choice do we have?"

Boube wanted to move to a two- or three-room apartment with a bath and toilet, but his monthly expenses prevented him from doing so. "I want to leave here," he said, "but when I finish paying for food, gas, electricity, telephone, inventory, parking and car insurance, and family expenses, there's not much left."

Boube was finally able to move to his current two-room apartment, which is on Lenox Avenue between 126th and 127th Streets in January 1995. According to Boube, a local entrepreneur who respects Africans owns the building. In March 1995, Boube pointed out a man dressed in a black leather jacket, dress slacks, and a black Stetson who was getting into his Jaguar. "That's my landlord," Boube said. "He likes Africans." He explained that when he moved in two months earlier, he had had some cash-flow problems. He needed to get his phone turned back on and had been having difficulty paying his rent. The new landlord, however, was willing to accept four hundred dollars per month for a space considerably larger than his previous apartment. Sensing that Boube was hard working and would turn out to be a reliable tenant, he allowed him stay rent-free for the first month. Sometimes, he let Boube use his phone. "And he is always respectful," Boube stated emphatically. "He's a fine man."

Boube's new apartment was on the third floor of a tenement in an apartment that looks out on Lenox Avenue. The stairwell was dark and creaky and smelled fetid even in winter. Mailboxes were located on the first floor, but because mail was so frequently stolen, Boube had a post office box at the local post office. Above the mailboxes was a sign: "Anyone caught throwing garbage out the window will be punished." In March 1995, Boube's neighbors included his compatriots Sala Fari and Issifi Mayaki, who shared an apartment; several African American men; and a young white man who studied at Columbia University. His new apartment consisted of two rooms, perhaps seven feet by twenty feet. One room remained empty. Boube planned to make this space his salon but had not yet bought furniture for it. He installed a curtain to cordon off the other room. A new double bed and a chest of drawers had been placed in the front room near the window. Boube had adorned the bedroom walls with print portraits of African American women. His salon consisted of his old plastic deck chairs and several low coffee tables. A cheap cotton carpet with Oriental designs stretched between the chairs and tables. On one table was a large boom box. The second table supported a television and VCR. He explained that he had bought a multisystem television so he could play PAL as well as VHS videocassettes. Many of the French and African videocassettes, he said, could be viewed only on the PAL system. "I used to have a much larger TV. I paid $1,400 for it," he mentioned, "but I sent it to my older brother in Abidjan." In the small foyer that separates the two front rooms from the kitchen and bathroom, Boube had hung three images on the wall: a poster of the Dome of the Rock in Jerusalem, a print of a beautiful African American woman, and a picture of Jesus.

> "Why the picture of Jesus?" we asked during an interview in his apartment one afternoon. He replied, "A woman from Canada gave it to me. We spent one whole day talking at the market and she gave me the photo. She thought that because I am Muslim I might refuse the picture. But I like all religions."

Although Boube had tired quickly of his experience in a SRO hotel in Chelsea, many West Africans are forced to remain in SROs despite their deplorable living conditions. Perhaps the best known African "village" in New York City is the Park View Hotel at 55 West 110th Street. Francophone West Africans who live there call it the Cent Dix (the 110th). The building is in a state of advanced disrepair. In 1994, city hall cited it for a variety of code violations that included the presence of leaks, urine, feces, roaches, trash, and garbage in public access. The cracked and peeling plaster walls that lined the hallways have attracted numerous drug dealers and other hustlers.

[...]

In 1992, the building's owner, Joe Cooper, said that perhaps three-quarters of the residents were West Africans. In 1995, however, less than half the occupants remained West African. Cooper said the deteriorating conditions compelled West Africans who had the funds to either leave the building or return to West Africa. The owner complained that the more recent occupants in the hotel were destructive. As soon as he fixed something, he asserted, someone destroyed the repairs or created new problems. Despite Cooper's claims, however, the local police say that crime is not widespread at the Park View.

The juxtaposition of African and African Americans has led to a number of social tensions. Several Africans said they had been disappointed to encounter hostility from blacks in the neighborhood. They reported with bitterness that they were sometimes accused of selling the Americans' ancestors into slavery. Some of "the Africans, on the other hand, believed that African Americans were unwilling to work hard." Many African Americans at the Park View, in contrast, praised their African neighbors, saying that they were friendly, respectful, and hard working.

If the West African residents at the Cent Dix hate the building in which they live, why have so many remained there? They stay, in part, because they do not know where else to go and because, despite everything, the Cent Dix offers something very essential: fellowship. Charles Kone, from the Ivory Coast, summed up this painful mix: "It's misery... There's no security, no maintenance. But I knew that when I got to the airport, there was a place I could go. It's like a corner of Africa." In fact, at the Park View Hotel, West Africans have even set up convenience stores and established communal kitchens. From the vantage of many West African residents, it has become a "vertical village."

These vertical villages, moreover, are well known in far-away West Africa. Experienced West African businessmen, who routinely travel to New York City, instruct first-time travelers to look for African taxi drivers on their arrival at John F. Kennedy Airport. These drivers, so it is said, know to take the new arrivals to one of several SRO hotels in Manhattan. When Boube Mounkaila first came to America in September 1990, he found a West African taxi driver, a Malian, who took him to a SRO hotel in Chelsea.

The idea of West African social cooperation, a theme deeply embedded in the cultures of most West African societies, is not limited to shabby, run-down SRO hotels. Islamic practice is centered on cooperative economics and the establishment and reinforcement of fellowship in a community of believers. West African traders also have constructed elaborate personal and professional networks that create a sense of fellowship. Many West African traders at the Malcolm Shabazz Harlem market share information on the best product suppliers, exchange goods, and recommend clients to their colleagues.

CULTURAL ISOLATION

Although the vast majority of West African street vendors in New York City have expressed profound appreciation for the economic opportunities they enjoy and exploit in the United States, they have invariably complained of loneliness, sociocultural isolation, and alienation from mainstream American social customs. These conditions, moreover, seem to have an impact on the subjective well-being of men like Moussa Boureima, Boube Mounkaila, and Issifi Mayaki.

[...]

Immigration usually reinforces social isolation. Intensified by cultural difference, feelings of isolation from the larger sociocultural environment can have a significant impact on physical and psychological well-being. Isolation limits the range of activities and interactions in which people can participate; it also reduces feelings of control and competence.

[...]

This lack of control compels Moussa Boureima, who is sick, to avoid hospitals; it convinces Boube Mounkaila that he can do little to resolve regulatory dilemmas provoked by the city of New York or the INS.

Sustaining such social and emotional support systems as family may diminish some of the negative effects of immigration. One of the greatest detriments to feelings of well-being among many West African street traders in New York City is, indeed, the absence of family. Constructed as lineages, their families are usually their primary source of emotional and social support. Caught in regulatory limbo, Issifi Mayaki, a principal figure in our study, is unable to return to West Africa to see his family, whom he misses and longs for. He says this situation frustrates him and sometimes makes him mean spirited.

[...]

For most West Africans, the ideal, if not the reality, of a cohesive family that lives and works together is paramount. This ideal, however remote, has survived regional, national, and international family dispersion. It compels men like Moussa Boureima, Issifi Mayaki, and Boube Mounkaila to phone regularly their kin in West Africa; it compels them to send as much money as possible to help support their wives, their children, and their aging parents, aunts, uncles, and cousins.

The absence of an extended family has several psychological ramifications for West African traders. Besides support, families provide a sense of trust and feelings of competence. As Issifi Mayaki has said, one can usually trust her or his blood kin. Generally speaking, the closer the blood ties, the greater the degree of trust. Absence of family therefore creates an absence of trust, which leads to a considerable amount of stress and anxiety. For young men, the absence of wives also means that they are in a kind of sexual and social limbo. They share profound cultural and social mores with their wives in whom they place great trust. In Niger, for example, marriage, which sometimes involves cousins, ties family relations in webs of mutual rights and obligations. Men expect their wives, even during their long absences, to remain faithful to them. To avoid opportunities for infidelity during long absences, long-distance traders often insist that their wives live in the family compound, surrounded by observant relatives who not only enforce codes of sexual fidelity but also help to raise the family's children. Many of the men, on the other hand, believe it is their inalienable right to have sexual relationships with other women – especially if they are traveling. As Muslims, moreover, they have the right, if they so choose, to marry up to four women, although this practice is increasingly rare. These are some of the cultural assumptions that many lonely and isolated West African traders bring to social/sexual relationships with the women they encounter in New York City. To say the least, these assumptions clash violently with contemporary social/sexual sensibilities in America.

El Hadj Moru Sifi, like many Nigerians in New York City, talked of being socially and culturally isolated during his time in America. A rotund man well into his fifties who hailed from Dosso in western Niger, El Hadj Moru did not like the food, detested what he considered American duplicity, and distrusted non-Africans. Between 1992 and 1994, he sold sunglasses on 125th Street in Harlem. Work and sleep constituted much of his life. El Hadj supported two devoted wives in Niger. "Our women," he said in August 1994, just prior to his departure,

> "show respect for their men. They also know how to cook real food. None of these Burger King and Big Macs. They make rice, gumbo sauce with hot pepper, and fresh and clean meat. That's what I miss. I want to sit outside with my friends and kin and eat from a common bowl. Then I want to talk and talk into the night. I want to be in a place that has real Muslim fellowship."

During his two years in New York City, El Hadj said that he had remained celibate – by choice. He did not trust the women he met. The women, he said, often took drugs, slept with men, and sometimes even gave birth to drug-addicted babies. "Some of these women even have AIDS. Soon, I will be in Niger in my own house surrounded by my wives and children. I will eat and talk well again." El Hadj Moru's attitudes are not uncommon among West African traders in New York City. Many of his "brothers" have also chosen to remain celibate.

Boube Mounkaila has been anything but celibate during his time in New York City. Like his brother traders, he misses his family, including a wife whom he has not seen in eight years and a daughter born several months after his departure for America whom he has never seen. Sometimes, when he thinks of his family, says Boube, "my heart is spoiled. That's when I listen to the *kountigi* [one stringed-lute] music."

From the time he arrived in New York City in 1990 as a twenty-eight-year-old undocumented immigrant, Boube has attracted the attention of many women. He is a tall, good-looking man who can be charming. He also has become fluent in what he calls "street English." Because he sells handbags, most of his clients are women, old and young. On any given day, a young woman might be sitting in Boube's stall waiting patiently for him. In speaking of Boube, some of his compatriots shake their heads and say, "Ah Boube, he likes the women too much."

Boube's domestic circumstances are exceptional among West African traders in New York City. For most of the traders, life is much less dramatic; it follows the course of a man like El Hadj Moru Sifi – one works, eats, and sleeps, with occasional interludes or with longstanding relationships with one woman. Issifi Mayaki's situation has been more typical. Issifi is a forty-year-old handsome and well-dressed man who speaks good English. Between 1994 and 1997, Issifi had a girlfriend, a social worker who was a single African American woman with a ten-year-old daughter. Issifi met his girlfriend when he sold African print cloth on 125th Street in 1994. She expressed interest in him. He told her of his wife and children in Africa. She said that was okay; she appreciated his forthrightness. They began to see one another but maintained separate residences.

When Issifi began to travel to festivals far from New York City, his relationship with the woman began to unravel. She did not like the idea of him traveling to festivals. She became jealous of his wife in Africa. When he told her about plans to travel home to see his family, she did not want him to go. She did not want to share him with anyone. Issifi began to believe that American women wished to totally consume their men, which, he said, was not the African way. This cultural clash became the source of contention, and eventually they drifted apart.

Other traders have made other domestic arrangements. Abdou Harouna, who like El Hadj Sifi comes from Dosso, Republic of Niger, is not a trader but a "gypsy" cab driver. Abdou came to New York City in 1992. In 1994, he married an African American woman, not simply because he wanted a way to obtain immigration papers but because he had fallen in love with Alice, who is a primary school teacher. They now have a daughter and live in Harlem. "Alice," Abdou said, "has a pure heart. She is a good person, and I'm a lucky man."

One of the Nigerian traders, Sidi Sansanne, has two families: one in the South Bronx and another in Niamey, Niger. In his thirties, Sidi has become a prosperous merchant who runs a profitable import–export business, which requires him to travel between Niamey and New York City seven to ten times a year. Sidi is perhaps the ideal model of West African trader success. He came to the United States in 1989 and sold goods on the streets of midtown Manhattan. He invested wisely and realized that the American market for Africana was immense. He saved his money and went to Niger to make contact with craft artisans. He began to import to the United States homespun West African cloth, traditional wool blankets, leather sacks, bags, and attaché cases, as well as silver jewelry.

In time, he established a family in New York City, obtained his Employment Authorization Permit, and in 1994 became a permanent resident – a green-card holder. As a permanent resident, Sidi has been able to travel between the United States and West Africa without restriction. Because he has traveled to and from Africa so frequently, Sidi became a private courier. For a small fee, he has taken to Africa important letters or money earmarked for the families of various traders. From Africa, he has carried letters and small gifts to his compatriots in New York City. The freedom to travel has also enabled Sidi to find new craft ateliers in Niger. During his six-week sojourns in Niger, he, of course, has tended to his other family.

This pattern is a transnational version of West African polygamous marriage practices. In western Niger, for example, prosperous itinerant traders establish residences in the major market towns of their trading circuit. In this way, they attempt to pay equal attention to their wives and children and minimize the inevitable disputes that are triggered when cowives live in one compound.

FELLOWSHIP AND COMMUNITY

[...]

For the West African traders at the Malcolm Shabazz Harlem market and elsewhere in New York City, a central component of community is Islam. In Islam, any adherent is a member of the community of believers. Islam unquestionably structures the everyday lives of the traders and keeps alive their sense of identity in what, for most of them, remains an alien and strange place. During six years of conversations with West African traders, the subject of Islam was invariably raised, especially when the conversation broached the subject of the quality of life in the United States. They have said that in the face of social deterioration in New York City, Islam has made them strong; its discipline and values, they have said, have empowered them to cope with social isolation in America. It has enabled them to resist divisive forces that, according to them, ruin American families. But the greatest buffer to cultural dislocation is the perception, held by almost all the traders, that Islam makes them emotionally and morally superior to most Americans.

El Hadj Harouna Souley is a Nigerian in his forties. He made the expensive pilgrimage to Mecca when he was thirty-four, which is an indication of his considerable success in commerce. El Hadj Harouna embodies the aforementioned sense of Islamic moral superiority. Between 1994 and 1997, he sold T-shirts, baseball caps, and sweatshirts from shelves stuffed between two storefronts on Canal Street in Lower Manhattan. Like most West African traders, he is a member of a large family. Although his parents are dead, he has one wife, fourteen children, four brothers, five sisters, and scores of nieces and nephews.

On a rainy afternoon in December 1995, El Hadj Harouna sat under an awning on the steps of Taj Mahal, a radio and electronics store on Canal Street near West Broadway. He pointed out two street hawkers, both African Americans, employed by the owners of Taj Mahal. "You see those men there," El Hadj Harouna said, referring to the hawkers. "They only know their mother. Sometimes they don't even know who their father is. That's the way it often is in America. Families are not unified. Look at him," he said, referring to the older of the two hawkers.

> "He's from Georgia. His family sends him money every month, and as long as I have known him he has not returned there to visit them. Why do some people here not honor their parents? Why don't families stick together – at least in spirit? I want to get back to my family compound where we can all live together,"

El Hadj Harouna stated emphatically. He continued,

> "Can parents here depend on their children to take care of them when they are old? I don't know. I've seen children who sit at home and eat their parents' money, but they think that they owe their parents no obligation. My children phone me every week and ask me to come home. When I am old, even if I have no money, my children will look after me. I will do no work. I will eat, sleep and talk with my friends."

El Hadj Harouna continued his conversation but now concentrated on religion:

> "My Muslim discipline gives me great strength to withstand America. I have been to Mecca. I give to the poor. I rise before dawn so that I can pray five times a day, every day. I fast during Ramadan. I avoid pork and alcohol. I honor the memory of my father and mother. I respect my wife. And even if I lose all my money, if I am able, Inshallah, to live with my family, I will be truly blessed."

West African community structures in New York City take on several forms that are more concrete than the "community of believers." Personal networks in which kinship, ethnicity, and nationality affect the density of contact and degree of trust and cooperation are baseline community forms. In addition to these personal networks, there are translocal communities based on national origin. These are formal associations like L'Association des Nigeriens de New York, L'Association des Maliens aux USA, L'Association des Senegalais aux USA, and the Club des Femmes d'Affaires Africaines de New York (a New York African businesswomen's association). These associations are usually connected to, if not organized by, the diplomatic missions of the various Francophone African countries. Meetings are held once a month in the evenings, usually at a particular nation's United Nations Mission, at which issues of mutual concern are discussed. The associations hold receptions for major Muslim and national holidays. They collect funds to help defray a compatriot's unexpected medical expenses. In the case of a compatriot's death, they also contribute funds to ship the body back to West Africa for burial. L'Association des Nigeriens, for example, has raised money to buy food for hungry people in Niger.

It would be easy and perhaps facile to suggest that these West African communities – formal and informal, economic and personal, translocal or imagined – supply community adherents with financial and emotional support. Such support, it could be argued, also provides social harmony and a sense of belonging that protects members from the disintegrative stresses of cultural alienation. On one level, this statement is most certainly true. Belonging to the community of the faithful provides a religiously sanctioned set of explanations for the West African's situation in America. As participating members in the Association des Nigeriens aux USA, Nigerians engage in a mutually reinforcing set of rights and obligations based on mutual citizenship. This organization represents the interests of Nigerians in New York City. Participation in personal networks yields both economic benefits and, in some cases, the concrete fellowship desired by most West African traders in New York City.

Closer inspection of these community forms, however, reveals a more complex scenario. Although West African traders speak highly of their various "national" associations, their participation in the regular activities of the organizations – the monthly meetings – is infrequent. There are a few traders, of course, who are active members, but the majority of the traders have neither the time nor the inclination to attend association meetings or events. The meetings are held in the evenings at the Nigerian Mission on East 44th Street between 1st and 2nd Avenues. Many traders do not want to travel there from Harlem or the South Bronx after a long day at the market.

More important, the presence of the association in New York City brings into relief a primary tension in Nigerian society – one that exists between members of the Niger's educated elite and its peasants. In western Niger, peasants often express a distrust of the literate civil servants, whom they sometimes refer to as *anasarra*, which can mean, depending on the context, "European," "non-Muslim," or "white man." Less educated Nigerians, including village traders, sometimes say that civil servants who command state power, having learned the European's language and ways, have become foreigners in their own country. In Niger, this strong statement may well be a means of articulating class differences. A similar distrust has been expressed in New York City. In February 1994, a former Nigerian civil servant and no friend of the government of Niger, who sold goods on 125th Street on weekends only, claimed that the Association des Nigeriens had deceived the merchants. "There is a clear division," he said, "between educated and uneducated Nigerians in New York City. The Association recently collected money from the street merchants and stated that the money went to help people in Niger. In fact," he said, "the money helped to pay the electric bill at the Nigerian Mission to the UN and the traders didn't know."

By the same token, participation in economic networks can produce negative as well as positive results. In Issifi Mayaki's case, his participation in a transnational network of African art traders, one based on the trust of cooperative economics, led to betrayal and the theft of his inventory. Boube Mounkaila lost the entire contents of his Econoline Van, which had been parked in a secure, fenced-off space in Harlem. The complicity of one of his economic associates enabled thieves to enter the guarded space and steal his goods. Neither Issifi nor Boube reported these thefts to the police, whom they distrust.

Membership in the community of the faithful, the community of Muslims, creates a spiritual bond and provides a source of support as well as a buffer against the stress of city life in New York. As we have mentioned, Islam, like any religion, provides explanations to the traders about the absurdities of life. It supplies an always-ready set of explanations for problems encountered by Muslims in societies in which Islam is not a major sociopolitical force. For many West African traders in New York City, Islam, as a way of life, is morally superior to other faiths practiced in the United States. And yet, being a member of the community of the faithful does not dissipate a West African trader's financial difficulties, nor does it eliminate the stress of potential illness or existential doubts brought on by cultural alienation.

[. . .]

COMMUNITY AND COMPETENCE

[. . .]

In this article, we have briefly considered the social, material, historical, and psychological dimensions of West African city life in New York. As we have seen, West Africans have skillfully used their traditions as Muslim traders to build personal and economic networks that result in a variety of communities. These communities, in turn, provide them the potential for economic security, social cohesion, and cultural stability in an alien environment.

Despite this rich set of resources, however, some traders have succeeded better than others have, which brings us to the issue of competence. The issue of competence influences the perception of control among West Africans in New York City. There is, for example, a wide diversity of linguistic competence among the traders. Men like Boube Mounkaila and Issifi Mayaki speak English well. Boube's linguistic competence makes him socially confident. His facility in English enables him to construct transnational exchange networks with Asians, African Americans, and Middle Easterners. Since transnational transactions in New York City are usually conducted in English, his linguistic competence has enabled him to expand his operations. Using his skills in English, Boube arranged to purchase a vehicle, buy automobile insurance, and obtain a driver's license. Boube also employs his considerable linguistic skills to charm his mostly African American customers. Mastery of English, in short, has increased Boube's profits and expanded considerably his social horizons. That expansion has made Boube a keen observer of shifting social and economic environments, which, in turn, increases further his business profits. The same can be said for Issifi Mayaki and scores of other West African street vendors.

Lack of competence in English, however, results in missed opportunities. Even though men like Moussa Boureima and his roommate, Idrissa Dan Inna, have been in New York City for more than three years, they speak little if any English. In 1994, they enrolled in a night school course sponsored by a church in Harlem but dropped out after one week. "I don't know," said Idrissa, who sold West African hats and bags on 125th Street at the time, "I just can't learn English. I don't have the head for it. I know it hurts my business. I can't really talk to the shoppers about the goods." When confronted with various financial, social, or personal problems, men like Moussa and Idrissa have to rely on more linguistically competent traders, which, in accordance with the findings of social psychologists, affects their self-image negatively and makes them even more socially isolated.

Although West African street vendors in New York City display various linguistic abilities, they must all confront the problem of cultural competence. Many seem to have mastered the culture of capitalism, but their lack of a more general cultural competence has cost them dearly. In this important domain, one of the key issues is that of trust. According to Islamic law, traders are expected to be completely honest and trustworthy in the dealings with suppliers, exchange partners, and customers. Among West African traders, who are mostly Muslims, trust is paramount. Most of the traders, the majority of whom come from families and ethnic groups long associated with long-distance trading in West Africa, adhere to the Muslim principles of economic transactions. Not surprisingly, their trust has often been betrayed. An exchange partner stole Issifi Mayaki's textile inventory. Betrayals cost Boube Mounkaila his inventory. And yet, these men had the competence to use community resources to rebound from these defeats and move along their paths in New York City. Other traders who have suffered similar setbacks have drifted into isolation, changed occupations, or returned to West Africa.

[. . .]

The communities that West Africans in New York have constructed do not, then, define their city lives; rather, they provide resources – economic, social, and cultural – that dilute the stress of living in an alien environment. More specifically, these communities enable many, but not all, West African traders to enhance their subjective feelings of well-being and control. The sense and reality of community, as we have suggested, is no panacea for the ills associated with state regulation, poverty, and sociocultural alienation. For West African traders, then, city life is molded by the congruence of historical, material, social, and psychological factors. These factors not only define a sense of community but also shape the cultural competencies that affect their adjustment to an alien environment.

The results of our study also underscore Ortner's (1997) contention that the notion of community – however problematic – is one worth retaining in the social sciences. Given a refined framework of "community," the social scientist is able to demonstrate how macro-forces (globalization, immigration, informal economies, and state regulation) affect the lives of individuals living in the fragmented transnational spaces that increasingly make up contemporary social worlds.

"Globalization and the Revalorizing of Ethnic Places in Immigration Gateway Cities"

from *Urban Affairs Review* (1998)

Jan Lin

Editors' Introduction

Jan Lin explores the changing disposition of ethnic places in the immigration gateway cities of the United States. Immigration cities have counterbalanced deindustrialization and urban decline by acting as gateways of labor, capital, commodity, and cultural exchange in the new global economy. Ethnic places are emblematic transnational spaces that both constitute and convey broader processes of economic and cultural globalization. Ethnic entrepreneurs, community activists, and artists have revalorized spaces in the zone-in-transition, places from which they were historically restricted, evicted, or displaced. These rejuvenated ethnic places serve as "polyglot honeypots" for urban managers pursuing growth machine strategies in the postindustrial symbolic economy. Contradictions and conflicts are presented by globalization as much as opportunities.

Lin conceptualizes immigration gateway cities as a subset of world cities, which serve as command centers in the global economy as well as facilitating multilateral trade flows between the U.S. and the Atlantic Rim, the Latin American and Caribbean Rim, and the Pacific Rim. Asian and Hispanic enterprises in the U.S. are over-concentrated in the transport and retail trade sectors. Globalization and ethnic commerce can be seen as leading edges of postindustrial urban growth, counterbalancing some of the decline brought by deindustrialization and runaway factories, explaining the continued vigor of some Frostbelt cities such as New York, Boston, and Chicago, as compared with non-immigration cities such as Buffalo, Philadelphia, and Detroit.

Immigrant enterprises have helped to revive industrial, warehousing, and retailing districts of the central city, which were declining with the departure of manufacturing and commercial activities to suburban locations. A Koreatown has emerged in the midtown area of New York City, for instance, and there is now a sign that identifies Broadway in Koreatown as "Korea Way." In newer Sunbelt cities, ethnic enterprises appear in the suburbs as well. In Los Angeles, the suburban Chinatown of the San Gabriel Valley is described by Wei Li as an "ethnoburb" (see "Anatomy of a New Ethnic Settlement: the Chinese Ethnoburb in Los Angeles," *Urban Studies* 35, 3 (March 1988): 479–502).

Ethnic places were shunned, excluded, or otherwise ignored during the early twentieth century. They were often deemed blighted areas and dangerous threats to public health or contagious arenas of vice and immorality. Chinatowns, Little Italies and Mexican barrios were regarded as obstacles to modernization and cultural assimilation, and often subject to slum clearance for urban renewal. Since the 1960s, however, immigrant enclaves have become more valued by urban planners and public officials as new purveyors of employment and tax revenue in U.S. central cities. Many cities have sought to promote this globalization process by erecting world trade centers, convention centers, and tourism complexes, and expanding port and airport facilities. These are new directions in the way that urban growth machines have adapted and prospered in the contemporary era of the world city. The growth of trade in immigrant goods and services is correlated with the growing significance of the culture industries in U.S. cities, or what Sharon Zukin (see Part Six, pp. 281–289) calls the "symbolic economy." This includes activities such as museums and the performing arts, the media, advertising, and tourism. Ethnicity has become more acceptable and saleable, and is increasingly celebrated in our multicultural global cities. Ethnicity, like other culture forms, may be produced, consumed, and traded in a variety of cuisinary, commodity, and communicative forms.

Lin did fieldwork and interviewing with a variety of ethnic museums and heritage societies, community activists, entrepreneurs, and urban planners in New York, Miami, Houston, and Los Angeles. He found a greater spirit of cultural insurgency and historical revisionism operating in New York and Los Angeles, while ethnic cultural and business interests in Houston and Miami worked a more conservative line of inclusion and patriotism. In both New York's Chinatown and Los Angeles' Little Tokyo, he found intragroup conflicts between overseas and local interests within the same ethnic community. Overseas investment can be a threat to local ethnic interests. There is a tricky political geography to ethnic place revival. The revalorization of the city can displace ethnic small businesses and poor residents through redevelopment, gentrification, and rising property taxes. There is a structurally similar process of revival of ethnic places in immigration gateway cities but this process differs in the way that it is locally constituted.

Jan Lin emigrated from Taiwan in 1966 and is now an American citizen. He is currently Associate Professor of Sociology at Occidental College. Previous to this, he taught at Amherst College in Massachusetts and at the University of Houston. He received a B.A. from Williams College, an M.Sc. in Sociology from the London School of Economics, and a Ph.D. in Sociology from the New School for Social Research in New York in 1992. His teaching and research interests are urban and community sociology, globalization and the sociology of the world-system, mass media and cultural studies, race and ethnic relations, and Asian Americans. He is the author of *Reconstructing Chinatown: Ethnic Enclave, Global Change* (Minneapolis: University of Minnesota Press, 1998), which received the Robert Park Award in 1999 for best book in community and urban sociology by the American Sociological Association. He is involved with a number of historic preservation, planning, and community improvement projects in Northeast Los Angeles and Chinatown. He was Principal Investigator for a three-year grant funded by the U.S. Department of Housing and Urban Development for the Northeast Los Angeles Community Outreach Partnership Center from 1999 to 2002.

The phenomenon of globalization in American cities is a complex set of transformations that suggest contrasting scenarios of promise and peril. These contradictions are manifest considering the vicissitudes of boom and bust buffeting U.S. localities and regions within the machinations of the increasingly interdependent and competitive global economy. Transnational capital, world trade, immigration, and cosmopolitan culture are all trends associated with the emergence of "world cities," which represent one vaunted route out of deindustrialization and urban decline. The dystopic underside to the dynamism of world cities has been observed, however, with reference to displacement, loss of local autonomy, socioeconomic polarization, and interracial conflict. Immigration gateway cities such as New York, Houston, Miami, and Los Angeles may be alternately identified as pluralistic "gorgeous mosaics" or Malthusian "Noah's arks." Globalization must be conceptualized, thus, as a powerful but fundamentally problematical process.

As these transformations have progressed, ethnic places are increasingly conspicuous as emblematic "transnational spaces," which both constitute and convey broader processes of economic and cultural globalization in the immigration gateway cities. The Chinatowns, Koreatowns, and Little Havanas of the postindustrial city have rejuvenated warehouse districts, retail corridors, and residential quarters of the zone-in-transition, reversing the obsolescence threatened with the decentralization of jobs and people to the urban periphery. Associated with the economic recovery of ethnic merchants and place entrepreneurs are the cultural reclamation efforts of a range of community-based artists, historians, and activists in undertakings such as public arts projects, cultural festivals, heritage preservation of certain symbolic or sacred cultural sites, and ethnic history museums, initiated often through acts of political contestation or community insurgency. These combined maneuvers mark the repossession of central urban economic and cultural spaces from which these ethnic actors were historically restricted, evicted, or displaced.

The new central-city ethnic places are primary purveyors of transnational commerce as well as culture, articulating closely with the world trade functions characteristic of global cities; furthermore, they may be employed by local state actors to project an image of the multiethnic city as an investment environment conducive to transnational corporate capital. Local state actors have found it advantageous to link ethnic commercial and cultural districts with economic devices such as world trade centers, convention center promotion, sports franchise expansions, arts promotion, and urban tourism. These schemes represent new frontiers of central-city growth amid the losses in manufacturing and commercial activity. Ethnic places, significantly, serve as "polyglot honeypots" in advancing these postindustrial functions in the transnational environment of the immigration gateway cities. Ethnic places (both commercial and cultural) thus may be seen alongside the producer services as leading edges of globalization.

Globalization presents a salubrious, yet slippery, terrain of new opportunities and potential hazards for ethnic entrepreneurs and culture producers, which I comparatively examine through a number of cases in the emergent global cities. Among the complications that emerge are intergroup conflicts over the character of ethnic place reclamation. Intragroup conflicts also may occur, especially between local and overseas factions within certain groups, such as the Chinese and the Japanese. More broadly, the repossession of ethnic commercial and cultural spaces may be critically conceptualized as a form of economic and symbolic revalorization that upgrades central urban space for higher, transnational uses antithetical to small-scale, local ethnic actors. Across the topography of urban space, ethnic places may authenticate "difference" while reinforcing class distinctions, serving as arenas for the investment and further accumulation of transnational capital.

[...]

Thus structural similarity in the process of globalization should not overdetermine the manner in which it takes place in the local context of American cities or ethnic enclaves. The onset of global processes since the 1960s has been differentially constituted in the various immigration gateway cities. Vociferous community insurgency has accompanied the reclamation of ethnic places in some cities, and revalorization efforts in other cities have involved greater cooperation between ethnic contenders and the local state. These contrasts reflect contingencies and variations in regional economies, the preceding history of local race and ethnic relations, traditions of community activism, and relations between local ethnics and overseas counterparts.

[...]

IMMIGRATION GATEWAYS, ETHNIC COMMERCE, AND THE ZONE-IN-TRANSITION

[...]

I conceptualize immigration gateway cities as a subset of world cities, which serve not only as command centers in the cross-border movement of capital and labor but also as critical nodes in processing flows of commodities and cultural products. The trade and transport firms concentrated in these cities facilitate the import and export of goods between the U.S. hinterlands and other global trading regions. Following this logic, we may observe that (1) New York and Boston are gateways to Atlantic Rim trade; (2) Miami, Dallas, and Houston are gateways to Latin American and Caribbean Rim trade; and (3) San Francisco and Los Angeles are gateways to Pacific Rim trade. Thus Los Angeles is dubbed the "capital of the Pacific Rim" and Miami the "capital of Latin America." Attention to the distributive and trade sectors thus augments one's understanding of economic globalization beyond the producer services as a leading edge of growth under postindustrialism.

[...]

The notion of globalization and ethnic commerce as leading edges of postindustrial urban growth revises an argument struck among urbanists in the 1970s that the real consequence of deindustrialization was an interregional job and population shift from northeastern (frostbelt) to southwestern (sunbelt) cities as manufacturing firms were lured by lower wages, lower land costs, and anti-union and preferential business climates of the Sunbelt. The refined argument counterbalances scenarios of decline and deindustrialization with the dynamism of globalization and immigration to explain the continued economic vigor of some older centers (e.g., New York City, Boston, and Chicago) in the 1990s. . . .

The function of ethnic enclaves in the global city may be more clearly understood through an urban spatial perspective that recognizes the role that ethnic enterprises have played in reviving industrial, warehousing, and retailing districts of the central city, which were declining with the departure of manufacturing and commercial activities to peripheral locations. Macy's and Bloomingdale's abandoned their downtown locations to anchor the new shopping malls and "gallerias" of the suburban cloverleafs. Kresge's retooled and entered the suburban periphery to become Kmart, but the once ubiquitous Woolworth five-and-dime stores clung persistently to their downtown locales, finally closing their last 400 stores in July 1997. Pakistani immigrants have filled some of this niche in the demand for low-end retail niche with the opening of 99-cent and dollar discount stores in New York City. But the new immigrant impacts on urban economies can be seen much more clearly across the geography of urban space.

New York City's midtown wholesaling district (from 14th to 34th Street) is a case in point. Previously dominated by white ethnic entrepreneurs (including Jewish and Italian immigrants), the district steadily lost economic vitality from the 1920s to the 1960s with the decentralization of manufacturing and retailing functions to the suburban and exurban periphery. Since the 1960s, however, new Asian, Latin American, and Caribbean entrants have filtered into the same built environment, renewing the economic dynamism of the district and giving it a new transnational atmosphere (a similar ethnic succession has occurred in the proximate Garment District). The New York Chinese Businessmen's Association purports to represent some 2,500 enterprises, mainly engaged in import–export wholesaling in the midtown area. A Korean business district of wholesalers, restaurants, and banks can be found in a rectangular area from 24th to 34th Street between Fifth and Sixth Avenues. In October 1995, it was officially recognized by the city of New York with the posting of "Koreatown" signage at the intersection of 32nd Street and Broadway. Underneath the Broadway street sign can be found an accompanying sign designating the corridor as "Korea Way."

New York's Chinatown has grown expansively since the 1960s from its location in lower Manhattan, succeeding Jewish and Italian outmovers in proximate blocks of the Lower East Side. Dominicans have established a commercial and residential presence in the Washington Heights district of upper Manhattan. South Asian Indians, though residentially decentralized, retain a commercial and cultural district in the Jackson Heights area of Queens, which serves as the "symbolic heart" of the dispersed community. Chinese, Koreans, and other new immigrant groups have succeeded white ethnic outmovers, establishing a multiethnic "Asiantown" in the Flushing area of Queens.

In Houston, Chinese and Mexican merchants have given new life to the deteriorating near-city warehousing and industrial districts to the east of the central business district (CBD). The downtown Chinatown wholesalers import food and restaurant products through the port of Houston for distribution in the metropolitan area as well as throughout the South and Midwest. A "Little Saigon" of Indochinese merchants has emerged in the disused midtown area to the immediate south of the CBD. New Asian and Latino commercial activity also has emerged in the suburban west and southwest economic corridors of the metropolis. Mexican activity can be found in Houston Heights, Chinese along Bellaire Boulevard, South Asians along Hillcroft, and Koreans in the Harwin wholesaling corridor as well as in retailing and restaurants on Long Point Road. The entrance of new ethnic merchants was expedited largely by the availability of low commercial rents and high vacancy rates during the 1980s, when Houston experienced a severe regional recession because of the tumble in petroleum prices associated with the oil glut on the world market. The emergence of ethnic commercial activity thus can be interpreted to have contributed toward the stabilization of the local economy.

In Los Angeles, the nation's largest Koreatown has emerged in an approximately 20 square-mile area west of downtown, which was formerly more than 90 percent white. Businesses in the area are primarily Korean, but Koreans make up less than 15 percent of the residents, who are primarily Hispanic (mainly Mexican). The old Chinatown north of downtown has been enhanced with the emergence of Indochinese merchants along the north Broadway corridor. Latino merchants now burgeon along south Broadway. East of the CBD, Little Tokyo has experienced some revival in the decades following the internment debacle. Beyond the central city, a suburban Chinatown, "Little Taipei," occupies Monterey Park. Southward in Orange County, Vietnamese merchants now proliferate along Bolsa Avenue in Westminster.

In Miami, Cubans settled initially in the old Riverside-Shenandoah section of Miami (between Flagler Street and the Tamiami Trail), a historically Jewish residential enclave. The Tamiami (Tampa-Miami) Trail, otherwise known as Highway 41 or SW 8th Street, was from the 1930s to the 1950s the major commercial spine for the Jewish community. Postwar prosperity led to residential outmovement of upwardly mobile Jewish-Americans into the Dade County suburbs, and the scene was set for the entrance of the Cubans in the 1960s, who redubbed SW 8th Street "Calle Ocho" (Eighth Street) as the surrounding district became known as Little Havana. Like the Jewish-American merchants who preceded them, the Cuban arrivals experienced discriminatory exclusions from the Anglo merchant establishment of Flagler Street, the main downtown commercial corridor. By the 1980s, Cubans themselves began filtering into the suburbs of Dade County, as Nicaraguans followed in their wake into the commercial and residential space of Little Havana. Though increasingly integrated residentially, the Cuban cultural and commercial presence remains conspicuous throughout the metropolis.

Although the new ethnic commercial enclaves are not exclusively located in central-city or near-city zones of the city – particularly in newer sunbelt automobile-centered cities such as Miami, Houston, and Los Angeles, where immigrant enterprises have penetrated the suburban commercial corridors of feeder roads and strip malls – some of the more prominent (Chinatowns, Koreatowns, and Little Havanas) still occupy central locations. I now turn to examine more specifically the historical transformations of the ethnic spaces of the zone-in-transition in the city center.

ETHNIC PLACES FROM URBAN VILLAGE TO GLOBAL VILLAGE

The massive immigrations of the industrial era (1840–1920) drew primarily from the "new stock" nonwhite Anglo-Saxon Protestant (WASP) European immigration (Catholic, Jewish). Industrial core cities of the Northeast and Midwest (NE–MW) drew the bulk of the foreign immigration to the United States. The Chicago School of Sociology viewed urban ethnic places (such as the Jewish "ghetto," Kleindeutschland, Little Italy, and Greektown) as "decompression chambers" for newly arrived immigrants, which facilitated their economic adaptation and cultural assimilation into American life. Ernest Burgess codified the human ecological supposition that the immigrant quarters of the zone-in-transition that surrounded the CBD eventually would wither away with the outward expansion of the CBD and the upward mobility of latter-generation ethnics into the outer-concentric residential zones of the suburbs. Harvard University scholar Oscar Handlin similarly inferred that the ethnic communities of the huddled masses were transitional phenomena by which the "uprooted" mitigated the shock of transoceanic passage.

The invasion–succession paradigm of the human ecologists privileged the invisible hand of the free market in determining urban land patterns and neglected political variables, including racial/ethnic discrimination and the interventions of the WASP-dominated state. Immigrant minorities were relegated to the zone-in-transition by the formal and de facto restrictions, exclusions, and covenants of Anglo powerbrokers as much as by the market barriers presented by high rents in the CBD. The restrictions experienced by nonwhite minorities (e.g., Asians, Latinos, African-Americans) were even greater than those experienced by white ethnic immigrants (e.g., Irish, Italians, Jews), especially in the cities outside of the NE–MW core region. Thus African-Americans and Latinos were restricted by Jim Crow or Jim Crow-type segregation laws and Asian-Americans by restrictive covenants and Alien Land Acts. Thus, in turn-of-the-century Los Angeles, Mexicans, African-Americans, and Chinese occupied fallow real estate around the Olvera Street Plaza north of the Anglo-controlled CBD centered on Broadway. The Mexican barrio became known as the Sonora or "Dogtown." The Chinese mixed with African-Americans on a corridor known as "Calle de los Negros" or "Nigger Alley." A Japanese colony appeared nearby at the intersection of East First and San Pedro Street. In Houston, Mexicans and African-Americans formed an approximate ring around the Anglo CBD. A Chinese merchant colony arose on the eastern periphery of the CBD, serving primarily African-Americans.

If ethnic settlements were shunned, excluded, or otherwise ignored during the industrial period of American urban growth, city managers and the federal government actively began bulldozing ethnic places under slum clearance policies of the interwar period, and more actively under urban renewal in the postwar period, to make room for expressway arterials, middle-class housing, or expansion of the CBD and government office buildings. Ethnic places were deemed unsanitary public health hazards, congested visual eyesores, and contagious mediums of vice and other social pathologies to middle-class urbanites. Chinatowns, Little Italies, and Mexican barrios were regarded as obstacles to modernization and cultural assimilation. In Manhattan's Lower East Side, riverfront tenements were cleared to make way for the East River Drive and public housing. The Cross-Bronx Expressway severed a huge Jewish tenement community. Houston's Chinatown was relocated to facilitate expansion of the CBD, and in Los Angeles, Chinatown was relocated to facilitate the construction of the Union Railway Station. These are some prominent cases in ethnic communities, but many observers have noted that the predominant victims of urban renewal programs were African-American communities.

[. . .]

The Hart–Cellar Immigration Act of 1965, passed during the liberal political environment of the civil rights era, overturned decades-long restrictive immigration quotas, auguring the arrival of a new wave of Latin American, Caribbean, and Asian immigration. The demographic, political, and cultural changes accompanying the civil rights movement and Hart–Cellar Act marked the passing of assimilation discourse as a battery of new conceptual paradigms emerged, which interpret ethnicity as an adaptable phenomenon that accompanies social change, rather than being a static, primordial status of premodernity. . . . Most recently, a fertile new terrain has arisen around concepts of globalization and transnationality.

Studies of ethnic transnational communities draw attention to the binational cultural and economic networks in which new immigrants interact. These frequent interactions between home and host societies are enabled by innovations in communications and transportation technologies, which have shrunk the barriers of geographical, economic, and cultural borders between constituent nations of the global system of states. A reciprocal circuitry of political and economic relations occurs through monetary remittances, seasonal labor migrations, circular migrations within the life cycle, dual residence, and binational investment practices. Dominican-Americans experience New York City like a "waiting room," as they forge livelihoods aqui y alla (here and there). Chinese cosmopolitans from China, Hong Kong, and Taiwan send their children (xiao liuxuesheng, "little foreign students," or more popularly "parachute kids") to study in U.S. schools and prospectively gain permanent residency while their parents (kongzhong feiren, "trapeze artists," or taikongren, "astronauts") frequently shuttle the transoceanic distances. . . .

These new immigration studies furrow some terrain remarkably similar to that cultivated by new perspectives in critical cultural studies, particularly Arjun Appadurai's concept of "ethnoscapes," which purports to release ethnicity like "a genie [previously] contained in the bottle of some sort of locality" (1990. Disjuncture and Difference in the Global Cultural Economy. *Public Culture* 2 (2), pg. 15). In privileging semiotic over material exchanges in international capitalism, Appadurai overdetermines the notion of "deterritorialization" through a reading of globalization as fractals-chaos-uncertainty, thereby obliterating the tangibility of "place" or "community" with an alternative cartography of rootless, diasporic networks of nomads, refugees, expatriates, tourists, and economic and artistic cosmopolitans. Preferring radical revisionism to iconoclasm, I would assert that traditional comprehensions of geographic places and territorial communities in the world system have not disappeared but have become problems, and the physical distances and temporal spans between places and localities have been increasingly compressed but not finally annihilated. Ethnic global villagers may be nearly freed from place-bounded restraints in their commercial and cultural pursuits, but ethnic spaces of the city are still highly salient as processors and distributors of ethnic products, durable and nondurable, in their cuisinary, commodity, and communicative forms (including mediums of print, cassette, disc, video, and live performance).

Thus, in the postindustrial environment of the world cities, American ethnic places as global villages are now more ethnically polyglot with nonwhite immigrants, who interact in sustained transnational networks of association, rather than provincial urban villages resistant to and marginalized by the pressures of cultural assimilation. Through the combined window of opportunity created by a domestic social upheaval in civil rights and external developments in global capitalism, ethnicity has a greater resilience and gravity in arenas of economic, political, and cultural interaction. But as discussed earlier, globalization may be as pernicious as it is auspicious, and the merchants, place entrepreneurs, and cultural practitioners have encountered risks as well as advantages.

TRANSNATIONAL CAPITAL, POLYGLOT HONEYPOTS, AND POSTINDUSTRIAL GROWTH COALITIONS

The promoters of growth in postindustrial cities have had to contend with a variety of economic and fiscal crises associated with postwar deindustrialization and suburbanization, dual trends that have deprived central cities of employment, residents, and a vital tax base. The fiscal impoverishments accompanying these structural transformations were further deepened with the curtailment of federal revenue-sharing programs under the Reagan administration. Increasingly left to their own devices, the localities responded in part by working in greater partnership with the private corporate sector to foster economic growth. In a number of cities, international capital has been a critical component of this new growth. Local planners and city managers have been active agents in promoting the globalization of metropolitan economies, facilitated through devices restructuring the central-city built environment, such as the erection of world trade centers and the enlargement of port and airport facilities, convention centers, and tourist complexes.

In Miami, a Downtown Action Committee began a publicity project in 1975 for a "New World Center," which envisioned Miami as center of the hemispheric past and future. One of the resulting projects, the Omni International Complex (containing a luxury hotel, shopping mall, theaters, restaurants), helped key Biscayne Boulevard's emergence as the "Fifth Avenue of the South". Biscayne Bay became not only an exclusive resort for the idle rich but also a vital hub for the transshipment of tourists and commodities in the Latin American and Caribbean Rim. The port of Miami became the largest cruise ship port in the world. A free trade zone was created near Miami International Airport. Fortune 500 companies began making Coral Cables the site of their Latin American headquarters. There was a massive metropolitan building boom in the 1980s.

In New York City, Chase Manhattan Bank president David Rockefeller formed a Downtown-Lower Manhattan Association in 1956, which commissioned studies recommending the razing and restructuring of the preindustrial built environment for global capital through the construction of a World Trade Center (WTC). The Port Authority of New York–New Jersey (which operates New York's port, airports, and major bridges and tunnels) assumed the WTC project in 1964, which, when completed between 1975 and 1980, became the world's largest office complex, helping to spur New York's recovery from the fiscal crisis of 1975. Now the material edifice and symbolic landmark for New York City's position as a command center and headquarters complex for global capitalism, the WTC has helped to offset losses in Fortune 500 corporation headquarters from the regional economy and has stimulated redevelopment in Lower

Manhattan, including the World Financial Center and high-income residential projects at Battery Park City.

These shifts have been paralleled in Houston with the fading power of a local Anglo power elite, the "8F crowd" (so named because of the Lamar Hotel suite at which they traditionally met), in the wake of the entrance of transnational oil and gas industry firms in the regional economy. The George R. Brown Convention Center, designed like a postmodern ocean liner to recognize Houston's importance as a port city, was built in 1987. The city's global aspirations were touted with the Bush administration's choice of Houston as the site for the 1990 Economic Summit of the advanced industrialized Group of Seven (G-7) nations. Los Angeles similarly experienced the steady entrance of the transnational corporate presence into its CBD, a process encouraged by long-time mayor Tom Bradley in the 1970s and 1980s.

Globalization is an incomplete characterization of the shifts taking place under postindustrialism if related changes are not identified. Urban growth machines, constituted of coalitions of place entrepreneurs (rentier capitalists) and public officials committed to local economic development, have been endemic to American cities since the era of westward expansion, but as Logan and Molotch observe, growth machines of the modern era are alliances of a more "multifaceted matrix" of interests that include the local media, utility companies, universities, arts institutions, professional sports, organized labor, and small retailers. . . .

Urban tourism efforts are predicated upon the strategy of assembling clusters of attractions or cultural "honeypots" near revivified railroad termini and hotel and convention center complexes, providing urban tourists, conventioneers, and business visitors with easy walking or transportation access. Having reached an advanced stage in European cities – which possess significant inventories of monuments, buildings, sites, and districts of architectural, historic, antiquitous, or sacred significance – urban tourism similarly has been fabricated in the American context through the much-vaunted prototypes of the Rouse Corporation's waterfront "festival marketplaces" and the Disney Corporation's heritage-facsimile theme parks. I focus here on those experiments in which strategies of economic globalization are linked with schemes to recover the urban multiethnic inheritance.

THE TRICKY POLITICAL GEOGRAPHY OF ETHNIC PLACE RECLAMATION

The linked civil rights and ethnic power movements of the 1950s and 1960s helped to initiate and reinforce parallel projects of community action and heritage reclamation in American race and ethnic places. Cultural reclamation efforts in ethnic communities were given further impetus during the years surrounding the American bicentennial activities of 1976, a period of cultural restoration during the nadir of the post-Watergate recessionary 1970s, which augmented the significance and respectability of ethnic heritage recovery. Manhattan's Lower East Side now contains a number of ethnic museums and preserved heritage sites, including the Chinatown History Museum, the Eldridge Street Synagogue, and the Lower East Side Tenement Museum. Cultural and community insurgency often was linked. Activists associated with the Chinatown Basement Workshop assembled an initial trove of artistic and material cultural artifacts that eventually were transferred to the Chinatown History Museum; this museum became a reality after agitations for preservation surrounded the proposed auction of a public school building that had fallen into city possession. The Little Italy Restoration Association unsuccessfully fought to acquire an abandoned police headquarters for an Italian cultural center; the spectacular Beaux-Arts-style edifice instead was converted to a luxury condominium complex.

In Los Angeles, opposition to a police center expansion and a municipal plan to demolish some historical sites in the process of widening East First Street in Little Tokyo spurred an activist movement in Little Tokyo, which eventually gained National Historic Landmark status for a city block that holds 13 buildings of particular merit, including a number of historic storefronts. The Hompa Hongwangi Buddhist Temple (built in 1925) was converted to the Japanese American National Museum. An "artistic sidewalk" wraps around the block, a multihued walkway of rose, white, and gray concrete afloat with images of tsutsumi, a Japanese custom of "wrappings" such as baskets, folded cloths, and suitcases. Quotes are also inscribed – warm memories of early neighborhood life in brass and harsh memories of internment in stainless steel. Sheila Levrant de Brettville, the designer of the Omoide no Shotokyo (Remembering Old Little Tokyo) walkway, said, "Buildings are like people. A portion of the story of the proprietors in each building is depicted. . . . To evoke their personal narrative is to evoke the larger community in which they lived." An antiquated housing barrack and watchtower (preserved and transferred from their original location at a Los Angeles processing center for Japanese-Americans awaiting relocation to wartime internment camps in the desert) sits starkly across the street surrounded by barbed wire.

Also in Los Angeles, Mexican historians and community activists have been seeking to repossess Chicano history at the Olvera Street Plaza. The plaza, which was the nucleus of Il Pueblo de Nuestra Senora la Reina de Los Angeles (the original agricultural settlement built by Mexican labor under the aegis of the Spanish crown in 1781), passed into Mexican rule in 1822 until being conquered by the United States in 1847. The site faded from significance as Anglo attentions were fixed on shifting CBD growth to Pershing Square south of the old center. Chinatown was razed to make way for the Union Railway Station in the 1930s, however, and other portions of the district were cleared to make way for highways and government building expansions in the post-World War II era. The site finally became El Pueblo de Los Angeles Historical Monument, under the Los Angeles Recreation and Parks Commission in the late 1980s under persistent Chicano pressure. The ideological struggle has continued, however, with the emergence of a vociferous intergroup conflict over the character of the heritage recovery effort. A restoration centered on the "prime historical period" of 1920–1932, when a plurality of ethnic groups existed at the plaza (but Mexican influence at the plaza was at its weakest), has been endorsed by the Euro-American-led Recreation and Parks Commission, but the Mexican Conservancy and other Chicano interests have favored a sequential history that recognizes the fundamental role of Native Americans and Mexicans in founding the settlement. The appointment of a diverse slate of board members representing all ethnic communities by Mayor Riordan somewhat resolved the conflict, but intergroup tensions persist.

Cuban heritage reclamation efforts in Miami focus on the "Freedom Tower" and the Tower Theater intersection on Calle Ocho in Little Havana. The so-called Freedom Tower, which is located on Biscayne Boulevard near harbor tourist amenities, was built in 1925 in the Spanish Renaissance architectural style, modeled after the Giralda Tower in Sevilla, Spain. For its first 32 years, it housed the offices of the Miami Herald. It later became the processing point for some 150,000 Cuban refugees airlifted to Miami on more than 3,000 "Freedom Flights" from 1965 to 1973. In the collective memory of Cuban emigres, the Freedom Tower occupies an urban iconographic position similar to the fusion of New York City's Ellis Island and Statue of Liberty. Saudi investors now hold the vacant building, but the Miami Office of Community Development and local preservationists have won historic designation for the site while they prepare proposals for its future disposition.

At a more advanced level of restoration is the Tower Theater site at the intersection of S.W. 8th Street (Calle Ocho) and S.W. 15th Avenue, the "heart" of the Little Havana commercial corridor. The theater is notable among Cuban-Americans as the first to begin screening American films dubbed in Spanish, giving many immigrants their most durable introduction to American culture. It is slated for conversion into a Latino film and performing arts center, with an interpretive museum of Cuban-American culture, and associated restaurants and retail shops. Also on the intersection is Maximo Gomez Park, a favorite spot for domino tournaments, where elderly Cubans are especially known to congregate, smoking cigars and engaging in political discourse and reminiscing about Cuba. Star-shaped plaques on the sidewalk commemorate renowned celebrities such as Gloria Estefan and the "Queen of Salsa," actress/singer Celia Cruz. City planners now refer to Little Havana as the "Latin Quarter" to recognize the greater diversity of members of the Latino diaspora, such as Central Americans, who now occupy the district. The S.W. 8th Street corridor is also the site of the Calle Ocho street festival; drawing up to

1 million people every year in March, the event is vaunted by its promoters as the "world's largest block party."

In Houston, Mexican merchant interests affiliated with the East End Chamber of Commerce have sought to promote the development of a promenade/bike trail and urban historical park along the length of the Buffalo Bayou from Allens Landing to the ship Turning Basin, to be marked at intervals with sites of natural, historical, and ethnic significance. Although the plan seems somewhat incongruous within the industrial environment of Houston's east side, proponents are quite serious about their proposal, brandishing the model of San Antonio's highly successful Riverwalk. Near the Our Lady of Guadalupe Church (the major spiritual and community center of the original Mexican settlement of the second ward of Houston) are a cluster of newer sites, including the palm-lined Guadalupe Plaza for civic gatherings and the partially constructed Mercado del Sol, intended as a Latino festival marketplace in an abandoned mattress factory.

The reclamation of ethnic place culture in Houston and Miami, although predicated upon an affirmative expression of ethnic pride, has less of the air of cultural insurgency and critical historical revisionism advanced by purveyors of ethnic heritage recovery in Los Angeles and New York. These differences extend somewhat from variations in the character of the preceding history, experienced by each group in their respective locales, and in the inventory of historians and community activists. To illustrate, the strongly anticommunist Cubans, who are largely refugees/immigrants of the more liberal postwar civil rights period, were the beneficiaries of significant U.S. government assistance during their relocation and settlement, despite some real initial discrimination on the part of Miami property owners and employers. In Houston (where Mexicans experienced conquest and dispossession), although place-based community development is strong in the east end Mexican community, place-based cultural restoration activities are still incipient. By contrast, Los Angeles efforts have emphasized the atrocities of Anglo conquest and the stark historical experiences of ethnic eviction, displacement, and internment. In New York City, heritage activities have centered on the salvaging of quotidian artifacts of tenement life and documenting the quiet but noble struggles of immigrant progenitors as pushcart peddlers, laundry workers, and garment workers in the promising but punishing crucible of the Lower East Side. The Chinatown History Museum recently changed its name to the Museum of the Chinese in the Americas in an attempt to move its scope and operations beyond local provincialities to encompass and promote new transnational, diasporic approaches to the study and dissemination of migration history.

The Japanese- and Chinese-American places of New York and Los Angeles also have been the arena of contentious struggles between local and overseas interests. Overseas interests gradually have made investment inroads into Little Tokyo in Los Angeles as the commercial and cultural reclamation of the district has proceeded in the past three decades, a trend that has been encouraged by the L.A. Community Redevelopment Agency (CRA). Little Tokyo community activists, fearing displacement of elderly residents and mom-and-pop merchants, launched an unsuccessful effort to block construction of a hotel (the New Otani) and shopping center (Weller Court) financed by overseas Japanese capital in the early 1970s. The "antieviction task force" decried public subsidies granted by the CRA to large investors via "tax increment financing" schemes, which reduced short-term land acquisitions costs, through the expected guarantee of long-term incremental increases in tax liability (opponents viewed the short-run subsidies as a "land grab" on the part of overseas capital). Local interests were placated in the ensuing years by long-awaited CRA projects supporting smaller merchants (Japanese Village Plaza) and senior housing needs (Little Tokyo Towers and Miyako Gardens Apartments). Overseas Japanese capital was, in fact, aggressively courted by the board of the Japanese-American National Museum in its recent capital campaign, which will finance a gargantuan new pavilion across the street. Akio Morita, founder and chairman of the Sony Corporation, dedicated one of his executives to raising an initial nest egg of some $9 million from the Keidenren of Japan (the major Japanese business federation). In revealing hyperbole, Morita has triumphantly saluted local Japanese-American communities for the "inroads" they have made for Japanese corporate investments in the United States.

New York's Chinatown has been the site of similar conflicts between local ethnics and their overseas counterparts. Encouraged by commercial dynamism in the district, the city in 1981 introduced revised zoning rules permitting high-density development near the entrance ramps of two bridges where there was a surplus of abandoned city-owned property (the Special Manhattan Bridge District). Community opposition quickly materialized, however, with the announcement of two major high-rise condominium projects to be financed by overseas Chinese capital. Protracted legal battles led to the revoking of one building permit, and the other project eventually was found culpable (with reference to the Mount Laurel doctrine) of inadequate "environmental impact review," where environment was given a wide latitude of meaning, including potential displacement of local residents and business and a negative impact on the community.

The opposition between local and overseas Chinese interests has not been so antithetical in downtown Houston, however, where the Houston Chinatown Council actively has sought backing of overseas investors in their efforts to market Chinatown as a tourist amenity for delegates at the nearby George R. Brown Convention Center. A six-block mixed-use development is envisioned with restaurants offering world cuisines (including Chinese, Vietnamese, Korean, Thai, Mexican, Italian, and Texas-style barbecue), a farmer's market, community center, theater for Chinese opera and other performances, and housing. Representatives of Shenzhen, China (Houston's sister city), will donate a gigantic Chinese gate with guardian lions.

THE FUTURE OF ETHNIC PLACES

Some observers may be too quick to proclaim the obsolescence of concepts such as place, community, and locality with the locomotive onrush of globalization, technological advance, and urban spatial change. Urban decentralization and the continuing growth of nonplace-based or electronic networks of commerce and communication – including catalog shopping, television, cable, video, and the Internet – suggest the continued diminishing salience and power of central urban places as marketplaces of commodities and cultures. But this article has found that central ethnic places are not only resilient but growing in significance, contributing to the cultural and commercial revalorization of the central spaces of immigration gateway cities.

The renewal of ethnic places in the zone-in-transition and central city has contributed to the rejuvenation of the economy as well as to the public culture of the city. As new marketplaces of cultural and economic products and sites of landmarks, representation, fetes, and artistic performance, these places play a major role in urban collective life, much as the agoras of ancient Greece. To the Greeks, the agora was not just the commercial marketplace but the "public hearth" or *hestia koine*, the center of the community:

[...]

A similar role is played by the Arab *suq*, the Latin American *mercado* or *feria*, the Chinese *ziyou shichang*, and the other various marketplaces, emporiums, bazaars, and public commercial places of the globe. A comparable communal and commercial role was played by Main Street in the many small towns of the American heartland.

The zone-in-transition truly merits its moniker in the postindustrial era as a spatial locus of new commercial and cultural transformations that, along with growth in the producer services complexes of the urban core, constitute the evolving spatial and economic apparatus of the evolving world city. The putative repossession and revalorizing of these central urban spaces from which ethnic actors were previously restricted or evicted is an outwardly auspicious process fraught with many latent dilemmas, revealing fundamental complexities regarding the links between space, capital, and power in the postindustrial immigration gateway cities.

The involvement of ethnic actors in the economic and cultural revalorization of ethnic places ultimately has incorporated them into the intricate dynamics of broader political–economic relations with local states and global capital. Community contenders are drawn into negotiation with state agencies, possibly becoming incorporated into the state apparatus itself, thus involving them in intergroup struggles within the state bureaucracy.

As polyglot honeypots, ethnic places may serve as multicultural tourist attractions and conduits of international trade and services, contributing both to the livelihood of ethnic contenders and municipal programs of globalization. But the revalorization of ethnic places also implicates them in broader strategies of gentrification and transnational capital accumulation, which ultimately may displace local ethnic residents and commercial merchants.

PART SIX

Culture and the Urban Economy

Plate 11 Jazz musician, New York City, 1958 by Dennis Stock (reproduced by permission of Magnum Photos).

Plate 12 Times Square, New York City, 1999 by Raymond Depardon (reproduced by permission of Magnum Photos).

INTRODUCTION TO PART SIX

Culture has always been a significant part of what cities are and do. The visual culture expressed in architectural styles, monuments, and the designs of parks as well as the less formal culture offered by street musicians and artists in neighborhood festivals and fairs contribute to how cities feel and are experienced. Until recently, however, the study of culture and its importance to the urban form and change was relatively circumscribed. Over the years, a number of sociocultural and symbolic interactionist sociologists campaigned to push the study of culture to the forefront of urban studies. These included Walter Firey, writing in 1945 (see this volume, Part Two, pp. 89–96), Anselm Strauss (*Images of the American City*. New York: Free Press of Glencoe, 1961), and Gerald Suttles ("The Cumulative Texture of Local Urban Culture," *American Journal of Sociology* 90, 2 (1984): 283–304). But the discipline's preoccupation with demographic and political–economic changes left very little room for "culture" as a primary focus of urban theory or research.

That changed with the precipitous decline in the manufacturing-based economies of cities that took place in the 1970s and 1980s. With the relentless pace of deindustrialization, older cities, both large and small, refashioned their economies from the *production of things* (from consumer durables, such as automobiles and appliances, to the light manufacturing of toys, clothes, and books) to the *production of spectacles* (events, leisure, and cultural activities). We live in an age where the production and consumption of symbolic and cultural objects can be as profitable as the production and consumption of durable commodities.

Interestingly, many urbanists found a renewed interest in culture through the prism of political economic urban theories. David Harvey's influential book, *The Condition of Postmodernity* (Oxford: Blackwell, 1989) outlined the connections between cultural changes (postmodernism) and political economic changes underway in western cities. Other efforts to integrate analyses of cultural production and consumption into urban theories include Sharon Zukin's and Mark Gottdiener's work in the 1990s (parts of which are included in this section). Significant changes in how urban capitalism works – namely, the shift toward the production of services and spectacles – have placed the analysis of symbols and imagery on the cutting edge of urban studies. The material development of the urban built environment is not ignored but the analytical focus has shifted to the production and interpretation of cultural symbols and themes that inundate the modern city.

But the analysis of culture within urban studies has moved beyond simply seeing culture as linked to political economic or class-based concerns. As both Sharon Zukin and Mark Gottdiener remind us in their readings, individuals and groups make sense of and experience the places where they shop, socialize and live, thereby shaping the urban environment. These experiences may be as diverse as the groups who inhabit the city. Thus, while certain powerful urban actors, such as planners and real estate developers, use culture to realize profits from urban development, their ability to shape how we experience the city is limited.

We can simply ignore the flurry of corporate images and themes that abound in consumption places like New York's Times Square or Chicago's Miracle Mile. Or we may associate them with distaste, revulsion, or other reactions and feelings unanticipated by advertisers and developers. Along with passivity, community resistance is another reaction to the manipulation of cultural themes and symbols to sell place. As Christopher Mele points out in his selection, individuals and groups may contest real estate developers' use of local cultural forms, such as art, music, and even styles of social protest, to package a community for high-end consumption. The redevelopment of New York's now highly popular SoHo neighborhood in the 1970s, for example, was largely due to real estate developers who capitalized on the area's artistic appeal and city policies that allowed manufacturing lofts to be rezoned residential. By the 1980s, the artists themselves had been displaced. On New York's Lower East Side a long history of political antagonism toward the real estate industry fostered a culture of insurgency among squatters, homesteaders, and low-income residents, many of whom were associated with the local artist community. As Mele shows, real estate developers sought to appropriate and "tame" this insurgent culture as a selling point for middle-class rental apartments. Community resistance tended to contribute to the corporate, sanitized version of the neighborhood's identity, showing culture is a powerful yet contested force in urban development.

Richard Florida's work on the creative class further exemplifies this shift in viewing culture as a determining (as opposed to dependent) factor in shaping the form of cities. Florida asks the question – why do certain cities in the US fare better in attracting talented professionals and creative people than others? In light of the shift toward a postindustrial economy since the 1970s (a full three decades ago), most cities now boast, in varying degrees, the infrastructure of lifestyle amenities (high-end restaurant and shopping districts, entertainment zones, and renovated parks, esplanades, and other leisure spaces). Despite increasing similarities, not all cities are reputed as exciting, progressive places to live. Florida's research shows that the ranking of cities as attractive to talented and creative people is based on both objective factors (demographic diversity, urban infrastructure) and subjective factors (open-mindedness and tolerance of social difference and forward-thinking toward new ideas). That is, a city's "cultural feel" matters. According to Florida, those cities with "backwards" or provincial place identities or reputations will fare poorly in the highly competitive endeavor to attract and keep talent. His work has interesting implications for urban development policy – how do cities foster the intangibles of tolerance and acceptance?

As testimony to the importance of cultural analysis, there has been an explosion of "cultural" subfields within urban studies, including tourism/leisure studies and consumption studies. These subfields have benefited greatly from the critical scholarship within cultural studies, especially the focus on the power of symbols and imagery, representation, and the importance of consumption to individual and group identity. For further information on the "cultural turn" in urban sociology see the introduction to John Eade and Christopher Mele, eds, *Understanding the City: Contemporary and Future Perspectives* (Oxford: Blackwell, 2002: 3–24). Additional sources of consumption and urban social theory include Scott Lash and John Urry's *Economies of Signs and Space* (London: Sage Publications, 1994), John Urry's *Consuming Places* (London: Routledge 1995), Celia Lury's *Consumer Culture* (London: Polity Press, 1996) and the edited volumes by Dennis R. Judd, ed., *The Infrastructure of Play: Building the Tourist City* (Armonk, NY: M. E. Sharpe, 2003), Michael Keith and Steve Pile, eds, *Place and the Politics of Identity* (London: Routledge 1993) and David Howes, ed., *Cross-Cultural Consumption: Global Markets, Local Realities* (London: Routledge 1996).

"Whose Culture? Whose City?"

from *The Cultures of Cities* (1995)

Sharon Zukin

Editors' Introduction

Sharon Zukin is a leading urban sociologist in the study of cities and culture. Her 1982 *Loft Living*, which examined New York City's SoHo neighborhood, is a landmark study of the intersection of culture and urban development. In it, she carefully presents the complementary and contradictory roles artists, tenants, manufacturers, real estate developers, and city officials play in the transforming of SoHo from a light manufacturing loft district in the 1960s to a trendy, increasingly upscale residential and commercial district. In the reading that follows, Zukin again addresses the interplay of various urban actors around issues of culture, which, she argues, has taken on greater significance in how cities are built and how we experience them.

Indeed, culture is the "motor of economic growth" for cities and forms the basis of what Zukin labels the "symbolic economy." The symbolic economy is comprised of two parallel production systems: the production of space, in which aesthetic ideals, cultural meanings, and themes are incorporated into the look and feel of buildings, streets, and parks and the production of symbols, in which more abstract cultural representations influence how particular spaces within cities should preferably be "consumed" or used and by whom. The latter generates a good deal of controversy: as more and more ostensibly "public" spaces become identified (and officially sanctioned) with particular, often commercially generated, themes, we are left to ask "whose culture? whose city?"

We can easily see the symbolic economy at work in urban places such as Boston's Faneuil Hall, New York's South Street Seaport, or Baltimore's Harborplace. Here, cultural themes – mainly gestures toward a romanticized, imaginary past of American industrial growth – are enlisted to define place and, more specifically, what we should do there (shop, eat) and who we should encounter (other shoppers, tourists). Such places, although carefully orchestrated in design and feel, are popular because they offer a respite from the homogeneity and bland uniformity of suburban spaces. Local government officials and business alliances have turned toward manufacturing new consumption spaces of urban diversity (albeit narrowly defined) or showcasing existing ones – ethnic neighborhoods, revitalized historic districts, artist enclaves – as a competitive economic advantage over suburbs and other cities.

Culture, then, is purposefully used by developers and city officials to frame urban space to attract new residential tenants, to entice high-end shoppers, or court tourists and visitors from around the globe. But the fusing of culture and space is not limited to governments, corporations, and the real estate industry. The arguably less powerful inhabitants of the city – the ordinary residents, community associations, and block clubs – use cultural representations, too, to stamp their identity on place and to exert their cultural presence in public spaces. Ethnic festivals and parades mark the city and provide a cultural roadmap to what its spaces mean for certain groups and users. Every summer the city of Toronto hosts Caribana, the largest Caribbean festival in North America. The festival celebrates the vibrant ties of this Canadian city's large immigrant population to the Caribbean. Brilliantly costumed masqueraders and dozens of trucks carrying live soca, calypso, steel pan, reggae and salsa artists enliven the city's streets. But the festival is by no means "local;" it attracts over a million participants annually, including hundreds of thousands of tourists, for whom Toronto "means" Caribana.

Finally, Zukin draws our attention to the increasing slippage between public and private spaces within the contemporary city. Historically, urban parks and streets have functioned as spaces where persons from different classes, ethnicities, and walks of life have intermingled and "rubbed elbows." Although always regulated and controlled, the identity of public spaces was seen as open and never exclusively defined by a single use or specific or preferred set of users. While parents and toddlers may "own" a corner of a city park on warm, sunny days, others – teenagers, lovers, or beer drinkers – lay claim to that same spot at different times of the day. As corporations increasingly sponsor public events and locally financed security forces police streets, the use of public spaces and their intended users are narrowed. Culture again becomes enlisted. Just as symbols, images, and other forms of representations may "define" shopping districts, restaurants, and theme parks, so they work in the (re)definition of public venues. Many urban plazas, waterfronts, and shopping streets have become "managed" by business associations, hence imposing their own identity on supposedly "public" space. The privatization (and militarization) of public space serves to reinforce the fear and conflict "between 'us' and 'them,' between security guards and criminals, [and] between elites and ethnic groups."

In addition to *Loft Living: Culture and Capital in Urban Change* (Baltimore, MD: The Johns Hopkins University Press, 1982), Sharon Zukin is the author of *Landscapes of Power: From Detroit to Disney World* (Berkeley, CA: University of California Press, 1991), *The Cultures of Cities* (from which this selection is excerpted) (Cambridge, MA: Blackwell, 1995), with Michael Sorkin, ed., *After the World Trade Center* (New York: Routledge, 2002), and *Point of Purchase: How Shopping Changed American Culture* (New York: Routledge, 2003). She is Broeklundian Professor of Sociology at Brooklyn College, CUNY.

Cities are often criticized because they represent the basest instincts of human society. They are built versions of Leviathan and Mammon, mapping the power of the bureaucratic machine or the social pressures of money. We who live in cities like to think of "culture" as the antidote to this crass vision. The Acropolis of the urban art museum or concert hall, the trendy art gallery and café, restaurants that fuse ethnic traditions into culinary logos – cultural activities are supposed to lift us out of the mire of our everyday lives and into the sacred spaces of ritualized pleasures.

Yet culture is also a powerful means of controlling cities. As a source of images and memories, it symbolizes "who belongs" in specific places. As a set of architectural themes, it plays a leading role in urban redevelopment strategies based on historic preservation or local "heritage." With the disappearance of local manufacturing industries and periodic crises in government and finance, culture is more and more the business of cities – the basis of their tourist attractions and their unique, competitive edge. The growth of cultural consumption (of art, food, fashion, music, tourism) and the industries that cater to it fuels the city's symbolic economy, its visible ability to produce both symbols and space.

In recent years, culture has also become a more explicit site of conflicts over social differences and urban fears. Large numbers of new immigrants and ethnic minorities have put pressure on public institutions, from schools to political parties, to deal with their individual demands. Such high culture institutions as art museums and symphony orchestras have been driven to expand and diversify their offerings to appeal to a broader public. These pressures, broadly speaking, are both ethnic and aesthetic. By creating policies and ideologies of "multiculturalism," they have forced public institutions to change.

On a different level, city boosters increasingly compete for tourist dollars and financial investments by bolstering the city's image as a center of cultural innovation, including restaurants, avant garde performances, and architectural design. These cultural strategies of redevelopment have fewer critics than multiculturalism. But they often pit the self-interest of real estate developers, politicians, and expansion minded cultural institutions against grassroots pressures from local communities.

At the same time, strangers mingling in public space and fears of violent crime have inspired the growth of private police forces, gated and barred communities, and a movement to design public spaces for maximum surveillance. These, too, are a source of contemporary urban culture. If one way of dealing with the material inequalities of city life has been to aestheticize diversity, another way has been to aestheticize fear.

Controlling the various cultures of cities suggests the possibility of controlling all sorts of urban ills, from violence and hate crime to economic decline. That this is an illusion has been amply shown by battles over multiculturalism and its warring factions – ethnic politics and urban riots. Yet the cultural power to create an image, to frame a vision, of the city has become more important as publics have become more mobile and diverse, and traditional institutions – both social classes and political parties – have become less relevant mechanisms of expressing identity. Those who create images stamp a collective identity. Whether they are media corporations like the Disney Company, art museums, or politicians, they are developing new spaces for public cultures. Significant public spaces of the late 19th and early 20th century – such as New York City's Central Park, the Broadway theater district, and the top of the Empire State Building – have been joined by Disney World, Bryant Park, and the entertaiment-based retail shops of Sony Plaza. By accepting these spaces without questioning their representations of urban life, we risk succumbing to a visually seductive, privatized public culture.

THE SYMBOLIC ECONOMY

[. . .]

Building a city depends on how people combine the traditional economic factors of land, labor, and capital. But it also depends on how they manipulate symbolic languages of exclusion and entitlement. The look and feel of cities reflect decisions about what – and who – should be visible and what should not, on concepts of order and disorder, and on uses of aesthetic power. In this primal sense, the city has always had a symbolic economy. Modern cities also owe their existence to a second, more abstract symbolic economy devised by "place entrepreneurs," officials and investors whose ability to deal with the symbols of growth yields "real" results in real estate development, new businesses, and jobs.

Related to this entrepreneurial activity is a third, traditional symbolic economy of city advocates and business elites who, through a combination of philanthropy, civic pride, and desire to establish their identity as a patrician class, build the majestic art museums, parks, and architectural complexes that represent a world-class city. What is new about the symbolic economy since the 1970s is its symbiosis of image and product, the scope and scale of selling images on a national and even a global level, and the role of the symbolic economy in speaking for, or representing, the city.

[. . .]

The entrepreneurial edge of the economy [has] shifted toward deal making and selling investments and toward those creative products that could not easily be reproduced elsewhere. Product design – creating the look of a thing – was said to show economic genius. Hollywood film studios and media empires were bought and sold and bought again. In the 1990s, with the harnessing of new computer-based technologies to marketing campaigns, the "information superhighway" promised to join companies to consumers in a Manichean embrace of technology and entertainment.

[. . .]

The growth of the symbolic economy in finance, media, and entertainment may not change the way entrepreneurs do business. But it has already forced the growth of towns and cities, created a vast new work force, and changed the way consumers and employees think. The facilities where these employees work – hotels, restaurants, expanses of new construction and undeveloped land – are more than just workplaces. They reshape geography and ecology; they are places of creation and transformation.

The Disney Company, for example, makes films and distributes them from Hollywood. It runs a television channel and sells commercial spinoffs, such as toys, books, and videos, from a national network of stores. Disney is also a real estate developer in Anaheim, Orlando, France, and Japan and the proposed developer of a theme park in Virginia and a hotel and theme park in Times Square. Moreover, as an employer, Disney has redefined work roles. Proposing a model for change in the emerging service economy, Disney has shifted from the white-collar worker to a new chameleon of "flexible" tasks. The planners at its corporate headquarters are "imaginers"; the costumed crowd-handlers at its theme parks are "cast members." Disney suggests that the symbolic economy is more than just the sum of the services it provides. The symbolic economy unifies material practices of finance, labor, art, performance, and design.

[...]

The symbolic economy recycles real estate as it does designer clothes. Visual display matters in American and European cities today, because the identities of places are established by sites of delectation. The sensual display of fruit at an urban farmers' market or gourmet food store puts a neighborhood "on the map" of visual delights and reclaims it for gentrification. A sidewalk cafe takes back the street from casual workers and homeless people.

[...]

Mass suburbanization since the 1950s has made it unreasonable to expect that most middle-class men and women will want to live in cities. But developing small places within the city as sites of visual delectation creates urban oases where everyone *appears* to be middle class. In the fronts of the restaurants or stores, at least, consumers are strolling, looking, eating, drinking, sometimes speaking English and sometimes not. In the back regions, an ethnic division of labor guarantees that immigrant workers are preparing food and cleaning up. This is not just a game of representations: developing the city's symbolic economy involves recycling workers, sorting people in housing markets, luring investment, and negotiating political claims for public goods and ethnic promotion. Cities from New York to Los Angeles and Miami seem to thrive by developing small districts around specific themes. Whether it is Times Square or el Calle Ocho, a commercial or an "ethnic" district, the narrative web spun by the symbolic economy around a specific place relies on a vision of cultural consumption and a social and an ethnic division of labor.

[...]

I see public culture as socially constructed on the micro-level. It is produced by the many social encounters that make up daily life in the streets, shops, and parks – the spaces in which we experience public life in cities. The right to be in these spaces, to use them in certain ways, to invest them with a sense of our selves and our communities – to claim them as ours and to be claimed in turn by them – make up a constantly changing public culture. People with economic and political power have the greatest opportunity to shape public culture by controlling the building of the city's public spaces in stone and concrete. Yet public space is inherently democratic. The question of who can occupy public space, and so define an image of the city, is open-ended.

Talking about the cultures of cities in purely visual terms does not do justice to the material practices of politics and economics that create a symbolic economy. But neither does a strictly political-economic approach suggest the subtle powers of visual and spatial strategies of social differentiation. The rise of the cities' symbolic economy is rooted in two long-term changes – the economic decline of cities compared to suburban and non-urban spaces and the expansion of abstract financial speculation – and in such short-term factors, dating from the 1970s and 1980s, as new mass immigration, the growth of cultural consumption, and the marketing of identity politics. We cannot speak about cities today without understanding how cities use culture as an economic base, how capitalizing on culture spills over into the privatization and militarization of public space, and how the power of culture is related to the aesthetics of fear.

CULTURE AS AN ECONOMIC BASE

[...]

Culture is intertwined with capital and identity in the city's production systems. From one point of view, cultural institutions establish a competitive advantage over other cities for attracting new businesses and corporate elites. Culture suggests the coherence and consistency of a brand name product. Like any commodity, "cultural" landscape has the possibility of generating other commodities.

[. . .]

In American and European cities during the 1970s, culture became more of an instrument in the entrepreneurial strategies of local governments and business alliances. In the shift to a post-postwar economy, who could build the biggest modern art museum suggested the vitality of the financial sector. Who could turn the waterfront from docklands rubble to parks and marinas suggested the possibilities for expansion of the managerial and professional corps. This was probably as rational a response as any to the unbeatable isolationist challenge of suburban industrial parks and office campuses. The city, such planners and developers as James Rouse believed, would counter the visual homogeneity of the suburbs by playing the card of aesthetic diversity.

[. . .]

Art museums, boutiques, restaurants, and other specialized sites of consumption create a social space for the exchange of ideas on which businesses thrive. While these can never be as private as a corporate dining room, urban consumption spaces allow for more social interaction among business elites. They are more democratic, accessible spaces than old-time businessmen's clubs. They open a window to the city – at least, to a rarified view of the city – and, to the extent they are written up in "lifestyle" magazines and consumer columns of the daily newspapers, they make ordinary people more aware of the elites' cultural consumption. Through the media, the elites' cultural preferences change what many ordinary people know about the city.

The high visibility of spokespersons, stars, and stylists for culture industries underlines the "sexy" quality of culture as a motor of economic growth. Not just in New York, Los Angeles, or Chicago, business leaders in a variety of low-profile, midsize cities are actively involved on the boards of trustees of cultural institutions because they believe that investing in the arts leads to more growth in other areas of the urban economy. They think a tourist economy develops the subjective image of place that "sells" a city to other corporate executives.

[. . .]

CULTURE AS A MEANS OF FRAMING SPACE

For several hundred years, visual representations of cities have "sold" urban growth. Images, from early maps to picture postcards, have not simply reflected real city spaces; instead, they have been imaginative reconstructions – from specific points of view – of a city's monumentality. The development of visual media in the 20th century made photography and movies the most important cultural means of framing urban space, at least until the 1970s. Since then, as the surrealism of King Kong shifted to that of *Blade Runner* and redevelopment came to focus on consumption activities, the material landscape itself – the buildings, parks, and streets – has become the city's most important visual representation. Historic preservation has been very important in this representation. Preserving old buildings and small sections of the city re-presents the scarce "monopoly" of the city's visible past. Such a monopoly has economic value in terms of tourist revenues and property values. Just an image of historic preservation, when taken out of context, has economic value. In Syracuse, New York, a crankshaft taken from a long-gone salt works was mounted as public sculpture to enhance a redevelopment project.

[. . .]

More common forms of visual re-presentation in all cities connect cultural activities and populist images in festivals, sports stadiums, and shopping centers. While these may simply be minimized as "loss leaders" supporting new office construction, they should also be understood as producing space for a symbolic economy.

[. . .]

Linking public culture to commercial cultures has important implications for social identity and social control. Preserving an ecology of images often takes a connoisseur's view of the past, re-reading the legible practices of social class discrimination and financial speculation by reshaping the city's collective memory. Boston's Faneuil Hall, South Street Seaport in New York, Harborplace in Baltimore, and London's Tobacco Wharf make the waterfront of older cities into a consumers' playground, far safer for tourists and cultural consumers than the closed worlds of wholesale fish and vegetable dealers and longshoremen. In such newer cities as Los Angeles or San Antonio, reclaiming the historic core, or the fictitious historic core, of the city for the middle classes puts the pueblo or the Alamo into an entirely different landscape from that of the surrounding inner city. On one level, there is a loss of authenticity, that is compensated for by a re-created historical narrative and a commodification of images; on another, men and women are simply displaced from public spaces they once considered theirs.

[...]

But incorporating new images into visual representations of the city can be democratic. It can integrate rather than segregate social and ethnic groups, and it can also help negotiate new group identities. In New York City, there is a big annual event organized by Caribbean immigrants, the West Indian-American Day Carnival parade, which is held every Labor Day on Eastern Parkway in Brooklyn. The parade has been instrumental in creating a pan-Caribbean identity among immigrants from the many small countries of that region. The parade also legitimizes the "gorgeous mosaic" of the ethnic population described by Mayor David N. Dinkins in 1989. The use of Eastern Parkway for a Caribbean festival reflects a geographical redistribution of ethnic groups – the Africanization of Brooklyn, the Caribbeanization of Crown Heights. More problematically, however, this cultural appropriation of public space supports the growing political identity of the Caribbean community and challenges the Lubavitcher Hassidim's appropriation of the same neighborhood. In Pasadena, California, African-American organizations have demanded representation on the nine-person commission that manages the annual Rose Parade, that city's big New Year's Day event. These cultural models of inclusion differ from the paradigm of legally imposed racial integration that eliminated segregated festivals and other symbolic activities in the 1950s and 1960s. By giving distinctive cultural groups access to the same public space, they incorporate separate visual images and cultural practices into the same public cultures.

Culture can also be used to frame, and humanize, the space of real estate development. Cultural producers who supply art (and sell "interpretation") are sought because they legitimize the appropriation of space. Office buildings are not just monumentalized by height and facades, or they are given a human face by video artists' screen installations and public concerts. Every well-designed downtown has a mixed-use shopping center and a nearby artists' quarter. Sometimes it seems that every derelict factory district or waterfront has been converted into one of those sites of visual delectation – a themed shopping space for seasonal produce, cooking equipment, restaurants, art galleries, and an aquarium. Urban redevelopment plans, from Lowell, Massachusetts, to downtown Philadelphia, San Francisco, and Los Angeles, focus on museums. Unsuccessful attempts to use cultural districts or aquariums to stop economic decline in Flint, Michigan and Camden, New Jersey – cities where there is no major employer – only emphasize the appeal of framing a space with a cultural institution when all other strategies of economic development fail.

Artists themselves have become a cultural means of framing space. They confirm the city's claim of continued cultural hegemony, in contrast to the suburbs and exurbs. Their presence – in studios, lofts, and galleries – puts a neighborhood on the road to gentrification. Ironically, this has happened since artists have become more self-conscious defenders of their own interests as artists and more involved in political organizations. Often they have been co-opted into property redevelopment projects as beneficiaries, both developers of an aesthetic mode of producing space (in public art, for example) and investors in a symbolic economy. There are, moreover, special connections between artists and corporate patrons. In such cities as New York and Los Angeles, the presence of artists documents a claim to these cities' status in the global hierarchy. The display of art, for public improvement or private gain, represents an abstraction of economic and social power. Among business elites, those from finance, insurance, and real estate are generally great patrons of both art museums and public art, as if to emphasize their prominence in the city's symbolic economy.

[...]

So the symbolic economy features two parallel production systems that are crucial to a city's material life: the *production of space*, with its synergy of capital investment and cultural meanings, and the *production of symbols*, which constructs both a currency of commercial exchange and a language of social identity. Every effort to rearrange space in the city is also an attempt at visual re-presentation. Raising property values, which remains a goal of most urban elites, requires imposing a new point of view. But negotiating whose point of view and the costs of imposing it create problems for public culture.

[...]

PUBLIC SPACE

[...]

It is important to understand the histories of symbolically central public spaces. The history of Central Park, for example, shows how, as definitions of who should have access to public space have changed, public cultures have steadily become more inclusive and democratic. From 1860 to 1880, the first uses of the park – for horse-back riders and carriages – rapidly yielded to sports activities and promenades for the mainly immigrant working class. Over the next 100 years, continued democratization of access to the park developed together with a language of political equality. In the whole country, it became more difficult to enforce outright segregation by race, sex, or age.

During the 1970s, public space, especially in cities, began to show the effects of movements to "deinstitutionalize" patients of mental hospitals without creating sufficient community facilities to support and house them. Streets became crowded with "others," some of whom clearly suffered from sickness and disorientation. By the early 1980s, the destruction of cheap housing in the centers of cities, particularly single-room-occupancy hotels, and the drastic decline in producing public housing, dramatically expanded the problem of homelessness. Public space, such as Central Park, became unintended public shelter. As had been true historically, the democratization of public space was entangled with the question of fear for physical security.

[...]

Business Improvement Districts follow a fairly new model in New York State and in smaller cities around the United States, that allows business and property owners in commercial districts to tax themselves voluntarily for maintenance and improvement of public areas and take these areas under their control. The concept originated in the 1970s as special assessment districts; in the 1980s, the name was changed to a more upbeat acronym, business improvement districts (BIDs). A BID can be incorporated in any commercial area. Because the city government has steadily reduced street cleaning and trash pickups in commercial streets since the fiscal crisis of 1975, there is a real incentive for business and property owners to take up the slack. A new law was required for such initiatives: unlike shopping malls, commercial streets are publicly owned, and local governments are responsible for their upkeep.

[...]

What kind of public culture is created under these conditions? Do urban BIDs create a Disney World in the streets, take the law into their own hands, and reward their entrepreneurial managers as richly as property values will allow? If elected public officials continue to urge the destruction of corrupt and bankrupt public institutions, I imagine a scenario of drastic privatization, with BIDs replacing the city government.

[...]

In their own way, under the guise of improving public spaces, BIDs nurture a visible social stratification. They channel investment into a central space, a space with both real and symbolic meaning for elites as well as other groups. The resources of the rich Manhattan BIDs far outstrip those even potentially available in other areas of the city, even if those areas set up BIDs. The rich BIDs' opportunity to exceed the constraints of the city's financial system confirms the fear that the prosperity of a few central spaces will stand in contrast to the impoverishment of the entire city.

BIDs can be equated with a return to civility, "an attempt to reclaim public space from the sense of menace that drives shoppers, and eventually store owners and citizens, to the suburbs." But rich BIDs can be criticized on the grounds of control, accountability, and vision. Public space that is no longer controlled by public agencies must inspire a liminal public culture open to all but governed by the private sector. Private management of public space does create some savings: saving money by hiring nonunion workers, saving time by removing design questions from the public arena. Because they choose an abstract aesthetic with no pretense of populism, private organizations avoid conflicts over representations of ethnic groups that public agencies encounter when they subsidize public art, including murals and statues.

Each area of the city gets a different form of visual consumption catering to a different constituency: culture functions as a mechanism of stratification. The public culture of midtown public space diffuses down through the poorer BIDs. It focuses on clean design, visible security, historic architectural features, and the sociability among strangers achieved by suburban shopping malls. Motifs of local identity are chosen by merchants and commercial property owners. Since most commercial property owners and merchants do not live in the area of their business or even in New York City, the sources of their vision of public culture may be eclectic: the nostalgically remembered city, European piazzas, suburban shopping malls, Disney World. In general, however, their vision of public space derives from commercial culture.

[. . .]

SECURITY, ETHNICITY, AND CULTURE

One of the most tangible threats to public culture comes from the politics of everyday fear. Physical assaults, random violence, hate crimes that target specific groups: the dangers of being in public spaces utterly destroy the principle of open access. Elderly men and women who live in cities commonly experience fear as a steady erosion of spaces and times available to them. An elderly Jewish politician who in the 1950s lived in Brownsville, a working-class Jewish neighborhood in Brooklyn where blacks began to move in greater numbers as whites moved out, told me, "My wife used to be able to come out to meet me at night, after a political meeting, and leave the kids in our apartment with the door unlocked." A Jewish woman remembers about that same era, "I used to go to concerts in Manhattan wearing a fur coat and come home on the subway at 1 a.m." There may be some exaggeration in these memories, but the point is clear. And it is not altogether different from the message behind crimes against black men who venture into mainly white areas of the city at night or attacks on authority figures such as police officers and firefighters who try to exercise that authority against street gangs, drug dealers, and gun-toting kids. Cities are not safe enough for people to participate in a public culture.

"Getting tough" on crime by building more prisons and imposing the death penalty are all too common answers to the politics of fear. Another answer is to privatize and militarize public space – making streets, parks, and even shops more secure but less free, or creating spaces, such as shopping malls and Disney World, that only *appear* to be public spaces because so many people use them for common purposes. It is not so easy, given a language of social equality, a tradition of civil rights, and a market economy, to enforce social distinctions in public space. The flight from "reality" that led to the privatization of public space in Disney World is an attempt to create a different, ultimately more menacing kind of public culture.

[. . .]

Gentrification, historic preservation, and other cultural strategies to enhance the visual appeal of urban spaces developed as major trends during the late 1960s and early 1970s. Yet these years were also a watershed in the institutionalization of urban fear. Voters and elites – a broadly conceived middle class in the United States – could have faced the choice of approving government policies to eliminate poverty, manage ethnic competition, and integrate everyone into common public institutions. Instead, they chose to buy protection, fueling the growth of the private security industry. This reaction was closely related to a perceived decline in public morality, an "elimination of almost all stabilizing authority" in urban public space.

[. . .]

In the past, those people who lived so close together they had to work out some etiquette for sharing, or dividing, public space were usually the poor. An exception that affected everyone was the system of racial segregation that worked by law in the south and by convention in many northern states until the 1960s, when – not surprisingly – perceptions of danger among whites increased. Like segregation, a traditional etiquette of public order of the urban poor involves dividing up territory by ethnic groups. This includes the system of "ordered segmentation" that the Chicago urban sociologist Gerald Suttles described a generation ago, at the very moment it was being outmoded by increased racial and ethnic mixing, ideologies of community empowerment, and the legitimization of ethnicity as a formal norm of political representation. Among city dwellers today, innumerable informal etiquettes for survival in public spaces flourish. The "streetwise" scrutiny of passersby described by the sociologist Elijah Anderson is one means for unarmed individuals to secure the streets. I think ethnicity – a cultural strategy for producing difference – is another, and it survives on the politics of fear by requiring people to keep their distance from certain aesthetic markers. These markers vary over time. Pants may be baggy or pegged, heads may be shaggy or shaved. Like fear itself, ethnicity becomes an aesthetic category.

[...]

For a brief moment in the late 1940s and early 1950s, working-class urban neighborhoods held the possibility of integrating white Americans and African-Americans in roughly the same social classes. This dream was laid to rest by movement to the suburbs, continued ethnic bias in employment, the decline of public services in expanding racial ghettos, criticism of integration movements for being associated with the Communist party, and fear of crime. Over the next 15 years, enough for a generation to grow up separate, the inner city developed its stereotyped image of "Otherness." The reality of minority groups' working-class life was demonized by a cultural view of the inner city "made up of four ideological domains: a physical environment of dilapidated houses, disused factories, and general dereliction; a romanticized notion of white working-class life with particular emphasis on the centrality of family life; a pathological image of black culture; and a stereotypical view of street culture."

By the 1980s, the development of a large black middle class with incomes more or less equal to white households' and the increase in immigrant groups raised a new possibility of developing ethnically and racially integrated cities. This time, however, there is a more explicit struggle over who will occupy the image of the city. Despite the real impoverishment of most urban populations, the larger issue is whether cities can again create an inclusive public culture. The forces of order have retreated into "small urban spaces," like privately managed public parks that can be refashioned to project an image of civility. Guardians of public institutions – teachers, cops – lack the time or inclination to *understand* the generalized ethnic Other.

[...]

When Disneyland recruited teenagers in South Central Los Angeles for summer jobs following the riots of 1992, it thrust into prominence a new confluence between the sources of contemporary public culture: a confluence between commercial culture and ethnic identity. Defining public culture in these terms recasts the way we view and describe the cultures of cities. Real cities are both material constructions, with human strengths and weaknesses, and symbolic projects developed by social representations, including affluence and technology, ethnicity and civility, local shopping streets and television news. Real cities are also macro-level struggles between major sources of change – global and local cultures, public stewardship and privatization, social diversity and homogeneity – and micro-level negotiations of power. Real cultures, for their part, are not torn by conflict between commercialism and ethnicity; they are made up of one-part corporate image selling and two-parts claims of group identity, and get their power from joining autobiography to hegemony – a powerful aesthetic fit with a collective lifestyle.

[...]

How do we connect what we experience in public space with ideologies and rhetorics of public culture? On the streets, the vernacular culture of the powerless provides a currency of economic exchange and a language of social revival. In other public spaces – grand plazas, waterfronts, and shopping streets reorganized by business improvement districts – another landscape incorporates vernacular culture or opposes it with its own image of identity and desire. Fear of reducing the distance between "us" and "them," between security guards and criminals, between elites and ethnic groups, makes culture a crucial weapon in reasserting order.

"Cities and the Creative Class"

from *City and Community* (2003)

Richard Florida

Editors' Introduction

We often tend to think of cities and culture in terms of museums, art galleries, and concert halls, as well as high-end shopping districts and bohemian artist enclaves. In this selection, Richard Florida asks us to view these cultural amenities and others as powerful attributes that attract a highly-desirable workforce he calls the "creative class." Florida is the author of the "creative capital thesis" which holds that a community of creative and talented people is fundamental to contemporary urban development and that there are underlying (and detectable) factors that favor certain cities ("creative centers") over others.

Cities have long been identified with diversity, creativity and innovation. But, as Florida argues, urban scholars and policymakers have typically and exclusively identified these assets with corporations and firms. This is apparent when one examines the tax abatements and other financial incentives municipal and state governments offer companies to relocate to or remain in their cities. When we turn our explanatory gaze to municipal efforts to attract talented and creative individuals (and to keep them as residents), we see certain theories might apply but each has shortcomings.

Agglomeration and cluster theories offer one explanation of why certain industries find cities desirable places to do business. Most cities have legal districts, for example, where law offices and courts houses are in close proximity. Legal transactions tend to require a good deal of face-to-face contact. Other industries, such as investment firms, tend to benefit in varying degrees from the efficiency and ease of geographic proximity.

Social capital theory also values the immediacy of close ties and connections among groups and individuals. Its major proponent, Robert Putnam, argued that a community's prosperity rests on strong ties and associations among its members. Yet, as Florida points out, the social and technological changes of the past twenty years make Putnam's call for a return to small-town community a romantic notion at best.

Florida finds greater utility in human capital theory, which posits that urban and regional growth is increasingly dependent on the presence of higher education and similar institutions to attract talented individuals. Many states and cities have funded university-corporate partnerships as economic incubators to enhance high-tech, well-paying employment.

Finally, Florida discusses his own "creative capital" theory of urban and regional development. He agrees partly with the human capital approach – infrastructure (good schools, good housing, etc.) is indeed important to attract talented and creative people. But positive lifestyle factors are also necessary to attract and maintain the creative class. This class is comprised of professionals who work in knowledge-based occupations in high-tech sectors, financial services, the legal and health-care professions, and business management.

In this age of telecommunications, professionals are able to live and work in a variety of cities or small towns. Why, then, do certain cities tend to attract a large share of the creative class? Florida argues that creative centers provide an environment where all forms of creativity – artistic and cultural, technological and economic – take place. Cities are misguided, Florida argues, in their emphasis upon infrastructural attractions, such as sports stadiums and tourism-and-entertainment districts. In his research, he found that members of the creative class desire "high-quality experiences, an openness to diversity of all kinds, and, above all else, the opportunity to validate their identities as creative people." The cities that are currently faring best have an abundance of technology, talent, and tolerance. Each is a necessary, Florida tells us, but by itself an insufficient condition. To attract creative and talented people and stimulate urban development, a place must have all three. The benefits of a concentrated creative class are many: growth in the well-paying employment sector, a higher tax base, reinvestment in the built environment, and population growth.

Florida is leading an effort to compel scholars to think more broadly about factors that contribute to successful cities – to move away from our late nineteenth and early twentieth century view of the city in terms of production of goods to seeing cities as centers of experience, lifestyle, amenities, and entertainment. This shift is not only apparent in cities such as San Francisco, Seattle, and Boston, but in older industrial cities such as Chicago and Philadelphia, which are moving to become centers of entertainment and cultural experiences.

Richard Florida is the Heinz Professor of Economic Development at Carnegie Mellon University. He is the author of *The Rise of the Creative Class: And How It's Transforming Work, Leisure, Community and Everyday Life* (New York: Basic Books, 2002) in which he identifies the underlying changes and the structural economic transformations behind dramatic changes in work and lifestyle that are fueling urban change. The book was awarded the Political Book Award for 2002 by the *Washington Monthly* and was listed as one of the ten most influential books of 2002 by the *Globe and Mail*. He is founder of Creativity Group, a consulting firm that works with corporations and governments around the world.

From the seminal work of Alfred Marshall to the 1920 studies by Robert Park to the pioneering writings of Jane Jacobs, cities have captured the imagination of sociologists, economists, and urbanists. For Park and especially for Jacobs, cities were cauldrons of diversity and difference, creativity and innovation. Yet over the last several decades, scholars have somehow forgotten this basic, underlying theme of urbanism. For the past two decades, I have conducted research on the social and economic functions of cities and regions. Generally speaking, the conventional wisdom in my field of regional development has been that companies, firms, and industries drive regional innovation and growth, and thus an almost exclusive focus in the literature on the location and, more recently, the clustering of firms and industries. From a policy perspective, this basic conceptual approach has undergirded policies that seek to spur growth by offering firms financial incentives and the like. More recently, scholars such as Robert Putnam have focused on the social functions of neighborhoods, communities, and cities, while others, such as the urban sociologist Terry Clark and the economist Edward Glasear, have turned their attention toward human capital, consumption, and cities as lifestyle and entertainment districts.

[. . .]

AGGLOMERATION AND CLUSTER THEORIES

Many researchers, sociologists, and academics have theorized on the continued importance of place in economic and social life. An increasingly influential view suggests that place remains important as a locus of economic activity because of the tendency of firms to cluster together. This view builds on the influential theories of the economist Alfred Marshall, who argued that firms cluster in "agglomerations" to gain productive efficiencies. The contemporary variant of this view, advanced by Harvard Business School professor Michael Porter, has many proponents in academia and in the practice of economic development. It is clear that similar firms tend to cluster. Examples of this sort of agglomeration include not only Detroit and Silicon Valley, but the *maquiladora* electronics and auto-parts districts in Mexico, the clustering of disk-drive makers in Singapore and of flat-panel-display producers in Japan, and the garment district and Broadway theater district in New York City.

The question is not whether firms cluster but why. Several answers have been offered. Some experts believe that clustering captures efficiencies generated from tight linkages between firms. Others say it has to do with the positive benefits of co-location, or what they call "spillovers." Still others claim it is because certain kinds of activity require face-to-face contact. But these are only partial answers. More importantly, companies cluster in order to draw from concentrations of talented people who power innovation and economic growth. The ability to rapidly mobilize talent from such a concentration of people is a tremendous source of competitive advantage for companies in our time-driven economy of the creative age.

THE SOCIAL CAPITAL PERSPECTIVE

An alternative view is based on Robert Putnam's social capital theory. From his perspective, regional economic growth is associated with tight-knit communities where people and firms form and share strong ties. In his widely read book *Bowling Alone*, he makes a compelling argument that many aspects of community life declined precipitously over the last half of the 20th century. Putnam gets his title from his finding that from 1980–1993, league bowling declined by 40 percent, whereas the number of individual bowlers rose by 10 percent. This, he argues, is just one indicator of a broader and more disturbing trend. Across the nation, people are less inclined to be part of civic groups: voter turnout is down, so is church attendance and union membership, and people are less and less inclined to volunteer. All of this stems from what Putnam sees as a long-term decline in social capital.

By this, he means that people have become increasingly disconnected from one another and from their communities. Putman finds this disengagement in the declining participation in churches, political parties, and recreational leagues, not to mention the loosening of familial bonds. Through painstakingly detailed empirical research, he documents the decline in social capital in civic and social life. For Putman, declining social capital means that society becomes less trustful and less civic-minded. Putnam believes a healthy, civic-minded community is essential to prosperity.

Although initially Putnam's theory resonated with me, my own research indicates a different trend. The people in my focus groups and interviews rarely wished for the kinds of community connectedness Putnam talks about. If anything, it appeared they were trying to get away from those kinds of environments. To a certain extent, participants acknowledged the importance of community, but they did not want it to be invasive or to prevent them from pursuing their own lives. Rather, they desired what I have come to term "quasi-anonymity." In the terms of modern sociology, these people prefer weak ties to strong.

This leads me to an even more basic observation. The kinds of communities that we both desire and that generate economic prosperity are very different than those of the past. Social structures that were important in earlier years now work against prosperity. Traditional notions of what it means to be a close, cohesive community and society tend to inhibit economic growth and innovation. Where strong ties among people were once important, weak ties are now more effective. Those social structures that historically embraced closeness may now appear restricting and invasive. These older communities are being exchanged for more inclusive and socially diverse arrangements. These trends are also what the statistics seem to bear out.

[. . .]

Historically, strong-tied communities were thought to be beneficial. However, there are some theorists that argue the disadvantages of such tight bonds. Indeed, social capital can and often does cut both ways: it can reinforce belonging and community, but it can just as easily shut out newcomers, raise barriers to entry, and retard innovation.

[. . .]

Places with dense ties and high levels of traditional social capital provide advantages to insiders and thus promote stability, while places with looser networks and weaker ties are more open to newcomers and thus promote novel combinations of resources and ideas.

HUMAN CAPITAL AND URBAN-REGIONAL GROWTH

Over the past decade or so, a potentially more powerful theory for city and regional growth has emerged. This theory postulates that people are the motor force behind regional growth. Its proponents thus refer to it as the "human capital" theory of regional development.

Economists and geographers have always accepted that economic growth is regional – that is driven by, and spreads from, regions, cities, or even neighborhoods. The traditional view, however, is that places grow either because they are located on transportation routes or because they have natural resources that encourage firms to locate there. According to this conventional view, the economic importance of a place is tied to the efficiency with which one can make things and do business. Governments employ this theory when they use tax breaks and highway construction to attract business. But these cost-related factors are no longer as crucial to success.

The proponents of the human capital theory argue that the key to regional growth lies not in reducing the costs of doing business, but in endowments of highly-educated and productive people. The human capital theory – like many theories of cities and urban areas – owes a debt to Jane Jacobs. Decades ago, Jacobs noted the ability of cities to attract creative people and thus spur economic growth. The Nobel-prize-winning economist Robert Lucas sees the productivity effects that come from the clustering of human capital as the critical factor in regional economic growth, referring to this as a "Jane Jacobs externality."

[...]

THE CREATIVE CAPITAL PERSPECTIVE

The human capital theory establishes that creative people are the driving force in regional economic growth. From that perspective, economic growth will occur in places that have highly educated people. But in treating human capital as a stock or endowment, this theory begs the question: Why do creative people cluster in certain places? In a world where people are highly mobile, why do they choose some cities over others and for what reasons?

Although economists and social scientists have paid a lot of attention to how companies decide where to locate, they have virtually ignored how people do so. This is the fundamental question I have tried to answer. In my interviews and focus groups, the same answer kept coming back: people said that economic *and* lifestyle considerations both matter, and so does the mix of both factors. In reality, people were not making the career decisions or geographic moves that the standard theories said they should: They were not slavishly following jobs to places. Instead, it appeared that highly educated individuals were drawn to places that were inclusive and diverse. Not only did my qualitative research indicate this trend, but the statistical analysis proved the same.

Gradually, I came to see my perspective, the creative capital theory, as distinct from the human capital theory. From my perspective, creative people power regional economic growth and these people prefer places that are innovative, diverse, and tolerant. My theory thus differs from the human capital theory in two respects: (1) it identifies a type of human capital, creative people, as being key to economic growth; and (2) it identifies the underlying factors that shape the location decisions of these people, instead of merely saying that regions are blessed with certain endowments of them.

To begin with, creative capital begins most fundamentally with the people I call the "creative class." The distinguishing characteristic of the creative class is that its members engage in work whose function is to "create meaningful new forms." The super-creative core of this new class includes scientists and engineers, university professors, poets and novelists, artists, entertainers, actors, designers, and architects, as well as the "thought leadership" of modern society: nonfiction writers, editors, cultural figures, think-tank researchers, analysts, and other opinion-makers. Members of this super-creative core produce new forms or designs that are readily transferable and broadly useful, such as designing a product that can be widely made, sold, and used; coming up with a theorem or strategy that can be applied in many cases; or composing music that can be performed again and again.

Beyond this core group, the creative class also includes "creative professionals" who work in a wide range of knowledge-based occupations in high-tech sectors, financial services, the legal and health-care professions, and business management. These people engage in creative problem-solving, drawing on complex bodies of knowledge to solve specific problems. Doing so typically requires a high degree of formal education and thus a high level of human capital. People who do this kind of work may sometimes come up with methods or products that turn out to be widely useful, but that is not part of the basic job description. What they are required to do regularly is think on their own. They apply or combine standard approaches in unique ways to fit the situation, exercise a great deal of judgment, and at times must independently try new ideas and innovations.

According to my estimates, the creative class now includes some 38.3 million Americans, roughly 30 percent of the entire U.S. workforce – up from just 10 percent at the turn of the 20th century and less than 20 percent as recently as 1980. However, it is important to point out that my theory recognizes creativity as a fundamental and intrinsic human characteristic. In a very real sense, all human beings are creative and all are potentially members of the creative class. It is just that 38 million people – roughly 30 percent of the workforce – are fortunate enough to be paid to use their creativity in their work.

In my research I have discovered a number of trends that are indicative of the new geography of creativity. These are some of the patterns of the creative class:

- The creative class is moving away from traditional corporate communities, working class centers, and even many Sunbelt regions to a set of places I call "creative centers."
- The creative centers tend to be the economic winners of our age. Not only do they have high concentrations of creative-class people, they have high concentrations of creative economic outcomes, in the form of innovations and high-tech industry growth. They also show strong signs of overall regional vitality, such as increases in regional employment and population.
- The creative centers are not thriving due to traditional economic reasons such as access to natural resources or transportation routes. Nor are they thriving because their local governments have gone bankrupt in the process of giving tax breaks and other incentives to lure business. They are succeeding largely because creative people want to live there. The companies follow the people – or, in many cases, are started by them. Creative centers provide the integrated ecosystem or habitat where all forms of creativity – artistic and cultural, technological and economic – can take root and flourish.
- Creative people are not moving to these places for traditional reasons. The physical attractions that most cities focus on – sports stadiums, freeways, urban malls, and tourism-and-entertainment districts that resemble theme parks – are irrelevant, insufficient, or actually unattractive to many creative-class people. What they look for in communities are abundant high-quality experiences, an openness to diversity of all kinds, and, above all else, the opportunity to validate their identities as creative people.

THE NEW GEOGRAPHY OF CREATIVITY

These shifts are giving rise to powerful migratory trends and an emerging new economic geography. In the leading creative centers, the creative class makes up more than 35 percent of the workforce, regions such as the greater Washington, DC, region, the Raleigh-Durham area, Boston, and Austin. But despite their considerable advantages, large regions have not cornered the market as creative-class locations. In fact, a number of smaller regions have some of the highest creative-class concentrations in the nation – notably college towns such as East Lansing, Michigan, and Madison, Wisconsin.

At the other end of the spectrum are regions that are being bypassed by the creative class. Among large regions, Las Vegas, Grand Rapids, and Memphis harbor the smallest concentrations of the creative class. Members of the creative class have nearly abandoned a wide range of smaller regions in the outskirts of the South and Midwest. In small metropolitan areas such as Victoria, Texas, and Jackson, Tennessee, the creative class comprises less than 15 percent of the workforce. The leading centers for the working class among large regions are Greensboro, North Carolina, and Memphis, Tennessee, where the working class makes up more than 30 percent of the workforce. Several smaller regions in the South and Midwest are veritable working-class enclaves with 40 to 50 percent or more of their workforce in the traditional industrial occupations. These places have some of the most minuscule concentrations of the creative class in the nation. They are symptomatic of a general lack of overlap between the major creative-class centers and those of the working class. Of the 26 large cities where the working class comprises more than one-quarter of the population, only one, Houston, ranks among the top 10 destinations for the creative class.

Las Vegas has the highest concentration of the service class among large cities, 58 percent, while West Palm Beach, Orlando, and Miami also have around half of their total workforce in the service class. These regions rank near the bottom of the list for the creative class. The service class makes up more than half the workforce in nearly 50 small and medium-size regions across the country. Few of them boast any significant concentrations of the creative class, save as vacationers, and offer little prospect for upward mobility. They include resort towns such as Honolulu and Cape Cod. But they also include places like Shreveport, Louisiana, and Pittsfield, Massachusetts. For these places that are not tourist destinations, the economic and social future is troubling to contemplate.

Places that are home to large concentrations of the creative class tend to rank highly as centers of innovation and high-tech industry. Three of the top five large creative-class regions are among the top five high-tech regions. Three of the top five large creative class regions are also among the top five most innovative regions (measured as patents granted per capita). And, the *same five* large regions that top the list on the Talent Index (measured as the percentage of people with a bachelor's degree or above) also have the highest creative-class concentration: Washington, DC, Boston, Austin, the Research Triangle, and San Francisco. The statistical correlations comparing creative-class locations to rates of patenting and high-tech industry are uniformly positive and statistically significant.

TECHNOLOGY, TALENT, AND TOLERANCE

The key to understanding the new economic geography of creativity and its effects on economic outcomes lies in what I call the 3Ts of economic development: *technology, talent, and tolerance*. Creativity and members of the creative class take root in places that possess all three of these critical factors. Each is a necessary but by itself insufficient condition. To attract creative people, generate innovation, and stimulate economic development, a place must have all three. I define tolerance as openness, inclusiveness, and diversity to all ethnicities, races, and walks of life. Talent is defined as those with a bachelor's degree and above. And technology is a function of both innovation and high-technology concentrations in a region. My focus group and interview results indicate that talented individuals are drawn to places that offer tolerant work and social environments. The statistical analysis validates not only the focus group results, but also indicates strong relationships between technology, tolerance, and talent.

The 3Ts explain why cities such as Baltimore, St. Louis, and Pittsburgh fail to grow despite their deep reservoirs of technology and world-class universities: they are unwilling to be sufficiently tolerant and open to attract and retain top creative talent. The interdependence of the 3Ts also explains why cities such as Miami and New Orleans do not make the grade even though they are lifestyle meccas: they lack the required technology base. The most successful places – the San Francisco Bay area, Boston, Washington, DC, Austin, and Seattle – put all 3Ts together. They are truly creative places.

My colleagues and I have conducted a great deal of statistical research to test the creative capital theory by looking at the way these 3Ts work together to power economic growth. We found that talent or creative capital is attracted to places that score high on our basic indicators of diversity – the Gay, Bohemian, and other indexes. It is not because high-tech industries are populated by great numbers of bohemians and gay people; rather, artists, musicians, gay people, and members of the creative class in general prefer places that are open and diverse. Such low entry barriers are especially important because, today, places grow not just through higher birth rates (in fact virtually all U.S. cities are declining on this measure), but by their ability to attract people from the outside.

As we have already seen, human capital theorists have shown that economic growth is closely associated with concentrations of highly-educated people. But few studies have specifically looked at the relationship between talent and technology, between clusters of educated and creative people and concentrations of innovation and high-tech industry. Using our measure of the creative class and the basic Talent Index, we examined these relationships for the 49 regions with more than one million people and for all 206 regions for which data are available. As well as some well-known technology centers, smaller college and university towns rank high on the Talent Index – places such as Santa Fe, Madison, Champaign-Urbana, State College, and Bloomington, Indiana. When I look at the subregional level, Ann Arbor (part of the Detroit region) and Boulder (part of the Denver region) rank first and third, respectively.

These findings show that both innovation and high-tech industry are strongly associated with locations of the creative class and of talent in general. Consider that 13 of the top 20 high-tech regions also rank among the top 20 creative-class centers, as do 14 of the top 20 regions for high-tech industry. Furthermore, an astounding 17 of the top 20 Talent Index regions also rank in the top 20 of the creative class. The statistical correlations between Talent Index and the creative-class centers are understandably among the strongest of any variables in my analysis because creative-class people tend to have high levels of education. But the correlations between the Talent Index and working-class regions are just the opposite – negative and highly significant – suggesting that working-class regions possess among the lowest levels of human capital.

Thus, the creative capital theory says that regional growth comes from the 3Ts of economic development, and to spur innovation and economic growth a region must have all three of them.

THE ROLE OF DIVERSITY

Economists have long argued that diversity is important to economic performance, but they have usually meant the diversity of firms or industries. The economist John Quigley, for instance, argues that regional economies benefit from the location of a diverse set of firms and industries. Jacobs long ago highlighted the role of diversity of both firms and people in powering innovation and city growth. As Jacobs saw it, great cities are places where people from virtually any background are welcome to turn their energy and ideas into innovations and wealth.

This raises an interesting question. Does living in an open and diverse environment help to make talented and creative people even more productive; or do its members simply cluster around one another and thus drive up these places' creativity only as a byproduct? I believe both are going on, but the former is more important. Places that are open and possess *low entry barriers* for people gain creativity advantage from their ability to attract people from a wide range of backgrounds. All else being equal, more open and diverse places are likely to attract greater numbers of talented and creative people – the sort of people who power innovation and growth.

LOW BARRIERS TO ENTRY

A large number of studies point to the role of immigrants in economic development. In *The Global Me,* the *Wall Street Journal* reporter Pascal Zachary argues that openness to immigration is the cornerstone of innovation and economic growth. He contends that America's successful economic performance is directly linked to its openness to innovative and energetic people from around the world, and attributes the decline of once prospering countries, such as Japan and Germany, to the homogeneity of their populations.

My team and I examined the relationships between immigration or percent foreign born and high-tech industry. Inspired by the Milken Institute study, we dubbed this the Melting Pot Index. The effect of openness to immigration on regions is mixed. Four out of the top 10 regions on the Melting Pot Index are also among the nation's top

10 high-technology areas; and seven of the top 10 are in the top 25 high-tech regions. The Melting Pot Index is positively associated with the Tech-Pole Index statistically. Clearly as University of California at Berkeley researcher Annalee Saxenian argues, immigration is associated with high-tech industry. However, immigration is not strongly associated with innovation. The Melting Pot Index is not statistically correlated with the Innovation Index, measured as rates of patenting. Although it is positively associated with population growth, it is not correlated with job growth. Furthermore, places that are open to immigration do not necessarily number among the leading creative-class centers. Even though 12 of the top 20 Melting Pot regions number in the top 20 centers for the creative class, there is no significant statistical relationship between the Melting Pot Index and the creative class.

THE GAY INDEX

Immigrants may be important to regional growth, but there are other types of diversity that prove even more important statistically. In the late 1990s, the Urban Institute's Gary Gates, along with the economists Dan Black, Seth Sanders, and Lowell Taylor, used information from the U.S. Census of Population to figure out where gay couples located. He discovered that particular cities were favorites among the gay population.

The U.S. Census Bureau collects detailed information on the American population, but until the 2000 Census it did not ask people to identify their sexual orientation. The 1990 Census allowed couples that were not married to identify as "unmarried partners," different from "roommates" or "unrelated adults." By determining which unmarried partners were of the same sex, Gates identified gay and lesbian couples. The Gay Index divides the percentage of coupled gay men and women in a region by the percentage of the population that lives there and thus permits a ranking of regions by their gay populations. Gates later updated the index to include the year 2000.

The results of our statistical analysis on the gay population are squarely in line with the creative capital theory. The Gay Index is a very strong predictor of a region's high tech industry concentration. Six of the top 10 1990 and five of the top 10 2000 Gay Index regions also rank among the nation's top 10 high-tech regions. In virtually all of our statistical analyses, the Gay Index did better any than other individual measure of diversity as a predictor of high-tech industry. Gays not only predict the concentration of high-tech industry, they also predict its growth. Four of the regions that rank in the top 10 for high-technology growth from 1990–1998 also rank in the top 10 on the Gay Index in both 1990 and 2000. In addition, the correlation between the Gay Index (measured in 1990) and the Tech-Pole Index calculated for 1990–2000 increases over time. This suggests that the benefits of diversity may actually compound.

There are several reasons why the Gay Index is a good measure for diversity. As a group, gays have been subject to a particularly high level of discrimination. Attempts by gays to integrate into the mainstream of society have met substantial opposition. To some extent, homosexuality represents the last frontier of diversity in our society, and thus a place that welcomes the gay community welcomes all kinds of people.

THE BOHEMIAN INDEX

As early as the 1920 studies by Robert Park, sociologists have observed the link between successful cities and the prevalence of bohemian culture. Working with my Carnegie Mellon team, I developed a new measure called the Bohemian Index, which uses Census occupation data to measure the number of writers, designers, musicians, actors, directors, painters, sculptors, photographers, and dancers in a region. The Bohemian Index is an improvement over traditional measures of amenities because it directly counts the producers of the amenities using reliable Census data. In addition to large regions, such as San Francisco, Boston, Seattle, and Los Angeles, smaller communities such as Boulder and Fort Collins, Colorado; Sarasota, Florida; Santa Barbara, California; and Madison, Wisconsin, rank rather highly when all regions are taken into account.

The Bohemian Index turns out to be an amazingly strong predictor of everything from a region's high-technology base to its overall population and employment growth. Five of the top

10 and 12 of the top 20 Bohemian Index regions number among the nation's top 20 high-technology regions. Eleven of the top 20 Bohemian Index regions number among the top 20 most innovative regions. The Bohemian Index is also a strong predictor of both regional employment and population growth. A region's Bohemian presence in 1990 predicts both its high-tech industry concentration and its employment and population growth between 1990 and 2000. This provides strong support for the view that places that provide a broad creative environment are the ones that flourish in the Creative Age.

TESTING THE THEORIES

Robert Cushing of the University of Texas has undertaken to systematically test the three major theories of regional growth: social capital, human capital, and creative capital. His findings are startling. In a nutshell, Cushing finds that social capital theory provides little explanation for regional growth. Both the human capital and creative capital theories are much better at accounting for such growth. Furthermore, he finds that creative communities and social capital communities are moving in opposite directions. Creative communities are centers of diversity, innovation, and economic growth; social capital communities are not.

Cushing went to great pains to replicate Putnam's data sources. He looked at the surveys conducted by a team that, under Putnam's direction, did extensive telephone interviewing in 40 cities to gauge the depth and breadth of social capital. Based on the data, Putnam measured 13 different kinds of social capital and gave each region a score for attributes like "political involvement," "civic leadership," "faith-based institutions," "protest politics," and "giving and volunteering." Using Putnam's own data, Cushing found very little evidence of a decline in volunteering. Rather, he found that volunteering was up in recent years. People were more likely to engage in volunteer activity in the late 1990s than they were in the 1970s. Volunteering by men was 5.8 percent higher in the five-year period 1993–1998 than it had been in the period 1975–1980. Volunteering by women was up by 7.6 percent. A variety of statistical tests confirmed these results, but Cushing did not stop there. He then combined this information on social capital trends with independent data on high-tech industry, innovation, human capital, and diversity. He added the Milken Institute's Tech-Pole Index, the Innovation Index, and measures of talent, diversity, and creativity (the Talent Index, the Gay Index, and the Bohemian Index). He grouped the regions according to the Tech-Pole Index and the Innovation Index (their ability to produce patents).

Cushing found that regions ranked high on the Milken Tech-Pole Index and Innovation Index ranked low on 11 of Putnam's 13 measures of social capital. High-tech regions scored below average on almost every measure of social capital. High-tech regions had less trust, less reliance on faith-based institutions, fewer clubs, less volunteering, less interest in traditional politics, and less civic leadership. The two measures of social capital where these regions excelled were "protest politics" and "diversity of friendships." Regions low on the Tech-Pole Index and the Innovation Index were exactly the opposite. They scored high on 11 of the 13 Putnam measures but below average on protest politics and diversity. Cushing then threw into the mix individual wages, income distribution, population growth, numbers of college-educated residents, and scientists and engineers. He found that the high-tech regions had higher incomes, more growth, more income inequality, and more scientists, engineers, and professions than their low-tech, but higher social capital counterparts. When Cushing compared the Gay and Bohemian Indexes to Putnam's measures of social capital in the 40 regions surveyed in 2000, the same basic pattern emerged: Regions high on these two diversity indexes were low on 11 of 13 of Putnam's categories of social capital. In Cushing's words, "conventional political involvement and social capital seem to relate negatively to technological development and higher economic growth." Based on this analysis, Cushing identified four distinct types of communities. The analysis is Cushing's; the labels are my own.

- *Classic Social Capital Communities.* These are the places that best fit the Putnam theory – places such as Bismarck, North Dakota; rural South Dakota; Baton Rouge, Louisiana; Birmingham, Alabama; and Greensboro, Charlotte, and

Winston-Salem, North Carolina. They score high on social capital and political involvement but low on diversity, innovation, and high-tech industry.

- *Organizational Age Communities.* These are older, corporate-dominated communities such as Cleveland, Detroit, Grand Rapids, and Kalamazoo. They have average social capital, higher-than-average political involvement, low levels of diversity, and low levels of innovation and high-tech industry. They score high on my Working Class Index. In my view, they represent the classic corporate centers of the organizational age.
- *Nerdistans.* These are fast-growing regions such as Silicon Valley, San Diego, Phoenix, Atlanta, Los Angeles, and Houston, lauded by some as models of rapid economic growth but seen by others as plagued with sprawl, pollution, and congestion. These regions have lots of high-tech industry, above-average diversity, low social capital, and low political involvement.
- *Creative Centers.* These large urban centers, such as San Francisco, Seattle, Boston, Chicago, Minneapolis, Denver, and Boulder, have high levels of innovation and high-tech industry and very high levels of diversity but lower than average levels of social capital and moderate levels of political involvement. These cities score highly on my Creativity Index and are repeatedly identified in my focus groups and interviews as desirable places to live and work. That's why I see them as representing the new creative mainstream.

In the winter of 2001, Cushing extended his analysis to include more than three decades of data for 100 regions. Again he based his analysis on Putnam's own data sources: the 30-year time series collected by DDB Worldwide, the advertising firm, on activities such as churchgoing, participation in clubs and committees, volunteer activity, and entertaining people at home. He used these data to group the regions into high and low social capital communities and found that social capital had little to do with regional economic growth. The high social capital communities showed a strong preference for "social isolation" and "security and stability" and grew the least – their defining attribute being a "close the gates" mentality. The low social capital communities had the highest rates of diversity and population growth.

Finally, Cushing undertook an objective and systematic comparison of the effect of the three theories – social capital, human capital, and creative capital – on economic growth. He built statistical models to determine the effect of these factors on population growth (a well-accepted measure of regional growth) between 1990 and 2000. To do so, he included separate measures of education and human capital; occupation, wages, and hours worked; poverty and income inequality; innovation and high-tech industry; and creativity and diversity for the period 1970–1990.

Again his results were striking. He found no evidence that social capital leads to regional economic growth; in fact the effects were negative. Both the human capital and creative capital models performed much better, according to his analysis. Turning first to the human capital approach, he found that while it did a good job of accounting for regional growth, "the interpretation is not as straightforward as the human capital approach might presume." Using creative occupations, bohemians, the Tech-Pole Index, and innovations as indicators of creative capital, he found the creative capital theory produced formidable results, with the predictive power of the Bohemian and Innovation Indexes being particularly high. Cushing concluded that the "creative capital model generates equally impressive results as the human capital model and perhaps better."

DIRECTIONS FOR FUTURE RESEARCH

The nature and function of the city is changing in ways and dimensions we could scarcely have expected even a decade or two ago. For much of the past century the city has been viewed as a center for physical production and trade, industrialization, and the agglomeration of finance, service, and retail activities. Our theories of the city are all based on the basic notion of the city as an arena for production and largely based on activities that take place in the city during the daylight hours. Similarly, our theories of community are largely based on notions of the tightly knit community of the past, a community defined by strong ties – a conceptual theme that has been revived by the work of Robert Putnam and widespread interest in social capital both inside academe and in public policy circles. But, as this article has tried to show, the past decade has seen a sweeping transformation in the nature and functions of cities and communities. My own field research, as well as the research of others, has shown the preference for weak ties and quasi-anonymity. Social capital is at best a limited theory of community – one that fits uneasily with many present day realities. Magically invoking it will not somehow recreate the stable communities characterized by strong ties and commitments of the past. On this score, the key is to understand the new kinds of communities – communities of interest – that are emerging in an era defined by weak ties and contingent commitments. Much more research is needed on these and related issues.

Our theories of cities, neighborhoods, and urban life are undergoing even more sweeping transformation. Sociologists such as Terry Clark, Richard Lloyd, and Leonard Nevarez have been dissecting the new reality of the city as a center for experience, lifestyle, amenities, and entertainment. This shift is not only noticeable in cites such as San Francisco, Seattle, and Boston, which have long been centers of culture and lifestyle, but in older industrial cities such as Chicago, which have been dramatically transformed into centers of entertainment, experience, and amenity. According to Clark, entertainment has replaced manufacturing and even services as Chicago's "number 1" industry. Understanding the city as an arena for consumption, for entertainment, and for amenities – a city that competes for people as well as for firms, a city of symbols and experiences, the city at night – is a huge research opportunity for sociology, geography, and related disciplines.

At the organizational level, there is a great need for research on the factors that motivate creative people and how organizations and workplaces can adapt. We are at the very infancy of organizational and workplace experiments on how to motivate creative people. Recent experiments with open office design, flexible schedules, and various accoutrements are only the very beginning. Research on the psychology of creativity by Teresa Amabile, Robert Sternberg, and others shows that creativity is an intrinsically motivated process and further suggests that the use of extrinsic rewards, such as financial incentives, may actually be counterproductive to motivating creative work. This suggests that both academic economists and professional managers have gone off in the wrong direction, particularly during the 1990s, with the use of stock options and other forms of equity compensation to motivate creative workers. A great deal more research is needed on the intrinsic factors that motivate creative workers and, even more importantly, on the characteristics and factors associated with organizations and workplaces that can best motivate and enhance creative work.

Turning now to larger macrosocietal questions. The past several decades have seen a dramatic shift in the underlying nature of advanced capitalist economies, from a traditional industrial-organizational system based on large factories and large corporate office towers, and premised on economies of scale and the extraction of physical labor, to newer, emergent systems based on knowledge, intellectual labor, and human creativity. Understanding the underlying dynamics of the system, the social structures on which it rests, the kinds of workplace transformations it is setting in motion, and its effects on community as well as city form, structure, and function is a tremendous opportunity for research. In *The Rise of the Creative Class* I try to identify some of these underlying changes and the structural transformations they have set in motion as workplaces, lifestyles, and communities all begin to adapt and evolve in light of these deep economic and social shifts. That work is my best first pass, but there is much, much more to do.

A final and critically important avenue for research is to begin to get a handle on the downsides, tensions, and contradictions of this new Creative Age – and there are many. One that I am exploring with my team is rising inequality. Our preliminary investigations suggest that inequality is increasing at both the inter-regional and intra-regional scales. At the inter-regional level, increased inequality appears to be a consequence of what we have come to call the "new great migration" as creative-class people relocate to roughly a dozen key creative regions nationwide. Other preliminary research, for example, by Robert Cushing at the University of Texas, suggests that Austin is importing high-skill creative-class people and exporting lower-skill individuals. The same pattern appears to hold for other creative centers. Inequality is also on the rise within regions. Preliminary research I have conducted with Kevin Stolarick indicates that inequality is highest in creative centers such as San Francisco and Austin. Then there is the question of the relationship between knowledge-based, creative capitalism and new types of workplace injury. At the turn of the century, during the explosion of industrial capitalism, there was great incidence and, later, great concern over physical injuries in the workplace. Eventually, after much examination and policy debate, there emerged mechanisms like OSHA to reduce physical injury in the workplace. In the Creative Age, when the mind itself becomes the mode of production, so to speak, the nature of workplace injury has changed to what I term "mental injury." Sociologists and social psychologists have much to offer in identifying the factors associated with the increasing incidence of anxiety disorder, depression, substance abuse, and other forms of mental injury, and their relationship to creative work. I often make an analogy to Charlie Chaplin's *Modern Times*: The creative-class worker racing frantically to keep up with e-mails, telephone calls, and other aspects of information overload resembles Chaplin's assembly-line worker frantically trying to keep up with the assembly line – but the creative-class worker has to do this on a 24/7, around-the-clock-basis. A great deal of research needs to be done on the incidence of mental injury and its relationship to new ways of working.

[. . .]

"Looking at Themed Environments"

from *The Theming of America* (1997)

Mark Gottdiener

Editors' Introduction

Contemporary cities engage in innovative efforts to (re)build their economies, especially in light of globalization processes that reward economically competitive spaces (see the selections by Friedmann and Sassen, this volume). In particular, the global restructuring of commodities production (manufacturing occurring off-shore with command and control decision-making done in key cities) since the 1970s has led many cities to adopt ambitious entertainment- and consumption-based economic development in their downtowns and waterfronts. Such development requires effective marketing of these spaces in order to attract tourists and locals to visit, shop, and play (i.e., consume).

In this selection, urban sociologist Mark Gottdiener addresses the development of post-industrial era spaces within cities and suburbs, focusing on the use of desirable symbols, themes, and motifs to create new environments of consumption. In a process he calls "theming," Gottdiener argues a two-fold material and symbolic production of the contemporary built environment. First, real estate developers and politicians have pushed for the development of shopping malls, cinema multiplexes, Main Streets, sports stadiums, concert venues, and restaurants – "containers of commodified human interaction." This infrastructure of consumption has replaced, in many older cities especially, the manufacturing infrastructure (factories, docks, and rail yards) of the past. Second, theming involves a complex cultural process of investing this new built environment with symbolic meaning and its consumers through the use of symbolic motifs.

In the selection that follows, Gottdiener explores the second, more complex aspect of theming. Theming involves the practice of signifying, in which symbols (a ship's anchor, a chile pepper, or a golden arch) convey an immediate, evident level of meaning and a more variable, complex level of meanings connected to specific cultural contexts. It is important to note that the second level of meanings can never be fully predicted or known; individuals interpret signs using their personal frames of reference (hence, a golden arch can mean many things in addition to appetizing, unappetizing, playfulness, or revulsion). Signs, then, are material objects that operate on many social levels.

Signs are used in the development of themed built environments to expose their users to the experience of consumption. Consumption includes not only the obvious act of purchasing but the more fascinating aspect of how individuals or groups use or interpret the constructed space by imputing some meaning or meanings to it. Consumption of a themed environment, Gottdiener argues, refers to the experience of individuals within a themed milieu. Thus, visitors to South Street Seaport or Times Square in New York City or Universal Studios or Disneyland in Southern California consume the environment itself as well as souvenirs, CDs, and a matinee or thrill rides and IMAX films. It is the intent of the developers to have symbols and motifs that are materially attached to the built environment of these spaces to heighten the visitors' experience (but it may not since an individual's interpretation of meaning, as already mentioned, is variable).

In the highly competitive global economy, the use of theming to lure prized tourists, strollers, and shoppers has become an important redevelopment strategy for real estate developers, planners, and corporations. The marriage of brand name themes, such as the Nike swoosh, and commercial spaces, such as Niketown, imparts considerable prestige or cache on a downtown shopping street. In Chicago, for example, the recent redevelopment of the Loop has meant the proliferation of themed chain stores and restaurants where many out-of-town visitors "feel right at home." In a digital age of mass media and a frenzy of corporate images, the use of motifs and symbols is fundamental to the production and consumption of the urban environment.

Mark Gottdiener is Professor of Sociology at the University at Buffalo. He is a leading proponent of the "new urban sociology" – a critical perspective largely influenced by European social theory that examines the roles of the state, the real estate industry, and social movements in the political, economic, and cultural production of urban space. He is author of several books, including *The Social Production of Urban Space* (Austin, TX: University of Texas Press, 1985), *The New Urban Sociology* (with Ray Hutchison) (New York: McGraw-Hill, 1999), *Postmodern Semiotics: Material Culture and The Forms of Postmodern Life* (Oxford: Blackwell, 1995), and *The Theming of America: Dreams, Visions and Commercial Spaces*, second edition (Boulder, CO: Westview Press, 2001).

THE INCREASING USE OF THEMES IN EVERYDAY LIFE

Since the end of World War II, our everyday environment has been altered in profound ways. Before the 1960s there was a clear distinction between the city and the country. Cities grew as compact, dense industrial environments usually laid out along right-angled grid lines. They possessed a central sector of office towers and an adjacent area of factories tied to rail spurs and roads. Residential areas exemplified the classic contrast of the "gold coast and slum." Wealthy, privileged areas of housing were juxtaposed and in close proximity with more modest, sometimes squalid, neighborhoods of industrial workers. Lying outside the limits of the city was another contrasting space – the countryside of farm fields, wooded acres, and occasional houses separated by open space. Cultural styles of life that were either urban (urbane) or rural reflected this dichotomy of different land uses, as did depictions of city and country dwellers in novels and films.

Before the 1960s, fundamental class differences between capitalists and workers organized the land use of the industrial city. Neighborhoods, for example, were working class and the local community reflected ethnic, racial, and religious solidarity dedicated to the task of raising families. The center of the city, in contrast, belonged to business. During this period the symbols that provided meaning to daily life were manifested more cognitively than materially. Buildings reflected their functions with a minimum of symbolic trappings. People in neighborhoods signified their culture through the sometimes subtle markings of churches and store signs using foreign languages. Symbolic marking was muted, and inhabitants had to be perceptive to pick out denotations of place that signified particular ethnic solidarities or preferred places of business.

Since the 1960s a new trend of symbolic differentiation within the built environment has appeared that contrasts graphically with the earlier period. The use of symbols and motifs more and more frequently characterizes the space of everyday life in both the city and the suburb. Signification involves not only a proliferation of signs and themes but also a constant reworking of built facades and interior spaces to incorporate overarching motifs in such a way that we increasingly are exposed to new environmental experiences when we consume. Now these new consumer spaces with their new modes of thematic representation organize daily life in an increasing variety of ways. Social activities have moved beyond the symbolic work of designating ethnic, religious, or economic status to an expanding repertoire of meaningful motifs. Whereas symbolic elements were muted in the settlement spaces of the early twentieth century, the trend is reversed in the new consumer spaces of today.

UNDERSTANDING THEMED ENVIRONMENTS

When I refer to a themed environment, I explicitly mean the material product of two social processes. First, I am talking about socially constructed, built environments – about large material forms that are designed to serve as containers for commodified human interaction (for example, malls). Second, I have in mind themed material forms that are products of a cultural process aimed at investing constructed spaces with symbolic meaning and at conveying that meaning to inhabitants and users through symbolic motifs. These motifs may take on a range of meanings according to the interpretations of the individuals who are exposed to them. The range of responses can include everything from no response at all – that is, a failure of the symbolic content to stimulate – to a negative response, or displeasure. Themed environments do not automatically provoke desire and pleasure in their users. Such spaces also can be sources of great irritation, as can be seen in the strong negative reactions of some Europeans to McDonald's franchises in their countries.

Another important distinction concerns the way I use the concepts of "production" and "consumption." The former refers to a social process of creation that often involves a group of individuals brought together within an organized, institutional context, such as real estate development. Consumption involves the way individuals or groups use or interpret the constructed space by imputing some meaning or meanings to it. These people may be customers, inhabitants, visitors, or clients, but they are all users of the space in some fashion. Consumption of a themed environment refers to the experience of individuals within a themed milieu, including the assumption of a particular orientation to space, based on the personal or group interpretation of its symbolic content. Built forms have the power to alter human behavior through meaning, and this response is also part of what I mean by the process of consumption in space.

Visitors to a themed park consume the environment itself as well as the rides and attractions. They adjust their behavior according to the stimuli they receive from the signals embedded in built forms. Motifs and symbols developed through the medium of the park's material forms may be highly stimulating, or conversely, hardly noticed. Always, however, the physical presence of individuals within a space involves their use or "consumption" of the material environment.

Most commentaries – for example, those about places like Disneyland – ignore the production process of the park and focus exclusively on impressionistic accounts of a visit there. I seek to correct this one-sided view by also emphasizing the process of production, and the role of that process in the larger economic organization of our society. I am especially interested in the intermixing of creation and use, as consumer experiences increasingly are packaged within themed milieus.

PRODUCTION IN CONSUMPTION

Despite the dichotomization of the interrelated concepts of production and consumption, I must caution the reader against making sharp distinctions between these two social processes. To begin with, as our society progressively shifts from an economy dependent on manufacturing to one in which service industries predominate, the jobs held by the bulk of the population are increasingly associated with thematic experiences. Museum exhibits, for example, more frequently concern the elaboration of themes than in the past; service labor in restaurants or recreational areas requires employee conformity to the symbolic decor through the wearing of costumes and the like; and retailing activity increasingly locates in motifed milieus such as malls. Thus, the world of work, or production, penetrates and merges with the world of consumption.

Second, it has become harder to isolate shopping and other leisure pursuits as activities defined by consumption alone. In the past, observers of mass cultural participation often did just that by painting the users or the audience as a group of *passive* consumers, conditioned by advertising to behave in the way producers wished. More recently, analysts of culture have recognized that the gross manipulation of people by advertising is an exaggeration. We must acknowledge the relative autonomy of individuals in the act of consumption, as they blend personal history, the self-actualization of their identities, group pressures of various kinds, and the powerful compulsions of the consumer society that pressure people to make certain choices in the marketplace. As individual identities become wrapped up in modes of self-expression and the fashioning of particular lifestyles in response to the great variety of market choices, there is a blurred line between production and consumption. More and more we view the pursuit of particular styles of life and the development of contemporary subjectivities through the use of material objects as a form of production itself. There is always an element of production in the act of consumption, just as there is also a corresponding aspect of use-value exploited by the production process. These intersecting, liminal activities of the economy are increasingly organized by overarching symbolic motifs within consumer milieus.

Michel de Certeau (1984), the author of an influential book on the subject, argues that consumers are always employing a creative strategy or series of strategies in their buying activities that bear a resemblance to the production process. They try to juggle desires, prices, stores, and modes of purchasing. They create uses for objects to fill specific needs. They modify commodities to suit their own lifestyle. These and other responses are common strategies of consumption, deployed by average household or family groups as they cope with economic adversity. In short, people are not the passive, media-manipulated masses often depicted by analysts of advertising. They are very often proactive in their attitudes toward commodities and shopping. Through the daily use of strategies, they "produce" an attitude and a form of coping behavior in their social role as consumer.

Another way of viewing the links between the processes of production and consumption focuses on the development of subjectivity and the emergence of the self within a consumer-oriented society. To be sure, images and desires produced by the advertising industry constantly prime people to consume. When individuals enter commercial realms, such as in a visit to a mall, the themed, retailing environment actualizes their consumer self. This process, however, is not a passive one, with individuals acting like marionettes, pulled back and forth by powerful consumer conditioning. Instead, people self-actualize within the commercial milieu, seeking ways of satisfying their desires and pursuing personal fulfillment through the market that express deeply held images of themselves. Granted, mass advertising conditions much of this actualization of a consumer identity, especially aided by the group force of conformity to fashion. Equally valid is the observation that self-actualization is destined to be disappointed in the alienated world of mass marketing. However, as observers interested in the innovative (production) aspects of buying argue, the fashioning of consumer identities is much less controlled by advertising manipulation than is often supposed, and the incredibly prolific abundance of commercial products does promise the satisfaction of many of our desires, whether these are manufactured for us or not.

As an example of these aspects of production in consumption, consider the activity of dressing in contemporary society. Group norms highly regulate socially acceptable dress, and this has always been so. The large fashion industry that orchestrates modes of appearance in modern society lately has targeted men as well as women both for periodic changes in style and for the production of desire. Despite the power of fashion, most individuals hold a very personalized conception of their dress patterns. They often seek self-actualization and pursue certain distinct lifestyles through the medium of appearance. People in our society spend a great deal of money on clothing – more than they could possibly need for purely protective purposes. They use these material objects symbolically in many ways to exploit social situations for their own advantage. Individuals may "dress for success" or to impress; they may seek approval of others – women, men, prospective in-laws, a possible employer; and they often seek identification with particular groups by dressing like them. Finally, people often mix and match objects of clothing and accessories on a daily basis in a creative effort to fashion a personal look or image. When considering the social process of dress as distinct from fashion dictated by the clothing industry, it becomes difficult to separate aspects of production and consumption because the two are so interrelated in the daily behavior of dressing.

A BRIEF NOTE ON SIGNS AND SYMBOLS

In the opening section of this chapter we have already begun to use concepts associated with the analysis of symbols in specific ways. This is necessary to discuss the phenomenon of the themed environment in comprehensive detail. The *sign* is defined conceptually as something that stands for something else, and, more technically, as a spoken or written word, a drawn figure, or a material object unified in the mind with a particular cultural concept. The sign is this unity of a word/object, known as a *signifier*, with a corresponding, culturally prescribed concept or meaning, known as a *signified*. Thus our minds attach the word *dog* or the drawn figure of a dog, as a signifier, to the idea of a "dog," that is, a domesticated canine species possessing certain behavioral characteristics. If we came from a culture where dogs were not routinely encountered, we might not know what the signifier *dog* means.

When dealing with objects that are signifiers of certain concepts, cultural meanings, or ideologies, we can consider them not only as "signs" but as *sign vehicles*. Signifying objects carry meanings with them. They may purposefully be constructed to convey meaning. Thus, Disneyland, as a theme park, is a large sign vehicle of the Disney ideology. This concept, however, has two aspects that are often a source of confusion. When we use a signifier to convey simple information, usually of a functional nature, we *denote* meaning. The word *train*, for example, denotes a mode of transportation or movement. Objects that denote a particular function are called *sign functions*. Every material form within a given culture is a sign of its function and denotes its use. When we approach a building that is a bank, we understand its meaning at the denotative level in terms of its sign function as a repository and transaction space for money.

Every signifier, every meaningful object, however, in addition, conveys another meaning that exists at the *connotative* level – that is, it *connotes* some association defined by social context and social process beyond its denotative sign function. The word *train*, which denotes transportation, also connotes old-fashioned travel, perhaps the nineteenth century by association, maybe a sort of romanticism of traveling, even mystery, exoticism, and intrigue, as in the Orient Express; or in another vein, slowness, noise, pollution, crowds, and the like. The bank building, which is the sign function for the activity of "banking," also *connotes* a variety of socially ascribed associations including wealth, power, success, future prospects, college educations, and savings for vacations or Christmas. In short, every sign not only denotes some social function and conveys a social meaning at the denotative level but also connotes a variety of associations that have meanings within specific cultural contexts. Thus, sign vehicles that are material objects operate on many social levels. Understanding this fact provides an appreciation for the rich cultural life of objects. It also means that we can understand how material objects are used as signs in social interaction. Rather than considering objects as signs – that is, as existing exclusively in the mind of the interpreter – we also consider signs as objects, as material forms that are used and manipulated by social actors for personal reasons or material gain. The former is associated with the act of consuming a themed environment. The latter concerns the way producers use signs as objects for distinct purposes, such as making money. Both processes operate simultaneously, depending on the point of reference.

Because signs perform double duty in social interaction (denoting and connoting), their interpretation is fraught with ambiguity. Furthermore, individuals decoding signs use their personal frame of reference, unless taught not to, and this may lead to the interpretation of a particular sign or discourse that was unintended by its producer, who may have come from another social context. For these reasons, the meaning attached to signs is always *polysemic*, that is, there are always several equally valid ways of interpreting a sign. Due to the presence of polysemy, the understanding of meaningful interaction is always problematic. Communication between individuals is a difficult task, especially because the *interpretation* of communication invariably differs from person to person even within the same receiving context. Recognition of cases of polysemy is also an important aspect of the interpretation of signs.

Societies with a polysemic culture accomplish the task of communication by adhering to particular symbolic *codes* that may also be called ideologies. Codes or ideologies are belief systems that organize meanings and interpretations into a single, unified sense. Sometimes we also use *semantic field* or the *universe of meaning* to signify the concept of code or ideology. Codes are subcultural phenomena shared by others, and they vary from one social group to another. A diverse society contains various subcultures that interpret experience and commodities according to their individual ideologies. For example, some people love country and western music so much that they dress up in cowboy hats and jeans and go to bars where they line dance to it. Others hate country and western music, preferring some other genre that is consistent with the ideology of their particular subculture. A diverse society contains populations that invoke a variety of subcultural contexts in the process of communication. Simultaneously, powerful forces in society, such as advertising, economics, the media, and politics, marshal unified social organization through the deployment of specific codes that limit ambiguity and direct our understanding toward specific meanings or values. It is this operation of institutions that channel meaning – for example, Hollywood, or advertising – that makes the new themed consumer environments so appealing to a variety of subcultural groups.Where would McDonald's be, if the corporation had not spent billions of dollars on advertising over the years?

REFERENCE

de Certeau, Michel. 1984. *The Practice of Everyday Life.* Berkeley: University of California Press.

"Globalization, Culture and Neighborhood Change"

from *Urban Affairs Review* (1996)

Christopher Mele

Editors' Introduction

Thus far, we have seen the relative importance of culture – particularly its production and consumption – to urban development. In the wake of industrial decline and in the face of rampant globalization, it is reasonable to assume that harnessing the "power of culture" to reshape cities is likely to continue. For Sharon Zukin and Mark Gottdiener, the use of cultural symbols and signs to develop new urban spaces provides cities a competitive advantage in their efforts to rebuild downtowns. In the selection that follows, Christopher Mele tracks a slight variation in this course of cultural-based urban redevelopment. In his study, the focus is not on "big development" projects, such as New York's Times Square or Baltimore's Inner Harbor, where state and local governments and large real estate investors are the key players in highly publicized development schemes. Instead, he addresses the importance of culture in the redevelopment of a neighborhood infamous for its stubborn resistance to renewal or gentrification, New York's Lower East Side. Here, the key stakeholders are individuals and small real estate holding companies who own individual apartment buildings, or tenements, and commercial spaces, such as bars and restaurants.

As Mele points out, in the efforts to gentrify the Lower East Side in the 1980s and 1990s, there were no master plans or grand redevelopment schemes at play nor were there big banks or state-financed redevelopment corporations to ensure the process went smoothly. Redevelopment was piecemeal: building by building, block by block. To make things more difficult for developers, the Lower East Side had a long history of grassroots political resistance to residential displacement and gentrification. Under these conditions, Mele argues, bringing culture into play becomes even more consequential to urban redevelopment. In poor, working class neighborhoods, successful redevelopment requires that the prevailing place identity be "reinvented" in order to attract middle-class consumers (renters).

Reinvention may take different forms and Mele discusses three efforts to redefine the identity of the Lower East Side in the early twentieth century and the recent past. First, there are efforts to simply eradicate an existing place identity by displacing a large part of the neighborhood's population. In short, real estate developers "plan around" the existing residents with the idea they will no longer live there. A second form of reinvention superimposes a new neighborhood identity atop an old one. In the 1960s, for example, newcomers began to refer to (and market) parts of the Lower East Side as the "East Village" – a more youthful, freewheeling identity that existed side-by-side the area's poor, downtrodden reputation. A more recent form of neighborhood reinvention is cultural appropriation, made possible according to Mele by globalization processes. Local cultural practices – resistance art, music, and even graffiti – once considered obstacles to middle-class redevelopment now become the basis for it. Such forms are appropriated, reworked, and packaged as marketing tools for selling "hip" or "artistic" neighborhoods such as San Francisco's Haight Ashbury or Philadelphia's South Street. As Mele concludes, "the promotion of specialized housing markets – the selling of bohemian mix – mirrors the corporate promotion of cultural forms geared toward specialized, or niche, consumer markets." Grit sells and, in turn, the tried and true methods deployed by squatters, artists, musicians, and grassroots activists to keep gentrification at bay become appealing "local culture."

Christopher Mele is Associate Professor of Sociology at the University at Buffalo. His work is in the area of urban and community sociology with an interest in political and legal regulation and social control of urban space. He is co-editor (with Teresa Miller) of *Civil Penalties, Social Consequences* (New York: Routledge, 2004) and (with John Eade) *Understanding the City: Contemporary and Future Perspectives* (Oxford: Blackwell, 2002). His empirical work is focused on New York City. He is author of *Selling the Lower East Side: Culture, Real Estate and Resistance in New York City* (Minneapolis: University of Minnesota Press, 2000).

On any given street corner on any given day in lower Manhattan's poorest neighborhood, the Lower East Side, Latino teenagers gather in twos and threes, passing time, smoking Luckys, and engaging in boy–girl talk. They live nearby with their families in apartments in century-old unrenovated tenements or in one of the hundreds of public housing units in endless rows of identical high-rises. Like most teenagers in urban environments, they assemble and produce local codes or styles of dress (East Coast "ghettocentric") and music (variations of acid house/jazz, hip-hop, and rap). Of interest to many cultural theorists is that such symbols of urban culture have migrated beyond these places of origin; they are televised worldwide on MTV (the music video network) and merchandised in chain outlets in suburban shopping malls across continents. Innovation and diversity, niche marketing, and widespread distribution are the hallmarks of an emerging global cultural economy. The symbolic culture of politically and economically disenfranchised social groups who live in the marginalized zones of the inner city is encouraged, cultivated, and ultimately, appropriated for the global market-place of culture.

Corporate expropriation and commodification of *difference* has stimulated the growth of the global cultural economy, but it has altered the localities – the alleys, street corners, and neighborhoods – where such culture is produced as well. Rap, hip-hop, and the myriad varieties of urban fashion are forms of everyday or common culture invented and disseminated to confer meaning to social life in an urban environment. Because they are grounded in and reflective of specifically urban environments, they are necessarily symbolic of real places that are detached economically and politically from mainstream (i.e., suburban) America. Global commodification of urban culture, however, demands that it no longer be representative of its agents, the collective process of its invention, or the environment from which it was spawned. The multinational corporate production, distribution, and marketing of subcultural forms transform them into consumption niches. Safe, sanitized imitations of not only the cultural forms but also their places of origin are served up for worldwide consumption.

Globalization has visceral effects upon urban form as well. In this article, I demonstrate how globalization has affected the struggle between place entrepreneurs and residents over community change. I present a case study of how globalization simultaneously creates new opportunities for urban entrepreneurs to restructure and market low-income communities and challenges traditional forms of local resistance and protest to neighborhood change. The economy of global appropriation and consumption of local cultural forms has altered the social and economic disposition of those neighborhoods in which such forms originate. As globalization of culture transforms how places are consumed, it has a similar effect, as shown in this article, on how places are produced or restructured by urban entrepreneurs such as real estate speculators and developers and financial institutions.

Using evidence from the Lower East Side of New York, I identify the pattern of neighborhood restructuring in the postwar decades. Until recently, the series of contests between place entrepreneurs and residents centered on the community's prevailing working-class and ethnic identity. Place entrepreneurs, in alliance with state urban policies, sought to eradicate both the physical and the sociocultural vestiges of the neighborhood's immigrant, working-class past. Residents often effectively mounted resistance to such efforts.

I demonstrate the effect that globalization has had on the postwar formula of neighborhood change. Responding to opportunities afforded by the increasingly global economy of cultural production and consumption, place entrepreneurs have abandoned their efforts to displace existing notions of place and local character. Efforts to reinvent neighborhood now involve appropriating, packaging, and marketing of the identity of marginalized communities as a means of accumulating profits in the local real estate market. Finally, I argue that this recent form of neighborhood reinvention challenges the efficacy of established notions of resistance to neighborhood change.

THEORIZING NEIGHBORHOOD REINVENTION

In most older U.S. cities, changes in neighborhoods are products of struggle between entrepreneurial efforts to capitalize on the changing political economy of cities and the continued resistance of working-class residents threatened with a loss of community. Place entrepreneurs have orchestrated several efforts to restructure the urban environment since World War II, often in conjunction with the policies of federal, state, and local governments. Civic boosterism, urban renewal, and more recently, gentrification are landmarks along what Jon Teaford (1990) called the "rough road to renaissance." Until the early 1980s, the efforts of capitalists to profit through manipulation of the urban housing market were frustrated by the dual phenomena of increasing suburbanization and the declining industrial economic base of older cities. Recently, however, the postindustrial economy has provided new opportunities in the urban housing market for place entrepreneurs.

Throughout the many and varying episodes of postwar restructuring, capital accumulation through urban space has propelled the efforts of entrepreneurs. Producers of urban space – the real estate industry, speculators, and developers – restructure neighborhoods through investment or disinvestment in the built environment. For place entrepreneurs or producers, housing exists chiefly as a commodity that is speculated upon, bought, sold, and developed for purposes of accumulation. Entrepreneurs invest in the undervalued commodity of low-income housing stock when structural shifts in the regional economy, such as the emergence of global corporate services, create an auspicious environment for reinvestment in devalorized inner-city neighborhoods.

Although the flow of investment is an important predictor of change in the disposition of residential housing, it alone cannot trigger the social and cultural changes that are required to upgrade poor and working-class neighborhoods. For investment in housing to profit, place entrepreneurs are compelled to create the most propitious neighborhood social and cultural conditions to attract a new and higher status of residents. In short, they must reinvent place.

Although urban accumulation is instrumental to capitalist control of the production of space, social and cultural reinvention is equally important to the control over the use and consumption of place. Both are intrinsically connected. Private ownership is translated as control over the use of urban space and is license to socially and culturally redefine those elements of neighborhood character that are indeed spatialized. The particular qualities that make place are often embedded in the use of particular buildings such as neighborhood bars, ethnic groceries, or notorious apartment buildings. Changes in ownership of buildings often entail changes in their use, thereby threatening to erase existing identity that is tied to a particular place.

Reinvention is a broader effort to control the consumption of space. First, the production of specific *urban images* or *city myths* attracts new investment in the built environment. In addition to developers, quasi-governmental and public institutions, such as chambers of commerce or, as in the case of New York, the Regional Plan Association (RPA), promulgate these myths and images. Invention of new meanings of place that are tied to "higher and better" uses attracts consumers who are willing and able to pay higher rents. Additionally, reinvention legitimizes the entire process of urban restructuring and masks the class conflict that is inherent in the restructuring of working-class neighborhoods. Neighborhood improvements, such as those made to parks, streets, and public buildings, are articulated as bettering the quality of life for all residents. Such improvements, however beneficial, are often used to justify and exculpate the social cost of residential displacement that is the consequence of redevelopment efforts. Finally, the desired image of a neighborhood as promulgated by public and private actors is contingent upon such processes as deindustrialization and, more recently, globalization.

In the following sections, the concept of reinvention is applied to a case study of the Lower East Side of New York from World War II to the present. As will be demonstrated, patterns of reinvention that existed in the decades following World War II have given way to recent and radical efforts to reinvent place that are connected to globalization.

REINVENTING NEIGHBORHOOD: DISPLACEMENT OF LOCAL CULTURES

Attempts by place entrepreneurs to redo the Lower East Side since World War II follow discernible patterns of neighborhood reinvention. Throughout the postwar history of urban restructuring, ventures designed to eliminate social and cultural features of historically working-class and minority neighborhoods – a vibrant streetscape, communal social life, and solidarity – dominated the blueprint of neighborhood reinvention. A number of alternative identities and images linked to the creation of a higher, more profitable use (white middle-class renters) for existing space were offered. Local residents resisted these efforts with varying degrees of success.

Like other poor neighborhoods in similar manufacturing cities, the Lower East Side suffered a crisis of identity brought on by the effects of postwar suburbanization and deindustrialization. By the end of World War II, the neighborhood's renowned status as home for working-class immigrants had already languished from earlier legislation that curtailed southern and eastern European immigration. Decentralization of key sectors of the area's largest employers – the garment and printing trades – that had begun in the 1920s accelerated after the war. The neighborhood's abundant tenements, which had landlords who once profited from housing the certain and steady flow of laborers, grew vacant. The local housing market flourished only during the citywide housing shortages caused by the war economy and, later, from the growing number of Puerto Rican workers seeking affordable housing.

[...]

During World War II, local elites constructed elaborate plans that would downplay the area's slum image and replace its aging infrastructure. It had long been the goal of business elites, planners, and landowners to capitalize on the Lower East Side's location adjacent to Wall Street and Midtown and to provide housing for white-collar workers increasingly drawn to suburbs.

For many inner-city neighborhoods faced with decline during the postwar years, the defining course of strategy was to beat suburbia at its own game. Without capital investments, older neighborhoods were ill equipped to reverse the suburban trend and compete effectively for middle-class residents. Likewise, their status as working-class enclaves needed revision. Implicit in reinvention efforts was the belief that eradicating glaring physical differences between city and suburb would erase the rapidly emerging social distinctions in class, income, race, occupation, and educational levels between the inner city and the surrounding suburbs.

[...]

In the "new" Lower East Side, images of the past would be either sanitized or eradicated. Proximity to work would continue to be vaunted, but middle-class consumers would be targeted. Poor residents, especially ethnic and racial minorities, would be displaced or confined to certain portions of the city, a practice augmented by federal, state, and local housing policy.

[...]

For the most part, the somewhat ambitious plans to modernize the Lower East Side failed. By and large, profits from suburbanization lured real estate capital away from inner-city neighborhoods. Tenant-based social movements, coupled with the continuation of rent control and stabilization policies, dampened possibilities for developers to eradicate both low-income tenants and the meaning of neighborhood that residents sought to preserve.

INVENTING AN EAST VILLAGE OUT OF THE LOWER EAST SIDE

With the exception of a brief flurry of revitalization in the late 1960s, few entrepreneurs and developers endeavored to resurrect the idea of middle-class reinvention of the Lower East Side. The investment strategies of the real estate industry remained geared toward suburban development, and like many inner-core neighborhoods in older cities, the Lower East Side succumbed to the decentralization of both middle-class residents and manufacturing industries. In the latter half of the 1950s and throughout the 1960s, the Lower East Side housed a vibrant Puerto Rican culture that became economically trapped by the loss of working-class jobs in lower Manhattan's light manufacturing district.

[. . .]

High poverty levels and rising crime converged in the 1970s to create an image of one of the city's least desirable and most dangerous places to live. Abandonment proceeded uninterrupted. As housing abandonment continued in the late 1970s, many of the poorest residents, most of whom were Puerto Ricans, were displaced. Many Puerto Rican residents left the neighborhood altogether, moving uptown to East Harlem or across the East River to Williamsburg in Brooklyn. Others moved into the public housing projects that lined the East River or doubled up with friends and families who stayed in the neighborhood. For remaining residents – a mix of Puerto Rican families, artists, and activists – the multiplier effect of abandonment caused living conditions to deteriorate further. There were increases in problems linked to drug abuse, such as street crime and "shooting galleries." The blocks farthest east – Avenues A, B, C, and D (known collectively as "Alphabet City") – suffered the worst.

The displacement of the Puerto Rican community in the East Village prepared the ground for a successive period of attempted redevelopment. As land values and property prices plummeted, reflecting the neighborhood's despair, a new generation of investors entered the arena, seeking new opportunities for accumulating real estate profits. Entrepreneurs triggered the gentrification of the Lower East Side in the 1980s. On the heels of a crisis of housing abandonment exacerbated by the city's fiscal crisis in the late 1970s, speculators purchased blocks of housing and later resold them to private developers. As the growth of New York's financial service economy trebled in the early 1980s, the trend in Manhattan real estate toward conversion and construction of upscale rental housing and luxury condominiums diminished the supply of middle-class rental units. Entrepreneurs viewed the Lower East Side as the last frontier for redevelopment. Developers set out to upgrade the neighborhood by displacing low-income (mostly Latino/Latina) residents and purging its stigma as rough, dangerous, and undesirable. In the 1980s, a revitalized East Village expanded eastward as developers hoped to gentrify the farthest reaches of the impoverished neighborhood, referred to as "Alphabet City."

The 1980s rush of real estate firms and brokers into the East Village housing market never achieved the wholesale gentrification of the type experienced on Manhattan's west side and in neighboring SoHo, however. By the end of the decade, it was apparent that prototypical white, upper-class professionals would never flock to the neighborhood in great numbers. Likewise, the built environment never surrendered to the "brownstoning" of tenements and cooperative and condominium conversions that inundated adjacent neighborhoods. The 1990s regional bust in real estate curtailed luxury conversions citywide. Consequently, the production of new middle-class housing was slowed.

The strategy of reinvestment and reinvention was thwarted not only by changes in the larger real estate market but also by community resistance to gentrification and displacement. Organized neighborhood housing coalitions (Cooper Square, Good Ole Lower East Side, and especially, the umbrella organization to which they belonged, the Lower East Side Planning Council) were effective in deterring wholesale redevelopment of apartments that were owned by the city as a result of earlier tax delinquencies. Mass demonstrations, fund-raising, and popular participation (sweat equity) in nonprofit housing development were part of their repertoire to gain recognition and public support. These tactics were effective in stemming the onslaught of real estate developers in the easternmost blocks of the East Village, "Alphabet City" – the area that contained the worst housing and the most city-owned units.

[. . .]

In addition to well-organized community resistance to private redevelopment, developers faced everyday resistance to their effort to reshape the negative image of the East Village. Popular resistance to what was referred to as the "yuppification" of the East Village was intense; graffiti urged the mugging of yuppies and the boycotting of upscale boutiques and groceries. Both mobilized resistance, such as the case of the Tompkins Square police riot in 1988, and everyday opposition to the bland, middle-class aesthetics of gentrification galvanized residents against developers' interest in the East Village as a commodity and not as a community. This mood dampened developers' plans. A series of demonstrations between 1988 and 1990 involving squatter rebellions and evictions of homeless people did not play well into their intentions of refashioning the East Village housing market. Some of the city's larger property holders and development firms, ones that had taken slight interest in the area a few years before, began exiting the local real estate game, leaving individual owners with multiple holdings and management companies and brokers the primary property owners. There was a realization that given the level of resistance in the East Village, its housing may never compete with that of SoHo or the West Village. In addition, the market could not bear it. The collapse of the overheated real estate market of the 1980s left a glut of luxury rentals and condominiums throughout the city.

REINVENTING NEIGHBORHOOD: APPROPRIATION OF LOCAL CULTURES

A distinct pattern emerges from the preceding example of postwar reinvention. The belief that successful restructuring of low-income neighborhoods depended upon total displacement of working-class, marginal community identities dominated the thinking of developers and planners alike. This notion of a singular path toward neighborhood redevelopment, however, is challenged by a globalized economy that accommodates (and, indeed, encourages) a variety of highly profitable and specialized places. The notion of the city has changed with globalization of the economy and so, too, has the notion of urban neighborhood change.

Global capitalism has rendered existing patterns of neighborhood reinvention ineffective and has created new possibilities for urban redevelopment. The increasing global interdependence of cultural and economic processes has caused contradictory pressures on the identity of neighborhoods and locales. On one hand, globalization is equated with the flattening of differences between places. Global capitalism has erased traditional hierarchical distinctions between places, such as strategic proximity to markets and resources. Those distinctions once formed the basis of urban accumulation and uneven urban development. As Michael Sorkin (1990) noted, space has now become departicularized or generic. On the other hand, the rapidity of economic and cultural transactions that has smoothed over certain historic distinctions between places creates both the need and the opportunity for new, unique local identities. The growth of globalism does not correlate with a corresponding decline of localism; on the contrary, many new forms of localism are cropping up. For place entrepreneurs, these local identities have become the foundation for new strategies of urban accumulation. Thus the near-universal push toward middle-class, suburban sameness that dominated postwar neighborhood reinvention of the urban landscape has been rendered obsolete.

[. . .]

The success of the multinational corporate model of global culture depends on a steady supply of emerging new cultural forms to be appropriated and commodified. As the global distribution of cultural forms is mastered by multinational corporations, the discovery or mining of these emerging new forms becomes more pressing. What was once discovery of rare, diverse, and new material is now production. In most media, the unpredictability of locating talent (e.g., by using talent scouts) has been replaced by the calculated efficiency and certainty of multinational corporate production (e.g. the music industry). In a global cultural economy in which product diversity (not necessarily originality) is foremost in creating new markets, the existence of places as incubators of new cultural forms takes on heightened importance.

The entertainment and culture industries thrive on spaces like the East Village, where new and potentially successful (profitable) cultural forms are spawned. As a niche-market incubator, the East Village has produced a *culture of insurgency*: forms that oppose and resist dominant or mainstream society. Throughout the 1980s, resistance to gentrification had become a focal point for new forms of art and fiction labeled the *downtown genre*. Art galleries promoting the local art scene materialized in the East Village during the 1980s. Public art, such as the many murals and graffiti, symbolizes the Lower East Side's identity of social and political resistance. In the 1970s, Puerto Rican political artists had painted several building-size murals depicting the struggles for Puerto Rican independence and the fight against speculation of the low-income housing market. Local artist Chico Garcia painted several memorials to friends lost to drugs and violence. Many of these remain.

In the 1980s, squatters, anarchists, and other performance propagandists used various media in resistance to encroaching gentrification by city and real estate interests. They regularly papered cinder-blocked entrances of abandoned buildings, utility poles, and any vertical flat surface with information and material related to gentrification ("Gentrification=Class War") and its consquences, including homelessness ("Your House is Mine") and residential displacement ("Mug a Yuppie"). Periodic updates on the latest activities of city agencies, police, and the real estate industry were communicated via graffiti and posters, turning the walls of neighborhood tenements into a public ledger of the struggle between residents and agents of gentrification.

[...]

The production of niche culture has affected the redevelopment of the landscape of the Lower East Side as well. For certain neighborhoods in global cities, the pattern of neighborhood reinvention has shifted toward co-optation rather than displacement of local character and identity. Rather than attempt to abolish the existing marginal cultural and social particularities of place, entrepreneurs have ventured to appropriate and transform them for purposes of further accumulation. Zones urban ecologists once labeled *bohemian mix*, like Haight-Ashbury in San Francisco, South Street in Philadelphia, or the East Village in New York, until recently have appeared somewhat impervious to middle-class upgrading. Within the past few decades, however, place entrepreneurs have developed ways to make symbolic use of what is around them and turn it to their advantage. What place entrepreneurs once considered marginal becomes unique and marketable. Stalled in their efforts at middle-class gentrification, many landlords and developers have sought to attract profitable housing-consumer niches or submarkets.

[...]

Real estate developers have been successful in spatializing the cultural image of the East Village, embracing its most profitable elements and discarding others. Developers realized that there was no need to reinvent another SoHo or West Village out of the East Village. There was an existing and growing market for the East Village that needed only to be further cultivated and tapped. It became apparent to developers and landlords that the peculiarities of the East Village's physical and social landscape were no longer constraints but emerging opportunities made possible by New York's emerging status as information and entertainment producer. The promotion of specialized housing markets – the selling of bohemian mix – mirrors the corporate promotion of cultural forms geared toward specialized, or niche, consumer markets.

Conceding the limitations of easily displacing the existing low-income identity of the neighborhood to attract stereotypical gentrifiers, real estate developers have embraced the image of bohemia to target and attract diverse occupational, cultural, and economic submarkets of new residents willing to pay higher rents. Developers have courted single young adults who were drawn to the area's reputation. This new cohort includes students attending the three major universities near the East Village, gays and lesbians intent on inventing a space of their own, and creative professionals in several musical and artistic professions. All were generally disinterested in typically gentrified neighborhoods.

Students attending the many schools in lower Manhattan have been attracted to the area by its proximity to their campuses. Along Third Avenue, large dormitories for students of New York University, the New School for Social Research, and the Cooper Union have been built to house first-year students from outside the metropolitan area. The presence of three large dormitories housing hundreds of 18-year-old students has added to the flavor of the East Village. In addition to living in the East Village, these students socialize, shop, and recreate there. The dormitories also serve as a spring-board for future tenants in the East Village. Many, if not most, dormitory residents find apartments in the neighborhood once they leave the dormitories. Although this cohort was diverse, their shared traits as tenants fit well into the redevelopment aims of investors. As a residential cohort, they have had a higher-threshold level of tolerance for the many social problems that, despite changes that were occurring, continued to plague much of the Lower East Side. Given their limited economic resources and preferences for residing in alternative neighborhoods, these groups endured above-average levels of crime, noise, and drug-related problems. Members of these groups were also less concerned with the level of city services and conditions of local schools, amenities usually associated with gentrification. In the eyes of developers, they were prime tenants because they rented in areas that stereotypical gentrifiers avoided, and simultaneously, they contributed to the area's sense of youthfulness and alternative cultures. This type of reinvention discloses important clues about the flexibility of real estate capital in the transformation of urban residential space.

IMPLICATIONS OF GLOBALIZATION FOR COMMUNITY RESISTANCE

Opportunities afforded by the growing global economy have transformed the pattern of reinvention and change for older, working-class neighborhoods in select global cities. Globalization creates new opportunities for the restructuring of urban space. In what Sharon Zukin (1995) refers to as a "symbolic economy," the production of space is increasingly tethered to the production of cultural symbols that are either created (as in theme parks or shopping malls) or appropriated (as in historic districts). The postwar model of displacement of local identity and character has given way to cultural appropriation and commodification, made possible by the growth of the global culture industry. In the East Village, these changes are most apparent. Like previous shifts in the urban economy, the recent increasing dependence on information- and entertainment-based employment in New York City's urban economy (linked to its world-city position) does not determine the strategies of real estate capital accumulation; it creates new opportunities for accumulation. In the East Village, a single set of economic, cultural, and social symbols – such as landscape, local politics, drug culture, and community lore and custom – has formed the basis of two conflicting identities, one immediate and real for local residents and the other distant and contrived for global consumption.

[...]

Community action that may have been successful earlier in stonewalling attempts to displace community may not work as well in halting the appropriation of neighborhood identity. In this most recent episode of neighborhood reinvention, traditional forms of resistance to redevelopment feed into the area's global image as counterculture incubator. The more residents go out of their way to distance themselves from the superimposed global identity, the more, it appears, they contribute to it. New forms of resistance have appeared, however. Global networks of squatters and housing advocates share information and strategies for local campaigns to maintain affordable housing. Although globalization challenges time-honored forms of community mobilization, it does create new opportunities for grassroots resistance.

REFERENCES

Sorkin, M., Ed. 1990. *Variations on a Theme Park.* New York: Noonday Press.

Teaford, J. 1990. The *Rough Road to Renaissance: Urban Revitalization in America, 1940–1985.* Baltimore, MD: The Johns Hopkins University Press.

Zukin, S. 1995. *The Cultures of Cities.* New York and London: Blackwell.

SIX

PART SEVEN

Urban Exclusion and Social Resistance

Plate 13 Urban protest in New York City, January 28–30, 2002 by Larry Towell. During protests at the World Economic Forum, a protestor holds a "Make the Global Economy Work for Working Families" sign (reproduced by permission of Magnum Photos).

INTRODUCTION TO PART SEVEN

Spatial segregation is a feature of cities across the globe. In certain cities segregation is associated primarily with ethnic or racial groups, in others, class status. In Latin America, for example, urban spatial segregation is based primarily on socioeconomic status. In the U.S., as we have seen in the readings in Part Three, spatial segregation is based mostly on racial or ethnic divisions. Since its inception, the discipline of urban sociology has remained preoccupied with the underlying causes of segregation, from Park and Burgess's "naturalist" explanations to Wacquant and Wilson's accounting of the structural factors that create impacted ghettos.

In the first half of this section, we turn to the social, political, and cultural infrastructure that sustains and furthers the geography of separation. Mike Davis provides a visual recounting of social polarization in Los Angeles, a patchwork quilt city of ethnic and class enclaves. Neighborhoods of stark differences in resources and wealth sit cheek-by-jowl and their separation rests systematically upon political–electoral, economic, and social exclusion. As Davis makes clear, considerable political energy (and even greater sums of money) is expended to reproduce and maintain segregation. The epitome of formalized and sanctioned urban segregation – the fortified residential enclave – is explored in Teresa Caldeira's reading. Caldeira focuses on São Paulo, Brazil, where walled communities are status symbols for the wealthy and acquisitive middle classes. Picture landscaped and sprawling estates or condominium complexes hidden behind tall stucco walls painted in pastel colors, covered with vines, nearly totally disconnected from the surrounding cities. While its residents may enjoy its amenities and hyper-security, the fortified enclave raises some disturbing questions about the public life of cities. When Caldeira turns her eye to gated communities in Los Angeles she sees similar issues of status-enhancement and security consciousness, coupled with a fear of the Other. Both of these readings address the important fact that, regardless of our competing explanations of the causes of segregation, urban polarization and exclusion are actively (re)produced through the actions of architects, city planners, developers, the police, and residents.

Our final two readings address social action of a different sort – grassroots collective action among individuals and groups in the city. The urban sociologist Manuel Castells introduced the term "urban social movement" in the early 1970s, on the heels of extensive collective action among residents in cities across the globe. While Castells initially saw much hope in the efforts of citizens' movements to make headway in the face of economic and political power, he later viewed them as little more than symptoms of postindustrial capitalism. In his 1983 book, *The City and the Grassroots: A Cross-cultural Theory of Urban Social Movements* (London: Edward Arnold), Castells characterized movements "reactive utopias," essentially incapable of bringing about social change. Macro economic and political forces, he argued, largely determined the fates of cities.

Castells' gloomy prediction notwithstanding, the study of urban social movements has flourished – arguably more so as the social consequences of globalization are realized in cities around the world. The study of how individuals come together and organize and make demands and claims is fundamentally a sociological and, perhaps more importantly, hopeful intellectual enterprise.

Our reading selections in this section focus on the forms of activism that have gained prominence in the current period of global socio-economic restructuring of cities. The readings in Part Five enumerated the political, economic, and social consequences of globalization for cities, including the increased polarization and unevenness in wealth. Many urban social movements have formed in reaction to cuts in social programs that have diminished the quality of life. Social welfare spending, in the forms of subsidies for mass transportation, low-income housing, pollution reduction, public health care, and public safety, has been curtailed. In the era of global competitiveness, municipal governments provide a larger share of their revenues to tax abatements and financial incentives for companies and developers to remain or relocate in their cities.

Collective responses to urban social problems exacerbated by globalization vary across cities. Local circumstances provide some urban grassroots movements a space for collective mobilization within their communities and neighborhoods. The forms and strategies urban social movements adopt – operating food pantries, forming block associations, or organizing protests and rent strikes – differ from city to city. Despite these differences, however, most recent urban movements have mobilized in response to a common threat of economic restructuring, spatial segregation and urban poverty (in short, consequences of globalization). As the readings in the section point out, more and more people in very different and distant cities are facing identical challenges to everyday living – the struggle to secure safe housing, pay rent, and find affordable health care. Most urban social movements find their roots in community-based struggles for these "collective consumption" goods, as Castells referred to them.

What kinds of strategies and tactics might social movements deploy in the face of globalization, where the causes of local problems often appear distant and removed? Pierre Hamel, Henri Lustiger-Thaler and Margit Mayer claim contemporary urban social movements engage in three different types of struggle and collective action based on a calculation of local *and* global dimensions of authority, power, and resistance. The final reading by Bettina Kohler and Markus Wissen sees cities as key sites of grass-roots movements and resistance precisely because they exist in the intersection of global, national, and local processes. Kohler and Wissen draw on the themes introduced in the readings in Part Five, in which cities are both the strategic nodes within the flow of international capital and spaces in which globalization is articulated and "made real." Contemporary collective action takes a similar path; movements can be focused on local social problems while simultaneously "jumping scales" and making connections with a network of similar movements across the globe. Global capitalism, as we have seen, has reaped benefits from the mobility of capital and information flows, transnational forms of organization, and increasingly open borders. In their efforts to combat the social ills globalization has wreaked, urban movements, too, have harnessed transnational networks and information flows to their advantage.

"*Chinatown*, Part Two?: The 'Internationalization' of Downtown Los Angeles"

from *New Left Review* (1987)

Mike Davis

Editor's Introduction

Mike Davis is the Jeremiah of early twenty-first century Los Angeles, an iconoclastic writer and journalist who articulates a stark prophetic vision of the millennial city as urban dystopia. While Michael Dear is the theoretician of the L.A. School, Mike Davis is the eloquent scribe. His journalistic and critical voice has captured the popular reading imagination in books such as *City of Quartz: Excavating the Future in Los Angeles* (London: Verso, 1990), which has been translated into six languages and won the National Book Critics Circle Award. His cinematically informed view of Los Angeles assaults the traditional boosterist treacle of Southern California as an edenic land of sunshine and orange groves, theme parks, and palm-lined subdivisions. Eden has lost its garden in the Los Angeles that Davis inhabits, a deindustrializing and sprawling landscape that is troubled by crime, police brutality, intergroup violence, economic and social inequality, and environmental catastrophe. Los Angeles is a junkyard of dreams.

The Los Angeles of Mike Davis is a *Bladerunner* (1982) city, where the homeless, the riotous underclasses, and undocumented immigrants struggle with the Los Angeles Police Department (LAPD) just as fugitive "Nexus 6" replicants evade the helicopters of the Tyrell Corporation in the sprawling science fiction metropolis of the future. Ridley Scott's *Bladerunner* meets Roman Polanski's *Chinatown* in this article, which updates the popular myth of urban corruption in Los Angeles from the 1910s to the 1980s. The corrupt water barons and power brokers of the days of William Mulholland and the Department of Water and Power (DWP) have given way to a group that includes the Community Redevelopment Agency (CRA) and a variety of transnational corporate interests. The DWP-led swindle in water rights and land speculation has become a CRA-led scheme of tax increment financing and urban redevelopment that has included the razing of Bunker Hill, downtown renewal, and police sweeping of the homeless in Skid Row. Downtown has become a fortress and citadel for this transnational capital of the Pacific Rim.

Deindustrialization has plagued the greater Los Angeles region, starving Latino and African American communities of jobs and capital. The CRA, City Hall, and related interests have abandoned the minority communities of East and South Los Angeles through disinvestments and financial redlining, while spending heavily on Downtown redevelopmental subsidies, new infrastructure, and tax abatements for transnational corporate investment interests.

See John Walton, "Film History as Urban History: the Case of *Chinatown*," pp. 46–58 in *Cinema and the City: Film and Urban Societies in a Global Context*, edited by Mark Shiel and Tony Fitzmaurice (Oxford: Blackwell, 2001), for a discussion of the real story behind the "rape of the Owens River valley" that screenwriter Robert Towne adapted for Hollywood. For a related perspective on social polarization in downtown Los Angeles, see Don Parson, "The Search for a Center: the Recomposition of Race, Class and Space in Los Angeles," *International Journal of Urban and Regional Research* 17, 2 (June 1993): 232–240.

Mike Davis was born in Fontana in the Inland Empire of San Bernardino County east of Los Angeles in 1946 and grew up in Bostonia, a now lost hamlet east of San Diego. He has worked as a meat cutter and long distance truck driver. He was an activist with Students for a Democratic Society and the Communist Party in the 1960s. He entered UCLA as mature student in economics and history. After forays to Ireland and Scotland, he finished his Ph.D. qualifying exams in 1981, whereupon he went to London to work with *New Left Review*. He published *Prisoners of the American Dream: Politics and Economy in the U.S. Working Class* (London: Verso, 1986). He has taught at a variety of colleges and universities in the Southern California region, and held posts in urban theory at the Southern California Institute of Architecture and history at the State University of New York at Stony Brook.

His recent books include *Ecology of Fear: Los Angeles and the Imagination of Disaster* (New York: Metropolitan Books, 1998), *Late Victorian Holocausts: El Niño Famines and the Making of the Third World* (London: Verso, 1999), *Magical Urbanism: Latinos Reinvent the U.S. City* (London: Verso, 2002), and *Dead Cities: A Natural History* (New York: New Press, 2002). See Lewis MacAdams, "Jeremiah Among the Palms: The Lives and Dark Prophecies of Mike Davis," *Los Angeles Weekly*, November 26–December 3, 1998 for an insightful biography of Mike Davis.

JAKE GITTES: How much are you worth?
NOAH GROSS: I have no idea. How much do you want?
GITTES: I want to know how much you're worth – over ten million?
GROSS: Oh, my, yes.
GITTES: Then why are you doing it? How much better can you eat? What can you buy that you can't already afford?
GROSS: The future, Mr. Gittes, the future . . .
(Robert Towne, *Chinatown* script)

The shortest route between Heaven and Hell in contemporary America is probably Fifth Street in Downtown L.A. West of the refurbished Biltmore Hotel, and spilling across the moat of the Harbor Freeway, a post-1970 glass and steel skyscape advertises the landrush of Pacific Rim capital to the central city. Here, Japanese mega-developers, transnational bankers and billionaire corporate raiders plot the restructuring of the California economy. A few blocks east, across the no-man's-land of Pershing Square, Fifth Street metamorphoses into the 'Nickel': the notorious half-mile strip of blood-and-vomit-spewn concrete where several thousand homeless people – themselves trapped in the inner circle of Dante's inferno – have become pawns in a vast local power struggle. Intersecting these extremes of greed and immiseration is the axis of a third reality: *el gran Broadway*, the reverberant commercial centre of a burgeoning Spanish-speaking city-within-a-city, whose *barrios* (interpenetrating the ghetto to the south) now form a dense ring around the central business district. A ten-minute walk down Fifth Street thus passes through abrupt existential and class divides, a micro-tour of social polarization in the late-Reaganite era. Moreover this landscape – whether we recognize the location or not – has insinuated itself into the contemporary imagination. Because the Downtown skycity is so recent, and because of its proximity to the media factories of Hollywood, it figures prominently as a representation of the early twenty-first-century urbanism that is now emerging. It has become *de rigueur* for passing theorists and image-mongers, whether as critics or celebrants, to stop and comment on the architectural order and social topography that are coalescing out of the lava of development around Fifth Street.

For Fredric Jameson, in a seminal essay, the built environment of Los Angeles, especially its 'downtown renaissance,' is a paradigm of the 'postmodern' city where architecture and electronic image have fused into a single hyperspace. For the well-known urban designer James Sanders, director of the Bryant Park Project in Manhattan, the 'intense – even poetic – verticality' of Fifth and Grand is an expression of Los Angeles imperialism:

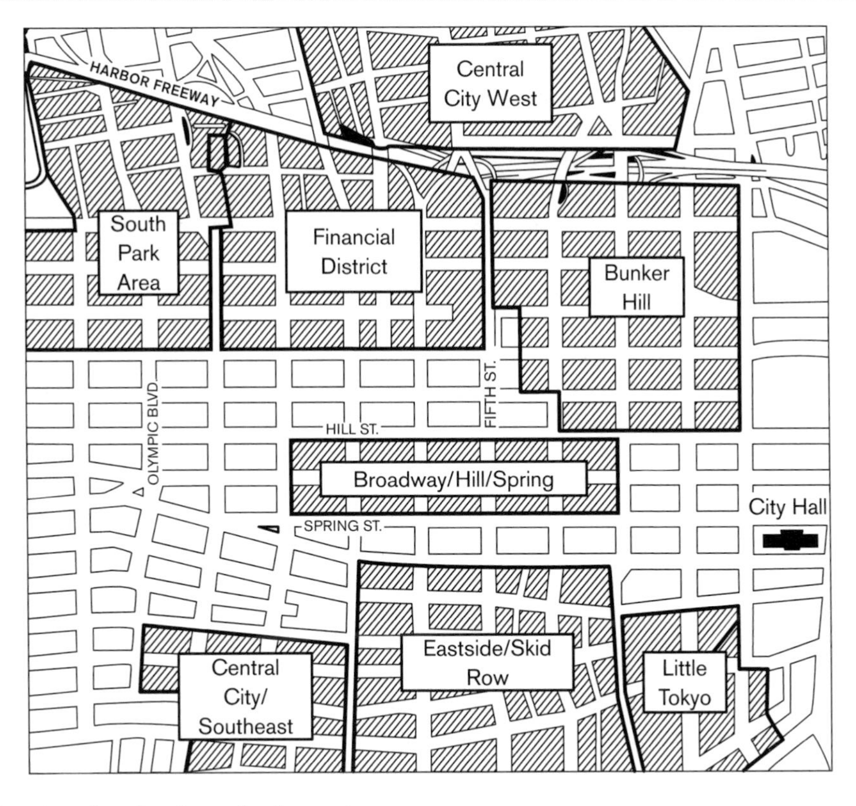

Figure 1 Downtown Los Angeles: redevelopment zones

"The 'new' downtown Los Angeles is pulling away international banking and finance, establishing a centre of great radiating lines of communication and trade for the Pacific Rim. As on the East Coast, where New York is grabbing the remaining marbles of the Atlantic economy community, so Los Angeles is setting itself up as the Pacific's economic capital. The two cities seem intent on carving the world into two great economic entities, with themselves as the centres."

Hollywood, meanwhile, has reached for different hyperboles. Younger directors have relentlessly exploited the social extremes of Downtown as a nightmare stage, a ground zero, for such contemporary apocalypses as *Repo Man, The Terminator, To Live and Die in L.A.*, and so on. This dystopian figuration acquires a dark grandeur in Ridley Scott's *Bladerunner* with its images of mile-high towers, ruled by interplanetary genetic-engineering conglomerates, rising above the poisonous congealed smog that drips acid rain upon thirty million inhabitants. None of the theories or visions on offer (with the partial exception of the racist 'yellow hordes' of *Bladerunner*), however, registers the presence, probably epochal, of an enlarged low-wage working class, living and working in the central city, and creating its own spatialized social world: networks of recreation, piety, reproduction and, ultimately, struggle. They fail to capture the growing tension – relayed through various mediations to the traditional L.A. working and middle classes – between international capital and international labour migration in the contested terrain of the inner city. For if L.A.'s Downtown is in any sense paradigmatic, it is because it condenses the intended and unintended spatial consequences of the political economy of post-Fordism: that is to say, the rise of new, globalized circuits of finance and luxury consumption amid the decline of much of the old mass-consumption and high-wage industrial economy. But there is no single, master logic of restructuring, rather the complex intersection of two separate macro-processes: one based on the overaccumulation of bank and real-estate capital (most recently, from the recycling of the East Asian trade surplus to California); the other arising from the reflux of low-wage manufacturing and labour-intensive services in the wake of unprecedented mass immigration from Mexico and Central America.

Within the larger systems of metropolitan Los Angeles and Southern California (separately, the ninth largest economy in the world), Downtown has become the privileged crucible where apparently infinite foreign capital and low-wage immigrant labour are first transformed into assets for the regional boom. But because Downtown is simultaneously a portal for capital and for immigration, and because the two functions remain concentrated in the same inner core of land development and infrastructure, there are growing contradictions. The yen-fueled momentum of highrise development cuts into the crowded work and residential spaces of the inner-city working poor: commercial overbuilding produces rampant underhousing. At the same time the uncoordinated dynamics of redevelopment and immigration, without investment in radically expanded welfare and physical infrastructures, are making powerful, if differential, impacts upon the living standards and residential positions of older working-class and middle-strata neighbourhoods from Boyle Heights to Venice and the San Fernando Valley. The political consequence is a far-reaching electoral realignment, excluding the working poor, as the old pro-development coalition under the figurehead of Mayor Bradley is attacked by a populist homeowners' rebellion orchestrated by his former Democratic allies on the white Westside. This is a complicated scenario, with sweeping assertions; a plot, perhaps, for *Chinatown*, Part Two. In the meantime, let me sketch, in bold outlines, the major action.

WHAT JAKE DISCOVERS ABOUT DOWNTOWN

The crisis of Downtown L.A. began in the same period in which Polanski's brilliant historical film noir is set, immediately after the great highrise building boom of 1923–24 that constructed the skyline as it remained until the 1960s. Commercial life in the centre began to wither as precocious automobilization (on a scale not achieved in Europe until the 1970s) gridlocked the Downtown traffic flow while the oil-rubber-paving lobby sabotaged the recapitalization of the city's once superb fixed-rail systems. A middle-class exodus to the Westside was followed by relocation of the large department stores and retail trade outward along Wilshire Boulevard. Depression and war filled Downtown tenements with an increasingly impoverished and shifting population; once aristocratic Bunker Hill near the civic centre became Raymond Chandler's notorious 'lost town, shabby town, crook town' with 'women with faces like stale beer . . . men with pulled-down hats.'

The New Deal hopes of local progressives and trade unionists that Downtown might be revivified with model public housing were vanquished in 1950 after a vicious, red-baiting mayoral campaign led by the *Los Angeles Times* and the traditional Downtown ruling circle. With a pliant city hall under Mayor Poulson, the *Times* and the Downtown Businessmen's Association, ventriloquizing their interests through the publicly unaccountable Community Redevelopment Agency (CRA), were able to ratify a masterplan (first conceived in 1931) to evict the nine thousand residents of Bunker Hill to make way for the first phase of ambitious commercial renewal. Redevelopment, however, was easier to fantasize about in the smoking rooms of the elite Jonathan Club than to implement in practice; it took almost a decade to clear Bunker Hill. Apart from the rearguard resistance of expropriated Downtown slumlords and the sporadic hostility of Valley taxpayers to costly improvements in the inner city (deftly exploited by the *poujadiste* demagoguery of Poulson's successor, Sam Yorty), the major obstacle to a vigorous recentralization of commercial development was the fragmentation of ruling-class interests in the L.A. Basin. Gone were the days when the Merchants and Manufacturers Association could mobilize the para-military unity of local business behind the Open Shop (invented in L.A. in the 1890s).

The first break in elite ranks had occurred in 1926 when the Hollywood movie moguls seceded from the Open Shop to establish their own sweetheart-union labour-relations system (the famous Studio Basic Agreement). Repelled by the country-club anti-semitism of L.A.'s WASP old money, and dealing with a different calculus of labour costs, the predominantly Jewish management of the entertainment industry evolved into a world apart. They endowed the state university system's UCLA not the private University of Southern California (USC); generally contributed to the national Democrats not Republicans; and focused their speculative energies on the development of Beverly Hills and the greater Westside. Meanwhile, the crucial aerospace industry – closely integrated, like Hollywood, with Wall Street – was even less historically entangled with the old inner city; its regional interests were defined by the great airport-and-manufacturing complexes outside the city limits. Finally, at the turn of the sixties, the Downtown renewal strategy was frontally challenged by Alcoa's (i.e., the Mellon Family's) announcement that it was building a huge highrise centre for the Westside on Twentieth Century Fox's old backlot in what is now Century City.

Ironically the CRA's Downtown plan was saved by the explosion of the local Black working class. The Watts Rebellion was, amongst other things, a protest against the racist 'Cotton Curtain' that excluded Blacks from the higher-wage jobs in the industrial belt east of Alameda Street, as well as against rampant police brutality, rackrenting and petty usury. The crisis of 1965, which continued to resonate through waves of innercity unrest and white backlash for almost a decade, was instrumentalized by redevelopment interests in two decisive ways. First and immediately, by raising the spectre of Downtown and USC engulfed by a militant Black population, the traditional corporate patrons of the CRA were able to galvanize broader ruling-class support for the renovation of Downtown, through the emergency Committee of 25 (later, the 'Community Committee') and an expanded Central City Association. The Community Committee, in particular, was the closest thing to an 'executive committee of the bourgeoisie' which Southern California had seen since the class wars of the 1930s. It broke precedent by including Jewish entertainment sector leaders (like Lew Wasserman of MCA), the CEOs of major aerospace corporations (like Roy Anderson of Lockheed and Tom Jones of Northrup), and a visible quotient of prominent Democrats. (USC and its alumni meanwhile used the crisis, with official city support, to evict poor households and impose a *cordon sanitaire* – parking lots, administrative buildings, shopping centres – between campus and the surrounding Black-Latino community.)

THE GROWTH COALITION

The second aftermath of the Watts uprising was the consolidation of a multiracial coalition, based on Jewish (10 per cent of the electorate) and Black votes under the dispensation of Downtown boosters led by Otis Chandler's 'liberalized' *Los Angeles Times*. After six years of blistering warfare against Mayor Yorty's white backlash the new coalition installed Thomas Bradley, a Black ex-cop and city councilman, as mayor in 1973. Bradley's regime, which over the years has drawn support from such landed powers as BankAmerica, the Irvine Ranch, and ARCO, enlarged the CRA mandate to encompass all of Downtown, opening a real-estate bonanza to commercial developers. At the same time his fourteen-year reign has been little short of catastrophic for inner-city residents. Black Southcentral Los Angeles has been reduced to a deindustrialized twilight zone, while Eastside Chicano-Latino neighbourhoods, unrepresented on the city council between 1963 and 1985, have been locked out of power. The key Bradley constituencies have been pacified with largely symbolic goods: for Blacks, a few celebrity politicians; for Jews, city hall's craven solicitation of Israel. The development interests, on the other hand, have received a plush welfare state all their own.

Because the city has avoided even the desultory levels of social service and patronage that have elsewhere been necessary supports of Black-led crisis-management, it has had still greater fiscal resources to subsidize urban renewal. The prime mechanism is 'tax increment financing,' which allows the CRA to function as an unelected, sovereign power, confiscating the tax increases from new development (the 'increment') to subsidize further development. As Black small businesses have been 'red-lined' out of existence by discriminatory bank credit practices, and as East L.A. continues to pay more taxes than it receives in services, Downtown redevelopers reap special low-interest loans, free infrastructural modernization, tax abatements and, above all, discounted prime land.

This is where we pick up the plot of Polanski's *Chinatown*: for the last twenty years big developers could be confident that their subsidized parcels – made available by the CRA at half the cost of land in Century City or downtown San Francisco – would triple or quadruple in market value within a few years of construction; the resulting tax increment being sluiced off for the next stage of land development. The CRA has quietly municipalized land speculation – just as in the early twentieth century aqueduct conspiracy upon which *Chinatown* was based, except on a vaster scale, with dirt and increments instead of dirt and water as the magic formula for super-profits.

With the majority of the city council routinely approving every request of the developers' lobby (or abdicating power to the CRA and the City Planning Department), it was not surprising that almost $2.5 billion of new investment flowed into Downtown in the decade after Bradley's election. Where there were just five new highrises above the old earthquake limit of thirteen floors in 1976, there are now forty-five. Increasingly the CRA operated a casino as players moved in and out of speculative positions, nearly a third of Downtown exchanging hands between 1976 and 1982. Ironically, as the ante has inexorably risen, many of the original champions of Downtown renewal, including large regional banks and oil companies with troubled cashflows, have had to cash in their equity and withdraw to the sidelines. As Volckerism first created a super-dollar and then weakened it, the volatile commercial real-estate markets around the country have favoured highly liquid investors and foreign capital. In 1979 the *Los Angeles Times* reported that a quarter of Downtown's major properties were foreign-owned; six years later the figure was revised to 75 percent (one authority claims 90 percent). The first wave of foreign investment in Downtown in the late 1970s, as in Manhattan, was led by Canadian real-estate capital, epitomized by Toronto-based Olympia and York. The Reichman clan who own Olympia and York collect skyscrapers like the mere rich collect rare stamps or Louis XIV furniture. Yet since 1984 they, along with the New York insurance companies and the British banks, have been swamped by a tsunami of East Asian finance and flight capital.

ZAITECH

What the Japanese call *zaitech*, the strategy of using financial technologies to shift cashflow from production to speculation, has radically restructured Downtown's investment portfolios and given a new impetus to sagging office construction. The liquid resources of other investors have been simply dwarfed by the sheer mass of the Japanese trade surplus which is rapidly finding its way from US Treasury bonds to prime real estate. The current super-yen puts the skyscrapers along Figueroa's 'Gold Coast' at a rummage-sale discount: a virtually unknown condominium developer from Tokyo, Shuwa Company Ltd., bought nearly $1 billion of L.A.'s new skyscape, including the twintowered Arco Plaza, in a single two-and-half-month shopping spree last year. Moreover, as local real estate analysts complain, the major Japanese companies are borrowing at very cheap rates, usually 5 percent or less. They borrow in Japan [in Shuwa's case, through ten L.A. branches of Japanese banks], deduct it from their taxes in Japan and convert it to dollars, invest in dollars in the United States.

The Japanese surge into Downtown real estate is coordinated with the wild stockmarket and property booms within Japan itself that are causing increasing alarm about the future of the Pacific Rim economy. As the super-yen and foreign protectionism depress industrial investment, most of the big Japanese corporations and trading firms are resorting to *zaitech* to keep themselves in the black. At the same time soaring stock values and a 100 percent annual rate of property appreciation in central Tokyo are inflating portfolios and pension trusts which seek new outlets overseas. One result is the impressive lineup of Mitsui Fudosan, Sumitomo, Dai-Ichi Life, Mitsubishi and a dozen other major Japanese players in a race to grab new Downtown development sites. Together with more shadowy Kuomintang capital from Hong Kong and the ASEAN region, they are positioning themselves to help push the frontiers of commercial speculation into the Spanish-speaking neighbourhoods west of the Harbor Freeway, as well as lobbying for the redevelopment of Skidrow as part of Little Tokyo.

"Fortified Enclaves: The New Urban Segregation"

from *Public Culture* (1996)

Teresa P. R. Caldeira

Editors' Introduction

In this selection, Teresa Caldeira examines fortified enclaves – what are sometimes referred to as "gated communities" – in São Paulo, Brazil and Los Angeles. Walled communities are an increasing presence in cities and suburbs around the world and Caldeira is interested in what these communities mean to the idea of the city as a space of opportunity, chance, flow and movement and, of course, social diversity.

In São Paulo, fortified enclaves are popular among the middle and upper classes for two different reasons. Walled or gated communities are popular among those who can afford them because they provide "total security" from the real and imagined dangers of the outside world. In addition to risk-conscious businessmen and their families and corporate elites, enclaves appeal to middle-classes fearful of street crime and the chaotic character of city streets. Enclaves provide a sense (perhaps illusion) of control and protection over one's surroundings. A second reason behind the proliferation of walled communities is status. Enclaves are physically separate and socially exclusive hence they confer a higher status among those who live within them (there is a prestige ranging among enclaves, as well). Given enclaves offer numerous amenities in private facilities (sport, leisure, etc.) and household and personal services, they are seen as spaces of privilege. As Caldeira also points out, class separation is a form of social distinction.

What effects do fortified enclaves have on the character of city life? As part of the built environment, enclaves contribute neither to public space nor freedom of mobility within the city. Walls, fences, and gates impede the flow of traffic, pedestrian and otherwise; enclaves serve no greater purpose for nonresidents who are forced to move around them. Second, the social, cultural, and economic life of enclaves is self-contained and inward-looking. Enclaves do not gesture – either architecturally or sociospatially – to the streets and neighborhoods around them. Caldeira writes: "In other words, the relationship they establish with the rest of the city and its public life is one of avoidance; they turn their backs on them."

Los Angeles shares with São Paulo a fascination and preoccupation with walled communities. Two memorable television images of middle-class homeowners during the 1992 riots were the walled communities with its residents safely intact and the residents of neighborhoods with street access erecting makeshift barricades and gates and manning the entrances with shotguns. Caldeira draws on Mike Davis's extensive work on the architecture of fear found in his book, *City of Quartz* (London: Verso, 1990). For Davis, walled communities are "class warfare at the level of the street." They tend to reinforce social inequality and spatial segregation and contribute nothing positive to the public life of the greater city. In fact, with their friendly, inviting names and pastel-colored walls, gated communities "normalize" inequality and segregation.

In the end, Caldeira finds very little to celebrate with walled communities. As anti-public spaces, they minimize the contact among persons of different class, ethnic, national and racial backgrounds and contribute to a greater fear of the "Other." Urban fear and its underlying causes are not addressed – they are concretized and memorialized. Whatever contact middle-classes have with the poor and ethnic and racial minorities is limited to consumption, she argues. Difference is encountered in brief exchanges at ethnic markets or restaurants or in contact with mostly minority maids and groundskeepers. Either way, contact is minimal, controlled, and on the terms of wealthier, more powerful whites.

Teresa Caldeira is Associate Professor of Anthropology at the University of California, Irvine. Her book, *City of Walls: Crime, Segregation, and Citizenship in São Paulo* (Berkeley, CA: University of California Press, 2000) won the American Ethnological Society Senior Book Prize in 2001. She has been a Visiting Scholar at the University of São Paulo and the Center for the Study of Violence (NEV). In addition to her research on urban segregation, she studies the impact of technological change on domestic spaces and generational relations in Brazil.

In the last few decades, the proliferation of fortified enclaves has created a new model of spatial segregation and transformed the quality of public life in many cities around the world. Fortified enclaves are privatized, enclosed, and monitored spaces for residence, consumption, leisure, and work. The fear of violence is one of their main justifications. They appeal to those who are abandoning the traditional public sphere of the streets to the poor, the "marginal," and the homeless. In cities fragmented by fortified enclaves, it is difficult to maintain the principles of openness and free circulation that have been among the most significant organizing values of modem cities. As a consequence, the character of public space and of citizens' participation in public life changes.

In order to sustain these arguments, this chapter analyzes the case of São Paulo, Brazil, and uses Los Angeles as a comparison. São Paulo is the largest metropolitan region (it has more than sixteen million inhabitants) of a society with one of the most inequitable distributions of wealth in the world. In São Paulo, social inequality is obvious. As a consequence, processes of spatial segregation are also particularly visible, expressed without disguise or subtlety. Sometimes, to look at an exaggerated form of a process is a way of throwing light onto some of its characteristics that might otherwise go unnoticed. It is like looking at a caricature. In fact, with its high walls and fences, armed guards, technologies of surveillance, and contrasts of ostentatious wealth and extreme poverty, contemporary São Paulo reveals with clarity a new pattern of segregation that is widespread in cities throughout the world, although generally in less severe and explicit forms.

[. . .]

BUILDING UP WALLS: SÃO PAULO'S RECENT TRANSFORMATIONS

The forms producing segregation in city space are historically variable. From the 1940s to the 1980s, a division between center and periphery organized the space of São Paulo, where great distances separated different social groups: the middle and upper classes lived in central and well-equipped neighborhoods and the poor lived in the precarious hinterland. In the last fifteen years, however, a combination of processes, some of them similar to those affecting other cities, deeply transformed the pattern of distribution of social groups and activities throughout the city. São Paulo continues to be a highly segregated city, but the way in which inequalities are inscribed into urban space has changed considerably.

[. . .]

São Paulo is today a city of walls. Physical barriers have been constructed everywhere – around houses, apartment buildings, parks, squares, office complexes, and schools. Apartment buildings and houses that used to be connected to the street by gardens are now everywhere separated by high fences and walls and guarded by electronic devices and armed security men. The new additions frequently look odd because they were improvised in spaces conceived without them, spaces designed to be open. However, these barriers are now fully integrated into new projects for individual houses, apartment buildings, shopping areas, and work spaces. A new aesthetic of security shapes all types of constructions and imposes its new logic of surveillance and distance as a means for displaying status, and it is changing the character of public life and public interactions.

[...]

Fortified enclaves represent a new alternative for the urban life of these middle and upper classes. As such, they are codified as something conferring high status. The construction of status symbols is a process that elaborates social distance and creates means for the assertion of social difference and inequality.

[...]

Closed condominiums are supposed to be separate worlds. Their advertisements propose a "total way of life" that would represent an alternative to the quality of life offered by the city and its deteriorated public space. Condominiums are distant, but they are supposed to be as independent and complete as possible to compensate for it; thus the emphasis on the common facilities they are supposed to have that transform them into sophisticated clubs. The facilities promised inside closed condominiums seem to be unlimited – from drugstores to tanning rooms, from bars and saunas to ballet rooms, from swimming pools to libraries.

In addition to common facilities, São Paulo's closed condominiums offer a wide range of services. Some of the services (excluding security) are psychologists and gymnastic teachers to manage children's recreation, classes of all sorts for all ages, organized sports, libraries, gardening, pet care, physicians, message centers, frozen food preparation, housekeeping administration, cooks, cleaning personnel, drivers, car washing, transportation, and servants to do the grocery shopping. If the list does not meet your dreams, do not worry, for "everything you might demand" can be made available. The expansion of domestic service is not a feature of Brazil alone. As Saskia Sassen has shown in the case of global cities, high-income gentrification requires an increase in low-wage jobs; yuppies and poor migrant workers depend on each other. In São Paulo, however, the intensive use of domestic labor is a continuation of an old pattern, although in recent years some relationships of labor have been altered, and this work has become more professional.

The multiplication of new services creates problems, including the spatial allocation of service areas. The solutions for this problem vary, but one of the most emblematic concerns the circulation areas. Despite many recent changes, the separation between two entrances – in buildings and in each individual apartment – and two elevators, one labeled "social" and the other "service," seems to be untouchable; different classes are not supposed to mix or interact in the public areas of the buildings. Sometimes the insistence on this distinction seems ridiculous, because the two elevators or doors are often placed side by side instead of being in separate areas. As space shrinks and the side-by-side solution spreads, the apartments that have totally separate areas of circulation advertise this fact with the phrase, "social hall independent from service hall." The idea is old: class separation as a form of distinction.

Another problem faced by the new developments is the control of a large number of servants. As the number of workers for each condominium increases, as many domestic jobs change their character, and as "creative services" proliferate for members of middle and upper classes who cannot do without them, so also do the mechanisms of control diversify. The "creative administrations" of the new enclaves in many cases take care of labor management and are in a position to impose strict forms of control that would create impossible daily relationships if adopted in the more personal interaction between domestic servants and the families who employ them. This more "professional" control is, therefore, a new service and is advertised as such. The basic method of control is direct and involves empowering some workers to control others. In various condominiums, both employees of the condominium and maids and cleaning workers of individual apartments (even those who live there) are required to show their identification tags to go in and out of the condominium. Often they and their personal belongings are searched when they leave work. Moreover, this control usually involves men exercising power over women.

The middle and upper classes are creating their dream of independence and freedom – both from the city and its mixture of classes and from everyday domestic tasks – on the basis of services from working-class people. They give guns to badly paid working-class guards to control their own movement in and out of their condominiums. They ask their badly paid "office boys" to solve all their bureaucratic problems, from paying their bills and standing in all types of lines to transporting incredible sums of money. They also ask their badly paid maids – who often live in the *favelas* on the other side of the condominium's wall – to wash and iron their clothes, make their beds, buy and prepare their food, and frequently care for their children all day long. In a context of increased fear of crime in which the poor are often associated with criminality, the upper classes fear contact and contamination, but they continue to depend on their servants. They can only be anxious about creating the most effective way of controlling these servants, with whom they have such ambiguous relationships of dependency and avoidance, intimacy and distrust.

Another feature of closed condominiums is isolation and distance from the city, a fact that is presented as offering the possibility of a better lifestyle. The latter is expressed, for example, by the location of the development in "nature" (green areas, parks, lakes), and in the use of phrases inspired by ecological discourses. However, it is clear in the advertisements that isolation means separation from those considered to be socially inferior, and that the key factor to assure this is security. This means fences and walls surrounding the condominium, guards on duty twenty-four hours a day controlling the entrances, and an array of facilities and services to ensure security – guardhouses with bathrooms and telephones, double doors in the garage, and armed guards patrolling the internal streets. "Total security" is crucial to "the new concept of residence." In sum, to relate security exclusively to crime is to fail to recognize all the meanings it is acquiring in various types of environments. The new systems of security not only provide protection from crime but also create segregated spaces in which the practice of exclusion is carefully and rigorously exercised.

[...]

ATTACKING MODERN PUBLIC SPACE

The new residential enclaves of the upper classes, associated with shopping malls, isolated office complexes, and other privately controlled environments represent a new form of organizing social differences and creating segregation in São Paulo and many other cities around the world. The characteristics of the Paulista enclaves that make their segregationist intentions viable may be summarized in four points. First, they use two instruments to create explicit separation: on the one hand, physical dividers such as fences and walls; on the other, large empty spaces creating distance and discouraging pedestrian circulation. Second, as if walls and distances were not enough, separation is guaranteed by private security systems: control and surveillance are conditions for internal social homogeneity and isolation. Third, the enclaves are private universes turned inward with designs and organization making no gestures toward the street. Fourth, the enclaves aim at being independent worlds that proscribe an exterior life, which is evaluated in negative terms. The enclaves are not subordinate either to public streets or to surrounding buildings and institutions. In other words, the relationship they establish with the rest of the city and its public life is one of avoidance: they turn their backs on them. Therefore, public streets become spaces in which the elite circulate by car and poor people circulate on foot or by public transportation. To walk on the public street is becoming a sign of class in many cities, an activity that the elite is abandoning. No longer using streets as spaces of sociability, the elite now want to prevent street life from entering their enclaves.

Private enclaves and the segregation they generate deny many of the basic elements that previously constituted the modern experience of public life: primacy of streets and their openness; free circulation of crowds and vehicles; impersonal and anonymous encounters of the pedestrian; unprogrammed public enjoyment and congregation in streets and squares; and the presence of people from different social backgrounds strolling and gazing at those passing by, looking at store windows, shopping, and sitting in cafes, joining political demonstrations, or using spaces especially designed for the entertainment of the masses (promenades, parks, stadiums, exhibitions).

[...]

Contemporary fortified enclaves use basically modernist instruments of planning, with some notable adaptations. First, the surrounding walls: unlike examples of modernist planning, such as Brasilia, where the residential areas were to have no fences or walls but to be delimited only by expressways, in São Paulo the walls are necessary to demarcate the private universes. However, this demarcation of private property is not supposed to create the same type of (nonmodernist) public space that characterizes the industrial city. Because the private universes are kept apart by voids (as in modernist design), they no longer generate street corridors. Moreover, pedestrian circulation is discouraged and shopping areas are kept away from the streets, again as in modernist design. The second adaptation occurs in the materials and forms of individual buildings. Here there are two possibilities. On the one hand, buildings may completely ignore the exterior walls, treating facades as their backs. On the other, plain modernist facades may be eliminated in favor of ornament, irregularity, and ostentatious materials that display the individuality and status of their owners. These buildings reject the glass and transparency of modernism and their disclosure of private life. In other words, internalization, privacy, and individuality are enhanced. Finally, sophisticated technologies of security assure the exclusivity of the already isolated buildings.

[. . .]

Instead of creating a space in which the distinctions between public and private disappear – making all space public as the modernists intended – the enclaves use modernist conventions to create spaces in which the private quality is enhanced beyond any doubt and in which the public, a shapeless void treated as residual, is deemed irrelevant. This was exactly the fate of modernist architecture and its "all public space" in Brasilia, a perversion of initial premises and intentions. The situation is just the opposite with the closed condominiums and other fortified enclaves of the 1980s and 1990s. Their aim is to segregate and to change the character of public life by bringing to private spaces constructed as socially homogeneous environments those activities that had been previously enacted in public spaces.

Today, in cities such as São Paulo, we find neither gestures toward openness and freedom of circulation regardless of differences nor a technocratic universalism aiming at erasing differences. Rather, we find a city space whose old modern urban design has been fragmented by the insertion of independent and well-delineated private enclaves (of modernist design) that pay no attention to an external overall ordination and are totally focused on their own internal organization. The fortified fragments are no longer meant to be subordinated to a total order kept together by ideologies of openness, commonality, or promises of incorporation. Heterogeneity is to be taken more seriously: fragments express radical inequalities, not simple differences. Stripped of the elements which in fact erased differences such as uniform and transparent facades, modernist architectural conventions used by the enclaves are helping to insure that different social worlds meet as infrequently as possible in city space – that is, that they belong to different spaces.

In sum, in a city of walls and enclaves such as São Paulo, public space undergoes a deep transformation. Felt as more dangerous, fractured by the new voids and enclaves, broken in its old alignments, privatized with chains closing streets, armed guards, guard dogs, guardhouses, and walled parks, public space in São Paulo is increasingly abandoned to those who do not have a chance to live, work, and shop in the new private, internalized, and fortified enclaves. As the spaces for the rich are enclosed and turned inside, the outside space is left for those who cannot afford to go in. A comparison with Los Angeles shows that this new type of segregation is not São Paulo's exclusive creation and suggests some of its consequences for the transformation of the public sphere.

SÃO PAULO, LOS ANGELES

Compared to São Paulo, Los Angeles has a more fragmented and dispersed urban structure. São Paulo still has a vivid downtown area and some central neighborhoods, concentrating commerce and office activities, which are shaped on the model of the corridor street and, in spite of all transformations, are still crowded during the day. Contemporary Los Angeles is "polynucleated and decentralized." And its renovated downtown, one of the city's economic and financial centers, does not have much street life: people's activities are contained in the corporate buildings and their under- and overpass connections to shopping, restaurants, and hotels. São Paulo's process of urban fragmentation by the construction of enclaves is more recent than Los Angeles's, but it has already changed the peripheral zones and the distribution of wealth and economic functions in ways similar to that of the metropolitan region of Los Angeles. According to Ed Soja, the latter is a multicentered region marked by a "peripheral urbanization," which is created by the expansion of high-technological, post-Fordist industrialization and marked by the presence of high-income residential developments, huge regional shopping centers, programmed environments for leisure (theme parks, Disneyland), links to major universities and the Department of Defense, and various enclaves of cheap labor, mostly immigrants. Although São Paulo lacks the high-technology industries found in Los Angeles, its tertiarization and distribution of services and commerce are starting to be organized according to the Los Angeles pattern.

Although we may say that São Paulo expresses Los Angeles's process of economic transformation and urban dispersion in a less explicit form, it is more explicit and exaggerated in the creation of separation and in the use of security procedures. Where rich neighborhoods such as Morumbi use high walls, iron fences, and armed guards, the West Side of Los Angeles uses mostly electronic alarms and small signs announcing "Armed Response." While São Paulo's elites clearly appropriate public spaces – closing public streets with chains and all sorts of physical obstacles, installing private armed guards to control circulation – Los Angeles elites still show more respect for public streets. However, walled communities appropriating public streets are already appearing in Los Angeles, and one can wonder if its more discreet pattern of separation and of surveillance is not in part associated with the fact that the poor are far from the West Side, while in Morumbi they live beside the enclaves. Another reason must surely be the fact that the Los Angeles Police Department – although considered one of the most biased and violent of the United States – still appears very effective and nonviolent if compared to São Paulo's police. São Paulo's upper classes explicitly rely on the services of an army of domestic servants and do not feel ashamed to transform the utilization of these services into status symbols, which in turn are incorporated in newspaper advertisements for enclaves. In West Los Angeles, although the domestic dependence on the services of immigrant maids, nannies, and gardeners seems to be increasing, the status associated with employing them has not yet become a matter for advertisement. In São Paulo, where the local government has been efficient in approving policies to help segregation, upper-class residents have not yet started any important social movement for this purpose. But in Los Angeles residents of expensive neighborhoods have been organizing powerful homeowner associations to lobby for zoning regulations that would maintain the isolation their neighborhoods now enjoy.

Despite the many differences between the two cases, it is also clear that in both Los Angeles and São Paulo conventions of modernist city planning and technologies of security are being used to create new forms of urban space and social segregation. In both cities, the elites are retreating to privatized environments, which they increasingly control, and are abandoning earlier types of urban space to the poor and their internal antagonisms. As might be expected given these common characteristics, in both cities we find debates involving planners and architects in which the new enclaves are frequently criticized but are also defended and theorized. In São Paulo, where modernism has been the dominant dogma in schools of architecture up to the present, the defense of walled constructions is recent and timid, using as arguments only practical reasons such as increasing rates of crime and of homelessness. Architects tend to talk about walls and security devices as an unavoidable evil. They talk to the press, but I could not find academic articles or books on the subject. In Los Angeles, however, the debate has already generated an important literature, and both criticism and praise of "defensible architecture" are already quite elaborated.

One person voicing the defense of the architectural style found in the new enclaves is Charles Jencks. He analyzes recent trends in Los Angeles architecture in relation to a diagnosis of the city's social configuration. In his view, Los Angeles's main problem is its heterogeneity, which inevitably generates chronic ethnic strife and explains episodes such as the 1992 uprising. Since he considers this heterogeneity as constitutive of Los Angeles's reality, and since his diagnosis of the economic situation is pessimistic, his expectation is that ethnic tension will increase, that the environment will become more defensive and that people will resort to nastier and more diverse measures of protection. Jencks sees the adoption of security devices as inevitable and as a matter of realism. Moreover, he discusses how this necessity is being transformed into art by styles that metamorphose hard-edged materials needed for security into "ambiguous signs of inventive beauty" and "keep out" and design facades with their backs to the street, camouflaging the contents of the houses. For him, the response to ethnic strife is "defensible architecture and riot realism." The "realism" lies in architects looking at "the dark side of division, conflict, and decay, and represent[ing] some unwelcome truths." Among the latter is the fact that heterogeneity and strife are here to stay, that the promises of the melting pot can no longer be fulfilled.

Jencks targets ethnic heterogeneity as the reason for Los Angeles's social conflicts and sees separation as a solution. He is not bothered by the fact that the intervention of architects and planners in the urban environment of Los Angeles reinforces social inequality and spatial segregation. He also does not interrogate the consequences of these creations for public space and political relationships. In fact, his admiration of the backside-to-the-street solution indicates a lack of concern with the maintenance of public streets as spaces that embed the values of openness and conviviality of the heterogeneous masses.

But Los Angeles's defensible architecture also has its critics. The most famous of them is Mike Davis, whose analysis I find illuminating, especially for its thinking about the transformations in the public sphere. For Davis, social inequality and spatial segregation are central characteristics of Los Angeles, and his expression "Fortress L.A." refers to the type of space being presently created in the city.

[. . .]

For Davis, the increasingly segregated and privatized Los Angeles is the result of a clear master plan of postliberal (i.e., Reagan–Bush Republican) elites, a theme he reiterates in his analysis of the 1992 riots. To talk of contemporary Los Angeles is, for Davis, to talk of a new "class war at the level of the built environment" and to demonstrate that "urban form is indeed following a repressive function in the political furrows of the Reagan–Bush era. Los Angeles, in its prefigurative mode, offers an especially disquieting catalogue of the emergent liaisons between architecture and the American police state."

Davis's writing is marked by an indignation fully supported by his wealth of evidence concerning Los Angeles. Nevertheless, sometimes he collapses complex social processes into a simplified scenario of warfare that his own rich description defies. Despite this tendency to look at social reality as the direct product of elite intentions, Davis elaborates a remarkable critique of social and spatial segregation and associates the emerging urban configuration with the crucial themes of social inequality and political options. For him, not only is there nothing inevitable about "fortress architecture," but it has, in fact, deep consequences for the way in which public space and public interactions are shaped.

My analysis of São Paulo's enclaves coincides with Davis's analysis of Los Angeles as far as the issue of public space is concerned. It is clear in both cases that the public order created by private enclaves of the "defensible" style has inequalities, isolation, and fragmentation as starting points. In this context, the fiction of the overall social contract and the ideals of universal rights and equality that legitimated the modern conception of public space vanish. We should ask, then, if there is already another political fiction organizing inequalities and differences at the societal level, and how best to conceive this new configuration as the old modern model loses its explanatory value. If social differences are brought to the center of the scene instead of being put aside by universalistic claims, then what kind of model for the public realm can we maintain? What kind of polity will correspond to the new fragmented public sphere? Is democracy still possible in this new public sphere?

PUBLIC SPHERE: INEQUALITIES AND BOUNDARIES

People attach meanings to the spaces where they live in flexible and varying ways, and the factors influencing these readings and uses are endless. However, cities are also material spaces with relative stability and rigidity that shape and bound people's lives and determine the types of encounters possible in public space. When walls are built up, they form the stage for public life regardless of the meanings people attach to them and regardless of the multiple "tactics" of resistance people use to appropriate urban space.

In this essay, I have been arguing that in cities where fortified enclaves produce spatial segregation, social inequalities become quite explicit. I have also been arguing that in these cities, residents' everyday interactions with people from other social groups diminish substantially, and public encounters primarily occur inside protected and relatively homogeneous groups. In the materiality of segregated spaces, in people's everyday trajectories, in their uses of public transportation, in their appropriations of streets and parks, and in their constructions of walls and defensive facades, social boundaries are rigidly constructed. Their crossing is under surveillance. When boundaries are crossed in this type of city, there is aggression, fear, and a feeling of unprotectedness; in a word, there is suspicion and danger. Residents from all social groups have a sense of exclusion and restriction. For some, the feeling of exclusion is obvious as they are denied access to various areas and are restricted to others. Affluent people who inhabit exclusive enclaves also feel restricted; their feelings of fear keep them away from regions and people identified in their mental maps of the city as dangerous.

Contemporary urban segregation is complementary to the issue of urban violence. On the one hand, the fear of crime is used to legitimate increasing measures of security and surveillance. On the other, the proliferation of everyday talk about crime becomes the context in which residents generate stereotypes as they label different social groups as dangerous and therefore as people to be feared and avoided. Everyday discussions about crime create rigid symbolic differences between social groups as they become aligned with either good or evil. In this sense, they contribute to a construction of inflexible separations as city walls do. Both enforce ungiving boundaries. In sum, one of the consequences of living in cities segregated by enclaves is that while heterogeneous contacts diminish, social differences are more rigidly perceived, and proximity to people from different groups is considered dangerous, thus emphasizing inequality and distance.

Nevertheless, the urban environment is not the only basis of people's experiences of social differences. In fact, there are other arenas in which differences tend to be experienced in almost opposite ways, offering an important counterpoint to the experience of the urban environment. This is the case of the perceptions of social difference forged through the intensification of communication networks and mass media (international news, documentaries about all types of lives and experiences), through mass movements of populations, through tourism, or through the consumption of ethnic products (food, clothes, films, music). In these contexts, boundaries between different social universes become more permeable and are constantly crossed as people have access to worlds not originally their own.

Thus, the perception and experience of social differences in contemporary cities may occur in quite distinct ways. Some tame social differences, allowing their appropriation by various types of consumers. Other experiences, such as those of emerging urban environments, characterized by fear and violence, magnify social differences and maintain distance and separateness. If the first type of experience may blur boundaries, the second type explicitly elaborates them. Both types of experience constitute the contemporary public sphere, but their consequences for public and political life are radically different. On the one hand, the softening of boundaries may still be related to the ideals of equality of the liberal-democratic polity and may serve as the basis of claims of incorporation. The tamed differences produced to be consumed do not threaten universalist ideals, and in their peculiar way put people into contact. On the other hand, the new urban morphologies of fear give new forms to inequality, keep groups apart, and inscribe a new sociability that runs against the ideals of the modern public and its democratic freedoms. When some people are denied access to certain areas and when different groups are not supposed to interact in public space, references to a universal principle of equality and freedom for social life are no longer possible, even as fiction. The consequences of the new separateness and restriction for public life are serious: contrary to what Jencks thinks, by making clear the extension of social inequalities and the lack of commonalities, defensible architecture and planning may only promote conflict instead of preventing it.

Among the conditions necessary for democracy is that people acknowledge those from different social groups as co-citizens, that is, as people having similar rights. If this is true, it is clear that contemporary cities that are segregated by fortified enclaves are not environments that generate conditions conducive to democracy. Rather, they foster inequality and the sense that different groups belong to separate universes and have irreconcilable claims. Cities of walls do not strengthen citizenship; rather, they contribute to its corrosion. Moreover, this effect does not depend either on the type of political regime or on the intentions of those in power, since the architecture of the enclaves by itself entails a certain social logic.

Discussions about cities such as Los Angeles, London, or Paris, that is, cities populated by people from the most diverse cultural origins, commonly invoke the theme of the limits of modern citizenship based on affiliation to a nation-state. One might rethink the parameters of citizenship in those cities and suggest that the criterion for participation in political life could be local residence rather than national citizenship. Moreover, it would be possible to argue that this local participation is increasingly necessary to make those cities livable and to improve the quality of life of the impoverished population, increasingly consisting of immigrants. The contrast between this alternative political vision and the reality of fortified cities allows for at least two conclusions, one pessimistic and one more optimistic.

The pessimist would say that the direction of new segregation and the extension of social separation already achieved would make impossible the engagement of a variety of social groups in a political life in which common goals and solutions would have to be negotiated. In this view, citizenship in cities of walls is meaningless. The optimistic interpretation, however, would consider that the change in the criteria for admission to political life, and the consequent change in status of a considerable part of the population, would generate a wider engagement in the search for solutions to common problems and would potentially bridge some distances. There are many reasons to be suspicious of such optimism: studies of homeowner associations in Los Angeles remind us how local democracy may be used as an instrument of segregation. However, the boom of social movements in São Paulo after the mid-1970s suggests a cautious optimism. Where excluded residents discover that they have rights to the city, they manage to transform their neighborhoods and to improve the quality of their lives. That fortified enclaves in part counteracted this process should not make us abandon this qualified optimism. The walls were not able to totally obstruct the exercise of citizenship, and poor residents continue to expand their rights.

NOTE

In Brazil in 1989, the proportion of income in the hands of the poorest 50 percent of the population was only 10.4 percent. At the same time, the richest 1 percent had 17.3 percent of the income. Data are from the National Research by Domicile Sample (PNAD) undertaken by the Brazilian Census Bureau. The distribution of wealth has become more inequitable since the early 1980s.

“Urban Social Movements – Local Thematics, Global Spaces”

from *Urban Movements in a Globalising World* (2000)

Pierre Hamel, Henri Lustiger-Thaler and Margit Mayer

Editors' Introduction

The authors featured in Part Five introduced the impact globalization has on the culture, economy and social life of cities. In this selection and the following, urban social movement scholars tackle the question of how globalization has affected grassroots political mobilization around issues of urban conflict. In this first article, Pierre Hamel, Henri Lustiger-Thaler, and Margit Mayer first provide an overview of how major globalization scholars account for urban social movements within their particular theories. Saskia Sassen's theory of globalization, for example, sets up a binary opposition within cities, with wealthy global elites on one hand and poor disenfranchised workers on the other. Opposition takes the form of resistance to “external” forces of globalization that produce negative effects at the local everyday level, especially in terms of social welfare.

The authors are largely dissatisfied with such accounts, however, and argue that the strategies and stated goals of contemporary urban social movements are more complex than Sassen and others have presented. “The cause–effect paradigms of local/global interrelations have largely been discounted for a dialectical, relational model of exchanges,” the authors contend, “problematising the way globalism forces us to rethink basic sociological constructs such as community, culture, the nation-state, and their asymmetric relations of power and authority.”

The first type of struggle involves what might be characterized as “bread and butter” urban issues: the reduction of traffic and congestion or efforts to reduce overdevelopment and sprawl in favor of “slow growth” development. The second type of struggle involves community-based organizations which aim to solve problems of poverty, unemployment, and public safety in disadvantaged neighborhoods. The third type of struggle involves what the authors call “new poor people's movements” which emerged in response to this erosion of the welfare state since the 1980s. These movements work directly with the urban poor, refugees and immigrants to address larger themes such as municipal racism and the regulation of public space (anti-homelessness initiatives).

Resistance to globalization is produced through the work of a variety of movement groups guided by the realities of current municipal and national politics. Globalization has altered the terrain of local politics and the more successful urban movements have learned to deal with the institutions of local urban power, including cooperating or compromising in order to make progress.

Pierre Hamel is Professor of Urban Planning and Sociology at the University of Montreal, Canada. He specializes in the study of urban social movements, local development, and city planning. He is author (with Henri Lustiger-Thaler and Louis Maheu) of “Is There a Role for Social Movements?” in Janet L. Abu-Lughod, ed., *Sociology for the Twenty-first Century: Continuities and Cutting Edges* (Chicago: University of Chicago

Press, 1999: 165–180) and "Collective Action and the Paradigm of Individualism" in Louis Maheu, ed., *Social Movements and Social Classes* (London: Sage, 1995).

Henri Lustiger-Thaler is a Professor of Sociology at Ramapo College in New Jersey. His research addresses how globalization has affected urban ghettos, particularly minorities and youth in New York City. He is coeditor (with Daniel Salee) of *Artful Practices: The Political Economy of Everyday Life* (Montreal: Black Rose Books, 1993) and editor of *Political Arrangements: Power and the City* (Montreal: Black Rose Books, 1992).

Margit Mayer is Professor in the Department of Political Sciences and the Institute of American Studies at the Free University of Berlin, Germany. She is a leading theorist of urban social movements. She is the author of "The Onward Sweep of Social Capital: Causes and Consequences for Understanding Cities, Communities and Urban Movements," *International Journal of Urban and Regional Research*, 27, 1, March 2003, coeditor (with Volker Eick and Jens Sambale) of *From Welfare to Work: Nonprofits and the Workfare State in Berlin and Los Angeles* (Berlin: FUB/JFK Institute and the Department of Political Science, 2003) and coeditor (with John Ely) of *The German Greens: Paradox Between Movement and Party* (Philadelphia: Temple University Press, 1998).

The study of urban social movements and the specificity of their collective actions have traditionally occupied a limited and marginal status in social movement theories and deliberations. This is due to a long-standing perception of urban movements in collective action theories both in Europe and the USA. Urban movements have usually been thought of as heavily vested in micro-, as opposed to macro-, social processes, underscoring the local characteristics of a specific urban political economy. This has led to a suspicion in the wider social sciences that urban movements are too readily interwoven in mediational processes or, worse, the humdrum of municipal politics, with its input/output policy factors. From a disciplinary standpoint, being so uniquely identified with the very essence of locality, as a form of collective action, has meant a strong association with community research studies, local growth and anti-growth machine coalitions, the stuff of the traditional as well as the new urban sociology and political science.

From this viewpoint, urban movements have not been traditionally perceived as representative of broader values, ideals or emancipatory possibilities, that have been standardly attributed to identity-based movements concerned with a claim on totality, such as the women's, environmental, civil rights, peace and human rights movements. Yet, even in the face of this implicit/explicit framing of action, urban movements are more and more recognised as very modern indeed in terms of how they articulate locality and what lies beyond. We might say that urban movements harbour a pervasive transfunctionality in terms of their corresponding state structures. They are not fearful of institutionalisation. In many cases they seek it out. They are corporatist and increasingly find expression in partnerships of all stripes and countenances, if the strategic environment is so predisposed. The latter, however, do not account for all their functions. They have another embedded characteristic: they are key to the social construction of conflict within the city. And this is often the case concerning processes within which urban movements are committed players. Their transfunctionality brings together service roles and conflict roles, as they pre-score the scripts and narratives of local politics.

This transfunctionality, and its fair measure of ambivalence, emerges from a rooted attachment to locality that is already beyond its own localisation, and embedded in regional and national levels of mediation, where the real action often is, in terms of the funding of community groups and other legislative concerns. Urban movements operate in a social space that we will refer to as the 'extra-local'. This is the domain of their collective actions, the realm where locality occurs elsewhere than in the local. One can argue that this has always been the case to some extent – community groups have always gravitated beyond city hall when it came to issues of 'well-being', or welfare – but never as pronounced as in this global epoch.

This curiously replays the classical 'urban question' that has preoccupied urban sociology for decades. What is the weight of the municipal in the larger structure of state and relations of domination and exploitation? Furthermore, we could argue that the future of urban movements will be more and more in the extra-local, the local beyond the local, if we take the deterritorialisation thesis seriously. But is this their only purchase on the phenomenon of globalisation? The socio-political characteristics of urban movements have been their abiding source of ambiguity as well as cultural identity. Globalisation is only making this characteristic more visible, reproblematising already dense relations of particularity and universality, a defining character of the new global configuration of social forces.

If the actions of urban movements ignite a wide spectrum of globalisation theses, it is because their extra-local field of collective action must take into account the political-economic, cultural and late-modern aspects of being an actor in the contemporary metropolis. Locality as well as extra-locality – what is already beyond the local in any given urban-based action – thus sit in a particular phenomenal relation to their global counterpart. If the recent literature on globalisation has instructed us in anything, it is surely an appreciation of the shift from an overly categorical perception of collective action to one that is more open-ended and dependent upon external factors as a recursive feature of structuration for interior scapes. The cause–effect paradigms of local/global interrelations have largely been discounted for a dialectical, relational model of exchanges, problematising the way globalism forces us to rethink basic sociological constructs such as community, culture, the nation-state, and their asymmetric relations of power and authority. From a theoretical perspective, broader-ranged movements have long harboured, at their core, a global dimension linking movement actors within different national contexts to other collective actors. The bridge between general social movements and their global strategies, while of interest and still largely unexplored in the literature, has been much better recognised than localised forms of action within a globalised context.

And there is good reason for this, as the linkages between the local and the global are not always self-evident or apparent. What is the global dimension of a local housing struggle aside from the pervasive effects of economic globalisation upon local real-estate markets? Is there more to be said than the obvious? What is the import of the post-colonial effect of culture and language in immigrant enclaves such as New York City's Washington Heights, which now harbours the largest Dominican population outside the Dominican Republic, creating globally extended families, economic exchanges, collapsed cultural borders? In the shifting post-colonial contexts of globalisation, are we witnessing ethnic movements such as the above that we have thus far only ascribed to the intensification of immigration patterns?

The global component of collective action, from the purview of urban movements, can be confirmed in the following way: if the global significance of a given practice can be detected and analysed as such within the urban everyday, we can say that there exists a basis for further exploring the regulative, cultural and political terrain of globalisation.

Curiously, urban movements have become an exemplary model for pinpointing these multitudinous effects. The interface of the global and the local/urban raises the following questions. How can we think about embedded local issues that evoke the personal, in a globalising context? What is the degree of autonomy, or relative autonomy, between the local and the global, mediated by the everyday concerns of urban dwellers? How have practices around the local economy, housing, quality of life and municipal politics extended themselves to other arenas of action in a context of intermeshing scapes, spheres and global expanses? How have global changes impacted upon developments that are so securely tied to the practices of everyday life that they become locked in the 'hidden transcripts' so evocatively suggested by anthropologist James C. Scott (1985)? Are these practices what they seem, or have they also been reconfigured, globalised, extended beyond themselves? What is the nature of community from the perspectival view of the urban dweller in a global landscape? If the local is but an aspect of globalisation, how have urban movements, in their service-related and grassroots democratic preoccupations, been reordered by global flows, where the democracies of nation-states are now subject to new transnational sites of power?

Arjun Appadurai (1990) has argued that it is useful to think of the global in terms of scapes and flows, rather than a specific place as such. This is a helpful insight when applied to the problem of locality in that it collapses the separation between the global and the local, and takes us beyond the unidimensionality of causality on these particular issues. If place can be understood in terms of mediational and relational flows, we can ask if globalisation has impacted on the urban arena of identity formation, trashing taken-for-granted place-oriented associations. Appadurai contrasts his idea of flows and scapes with the notion of relatively stable communities from which people act. But have these communities already been transfigured? The assumption of relatively stable communities in a context of disconnected flows is highly problematic given the perspectives from which people view their actions, rendering community an open question in the age of global restructuring.

Furthermore, if globalisation also refers to the time-space compression of multiple national contexts, it is important to understand how urban movements, as representatives of an identifiable community, a community interest, or increasingly unidentifiable entities, such as the shifting structure of rights and privileges of legal and illegal immigrants in the USA for instance, are doubly stratified by the globalisation process. Their 'real homes' as well as negotiated institutional homes (on the local and regional level) have been irrevocably altered by flows and disjunctives. One only has to examine inner-city immigrant communities to derive a sense of the power of flows and its impact on their social, cultural and political self-understandings. Urban social movements therefore need a theory of the global, if nothing more than to engender clarity about the local, as much as the globalisation debates are impoverished without a theory of collective action that seeks to understand urban movements as stratified actors within a local–global paradigm on all levels of the globalisation thesis. It is therefore impossible to speak about grassroots democracy, a defining characteristic of the progressive urban movement, without addressing the impact of globalisation, as both constraining and enabling phenomenon which connects the mundane of everyday life to discussions on the new forms of social and global citizenship detected in a locally inscribed practice.

Examining new contexts for global democracy may therefore emerge as a potentially useful research background for the contemporary study of urban social movements. If the nation-state is no longer at the apex of democratic theory, the former understood as the accountability of relations between those affected by a political decision and those that make the decision, a better understanding of the intermeshing of local, national, regional and international processes is called for. The question is therefore begged: how many of the activities of urban movements are framed by international pressures? Is the very idea of the local, which urban social movements defend, already an anachronism in a global civil society where loyalties to nation-states are no longer a central point of identity? Is the relationship between economic globalisation and the ability of political institutions to control it, creating a Hobbesian war of the cities, placing urban movements on the front line of the urban social economy, through local economic development strategies?

[. . .]

LOCAL-URBAN GLOBALITIES AND COLLECTIVE ACTION

From the above discussion, one could presume that, viewed from the perspective of urban social movements, debates on globalisation have ushered in a fundamental rethinking of some of the central premises of contemporary social relations of domination, exploitation and power. Certain authors argue that we are in a substantively new period of global transformation. The organisation of production and supply of goods and services on a global level, alongside direct foreign investment and international trade, is widely recognised as a core component of globalisation. The internationalisation of private and public technology domains has furthermore created an emerging techno-globalist discourse that has impregnated general cultural discourse. Indeed, globalisation has been most recognisable in public discourse through the reigning neo-liberal interpretation. However, changes in the global economy have had a structural as well as a discursive impact on problems faced by local economies, hence the refurbished roles of local advocates, NGOs and grassroots movements. The local impact of economic globalisation has been the foundation for their actions. Others sense that the present period gives too much to globalisation, most of which can be defined in terms of national forms of labour-market restructuring and post-Fordist practices. Is the globalisation effect on localities to be understood only through the feedback loop of changes in the reorganisation of production at the level of the local community? Is there a missing analytical dimension that can help us better understand the regulative forces which impact upon the collective actions of urban movements?

Anthony King (1990) has argued that while global consciousness may be new, the phenomena it points to most certainly are not. World cities, King states, did not suddenly emerge in the 1950s. They have long performed both national and international functions. Grassroots organisations have been mirroring these changes, by their claim-making, resistance, compromises and acquiescences. From a systemic view, globalisation, as Immanuel Wallerstein contends, has created new oppositional cultures (Wallerstein 1984). For Wallerstein, social movements are key indicators of change.

[...]

Scott Lash and John Urry (1994) take us more substantively into the arena of cultural definitions of globalism useful for an understanding of current global systems that ultimately diminish the significance of cities, workplaces and organisation, in the formation of the new subject of late modernity. Lash and Urry argue that it is not only individuals that are impacted by globalisation, i.e. the household as the link between the local and global, but that once available lifestyles based on where people live are not always immediately present. This kind of dislocation from community, a disinterring of class and ethnic relations, has its counterpart further up the class ladder in Ulf Hannerz's (1990) reference to global cosmopolitans, the 'new travelling middle classes' who privilege an identity based on mobility rather than territory. References to class, place and space are therefore a basis for understanding globalism as it impacts the urban cultural realm.

[...]

Globalisation is intrinsically linked to oppositional cultural activism. Post-national groups appeal to the power of localising economies from their objective life circumstances, as do groups in deindustrialising contexts. They are, in a sense, already in the extra-local, that is the national and the regional levels, while seeking to establish their own investment bank and local social economies. Resistance therefore becomes a multiple engagement as it occurs within increasingly flexible institutions. Globalisation as a process is intrinsically interrelated with its opposite, localisation as a transformative component of late modernity. So powerful is the idea of globalisation that it is emerging as a replacement for the concept of modernisation, as it traces the trajectory of cultural phenomena in a global capitalist economy.

[...]

URBAN MOVEMENTS IN THE GLOBAL FLOW

[...]

Saskia Sassen (1998) describes global cities as sites of a new frontier, arguing that the conflicts and battles that used to take place in colonies are now taking place in the major cities of the advanced world.

This is where today's post-colonial battles play themselves out: between the overvalorised new transnational professional class, which enacts the global corporate project, and which makes up the 'new city users' treating the city as transterritorial environment, on the one hand, and the devalued underpaid immigrant working class, which is providing the material condition of that corporate world of power, on the other. The concentration of these (transnational) 'servants' in global cities allows them a strategic position, a position they could never gain in small towns. This 'presence,' Sassen suggests, translates into the possibility of laying claims on the city, and allows these actors to unbundle the conditions of their powerlessness and to develop a form of politics that not only gives them visibility but that also has the potential of challenging the global elites' project.

By placing urban conflict and urban movements in the context of the emerging global grid of strategic places constituting a new 'geography of centrality' (that cuts across both national boundaries and the old north/south divide), this analysis is able to identify a systematic link between the functions and activities characteristic of these cities and the issues and actors of contemporary urban movements. The example of New York City's finance sector illustrates, for example, that 50 per cent of its workers are clerical and other manual workers (such as cleaners and janitors), who are made up of unskilled immigrants and workers. These manual service jobs as well as the new informal economy are as constitutive an element of the global economy as is the transnationalism from below which has been gaining presence in these cities.

While this approach is thus extremely helpful to our understanding of globalisation as a catalyst for emerging practices at the urban level, it remains silent in two important respects. It offers no analytic space for the role (different) political responses are playing in the emergence of this new urban frontier. And it treats the 'post-colonial struggle' which it highlights as a black box, instead of studying the actual work of those movement organisations that are producing the 'voice' now increasingly heard from the local level. More close-up analysis of the emerging new infrastructure of movement organisations produces a more complex picture than the binary opposition between an urban glamour zone and an urban war zone as implied by Sassen.

[. . .]

We might distinguish at least three different kinds of struggle delineating the contours of the movement terrain in the globalised city:

- those around the costs of striving towards the top of the global urban hierarchy;
- those dealing with the new phenomena of urban decay and marginality, increasingly in the form of routinised cooperation with the local state and mediated through a variety of revitalisation and/or (economic) development programmes; and,
- those reflecting the erosion of the local welfare state.

The thick infrastructure made up by these different kinds of movement is the result of the work of movement organisations who – through constant negotiation and conflict – produce very concrete and specific demands, and achieve specific victories, compromises and defeats, but in any case contribute to the concrete shape of the globalised city. Briefly looking at each of these areas of struggle, the concept of 'presence' of those challenging the corporate project of the international business class quickly gives way to one of ambiguity and tension, inviting a different set of questions as to the role and potential of these movements.

Struggles at the top of the global urban hierarchy

Striving to climb up the urban hierarchy, to become part of the cross-border network of cities that manage and coordinate the global economy, and becoming part of the 'strategic geography of centrality' (Sassen 1998) brings certain advantages for these cities. About 20 per cent of their residents do thrive economically, which is arguably an improvement over the far lower percentages of former urban elites. But in spite of the enormous concentrations of wealth in these cities, this striving has also inflicted costs. All of these cities exhibit higher than average rates of poverty and income inequality. Besides the costs for those who fall outside this strategic geography of power, there are the costs many more residents experience due to the gentrification, displacement, congestion and pollution that come with the grooming of the city for the top of the hierarchy.

Movements protesting these costs have emerged in a variety of forms. Some organise to protect their home environment – from congestion and traffic, development or projects they don't like to have in their backyards. Others fight urban growth policies and gentrification. Frequently such fights are triggered directly by the new development instruments of big-city politics, such as spectacular urban development projects, festivals, and the attraction of mega-events, sports entertainment complexes and theme-enhanced urban entertainment centres, all of which use the packaging and sale of urban place images. Movements have attacked the detrimental side effects of and the lack of democratic participation inherent in these strategies of restructuring the city and of raising funds, and they criticise the spatial and temporal concentration of such development projects. Often radical or autonomous movements, who consciously seize on the importance of image politics, have gained in the global competition of cities. They devise image-damaging actions to make their city less attractive to big investors and developers. So-called inner-city action weeks in the summers of 1997 and 1998 scandalised the costs of dressing up European cities for new city users. While these campaigns were carefully networked across German and Swiss cities, concentrating intense action in one week each summer, other actions have been more frequent and spontaneous, as, for example, the bicycle demonstrations taking place every last Friday of the month. This movement, which calls itself 'Critical Mass', first emerged in 1992 in San Francisco and has meanwhile spread to many other American and European cities. In Berlin, where 300 bicyclists are injured per year and one is killed every three weeks, about 500 cyclists cause a rush-hour traffic jam each last Friday of the month. Though protesting the concrete costs of competitive urban politics in concrete cities, this semi-spontaneous form of mobilisation (since no identifiable organisation gives notification of the demonstration) is a transnational form of mobilisation to the extent that it has been orchestrated through the Internet and thus spread across the Atlantic. So within this field of protest against the new competitive urban politics and its effects we find both movements defending privileged conditions or pursuing particularist interests, and movements with social justice concerns or the potential to unite different groups behind democratic planning demands.

Struggles over urban decay and revitalization

All cities, not just those striving for the top of the urban hierarchy, have had to tackle problems of disadvantaged neighbourhoods and urban renewal ever since central state funding has dried up while unemployment and new forms of social and spatial exclusion have intensified. In the process, community-based organisations that often had their origin in the urban revolts of the 1960s and 1970s have become part of municipal revitalisation and similar programmes.

Due to this opening of the local state, movements dealing with urban repair follow a very different logic from the less institutionalised ones just discussed. Nudged by a variety of philanthropic and state programmes, they have transformed into community-based and client-oriented institutions delivering polyvalent functions in the context of municipal social and economic development programmes. As the emphasis of poverty alleviation programmes has increasingly shifted from welfare to workfare, and economic and development programmes have shifted from brick-and-mortar approaches to micro-enterprise programmes, these community-based organisations have come to tackle problems of disadvantaged neighbourhoods and urban renewal with approaches that are collaborative (i.e. making use of public-private partnerships) and comprehensive (i.e. simultaneously engaged in economic development, employment generation, social services, housing, ecological projects and public safety). Importantly, they have all come to integrate economic development with poverty alleviation.

Since the 1990s, thanks to the spread of workfare and employment programmes, they have increasingly employed and trained welfare recipients not only to become cleaners and janitors but also administrative assistants, service representatives or phone operators for hotel chains, major brokerage houses and financial service companies. For example, Wildcat Service Corporation and Women's Housing and Economic Development Corporation in the Bronx train welfare recipients for what they have identified as the growing sectors of the local economy: in computer software, Business English and math and various life skills, like dealing with an unpleasant supervisor. The trainees then go to Salomon Smith Barney for a 16-week internship, after which the company can hire whomever it chooses.

While much of the movement literature emphasises the contestatory character of such grassroots organisations and the counterweight they pose to conventional ways of local economic development, these forms of institutionalisation also exhibit certain dangers – of cooptation, NGOisation, and the pursuit of insider interests. Increasingly, they contribute to forms of labour-market flexibilisation, the results of which are no longer under their control and may have little to do with their original political intentions. Only a few of the organisations participating in this routinised cooperation with the local state manage to keep public pressure on, as most of their energy needs to focus on private lobbying strategies to secure jobs and on the competition for dwindling funding sources. Increasingly, community-based development organisations and alternative renewal agents find themselves attacked by other movement actors who do not qualify for inclusion or who prefer different forms of political action.

Struggles over the erosion of the local welfare state

Social inequality and poverty rates intensify in cities competing for the top locations in the urban hierarchy, and at the same time traditional redistributive social policies are increasingly replaced by labour-market policies designed to promote flexibility, as well as by punitive measures designed to disenfranchise and exclude certain marginal groups. In response to this erosion of the local welfare state new poor people's movements have sprung up, accompanied by actions of civil rights groups working with refugees and immigrants, anti-racist initiatives and various advocacy organisations. While their protest activities tend to be episodic and spontaneous, local in nature, and disruptive in strategy, they do occasionally manage to block the efforts made by city governments and corporate elites to drive them out of downtowns. Especially when supported by more resource-rich advocacy groups, professional activists or church organisations, their struggles go beyond merely disrupting normal city government operations and manage to challenge the legitimacy of local policies of exclusion. Recently, the police raids to 'clean' European citadel plazas of the poor and marginalised have prompted campaigns by antiracist initiatives. This kind of 'external support' is also evident in the homeless newspapers that have become regular features in all the advanced cities' downtowns, and in the broad variety of projects and organisations servicing the marginalised, which have mushroomed everywhere. Many of these organisations, while focused on organising local support and services for the poor, engage in transnational exchange, actively using Internet sources and sharing experience across the borders.

Other ways of resisting are efforts at worksite organising amongst the precariously employed, low- or sub-minimum wage, often immigrant, workers. For example, workfare workers have been organised by new organisations such as WEP Workers Together! (WWT!) in New York since 1996 as well as old organisations such as ACORN (formed in 1970), which have managed to get unions and community organisations involved in joint demonstrations to protest the conditions of workfare. While a large part of workfare organising consists in the effort to symbolise that workfare workers are workers entitled to on-the-job protections and the full range of rights granted to 'normal' workers, and to break down the barriers between welfare recipients and workers, other worksite struggles revolve around groups almost as precarious, though regularly employed. The exploding demand for cleaners and janitors, which accompanied the boom in office buildings in globalising cities, has allowed the owners to rid themselves of contracts with unionised firms and contract with non-union janitorial firms that would hire immigrant workers. In reaction, the justice for janitors (J4J) campaign spread from Pittsburgh and Denver to Californian cities in the late 1980s. Today the union J4J addresses the problems of immigrant workers, offers legal help, files lawsuits, and mobilises the community to support the janitors around human and civil rights issues as well as the quality of life in the cities. Another illustration is Asian Immigrant Women Advocates (AIWA), which is a membership organisation of about 1,000 Asian-American immigrant women workers in the San Francisco area hotel, janitorial, garment and electronic industries. Groups such as these organise around the working conditions and survival needs of a growing precarious workforce, while resisting becoming social service organisations.

By contrast, groups that function as advocacy and service organisations for the poor have become incorporated into municipal strategies seeking to harness their reform energy. While of later and different origin than the community groups and self-help initiatives discussed above, the non-profit and voluntary sector groups servicing the new poor and marginalised are similarly becoming part of routine cooperation with the local state. With the support and direction provided by various state and foundation programmes, they produce affordable housing, commercial establishments, social services and training programmes. But they seek to go beyond merely 'mending' the disintegration processes characteristic of the globalised city by developing innovative strategies that explicitly acknowledge – and consolidate – the new divisions. For example, they might help recent immigrants find jobs and places to live by training them to find work in the growing informal sector as day labourers rather than channelling them into normal job-training programmes.

Because of such innovativeness, local knowledge and skills, such groups have come to be seen as ideal agents to mediate such services to poor and underprivileged urban residents. Like the community-based organisations discussed above, these grassroots organisations are uniquely disposed to organise and provide neighbourhood voice and social capital as well as job access and survival. However, while providing these goods and services, it becomes difficult, and increasingly rare, for these organisations to succeed in creating solidarity and empowerment, or to challenge the drawbacks of state programmes. Many are indeed becoming tied up with managing the housing and employment problems of groups whose exclusion by the 'normal' market mechanisms might otherwise threaten the social cohesion of the city.

Thus, there is indeed resistance to the global corporate project as enacted in our cities. It is produced through the work of a variety of movement groups, building on past legacies, and guided by the design of various funding programmes. Their institutional and economic infrastructure constitutes a kind of social economy, which is not only experienced as a challenge to mainstream institutions, but is increasingly valued for its employment potential and its capacities to process local conflict and create local social inclusion. Of course these infrastructures and the activities they support take on different shapes and bends: there are actual alternatives to exclusion, as examples in Quebec and France will illustrate, but there are also many ways of instrumentalising, in neo-liberal environments, the work of these groups, so as to tie them up with mere coping strategies. There are new forms of transnational organisation and networks strengthening grassroots groups across national borders, connecting cities in ways not envisioned by the global business elite. But there are also parochial and NIMBY (not in my backyard) movements that occasionally even take on anti-immigrant or racist overtones. Most of the current literature ignores these ambivalences in the urban movements active in today's cities and does not focus yet on their relation to the globalisation process. While there is widespread agreement that globalisation has markedly transformed the urban terrain in which these movements act, and while some of the relevant literature does capture the transformation of these movements into social and economic actors with transnational awareness, these changes are far from systematically explored nor analysed in the context of the global processes to which they relate.

GLOBALISATION AND COLLECTIVE ACTION

As highlighted in the previous section, the diversity of struggles that characterise urban movements is related to experiences of domination and exploitation encountered in everyday life and filtered through the grid of the new global order. Confronted with patterns of domination that local political elites and other social forces reproduce, urban movements deal with the institutions of local development and management from very different perspectives than more general movements. Thus, the determination to combat alienation or exclusion is counterbalanced by the necessity to cooperate or compromise with local actors and social forces. The ambivalence of their collective action is most visible through the relationships that movements build with institutions. In the globalised city, while the interrelations between social, economic and cultural dimensions of life are becoming more integrated, actors and systems of action are increasingly difficult to align. Institutions are facing crises due to their incapacity to embrace all the conflictual interactions they create in their response to the diversity of social demands. Actors and systems appear to be in a state of flux and non-correspondence. At the same time movements are torn between the defence of identity and the pragmatism or compromises commanded by the pursuit of their instrumental finalities.

[. . .]

Over the past few years, urban movement practices have rotated around the changing conditions of late modern societies, fighting against social injustice and social exclusion in cities that are becoming more and more globalised. In order to better understand their specific contribution to social change, researchers need to find a fruitful balance between system-centred and movement-centred modes of analysis.

[. . .]

In many respects, urban movements support the claim that the democratisation of civic life is inseparable from achieving 'decent' standards of living for all citizens. This is why in several cases these movements mobilise around issues of local economic development, in an effort to expand citizenship claims towards redistribution claims. In short, urban movements continue to wager on the possibilities offered by local democracy, trying to build compromises with other social agents or elaborating diverse strategies of resistance to unwanted models of social change.

REFERENCES

Appadurai, Arjun. 1990. "Disjuncture and Difference in the Global Economy." In Mike Featherstone, ed., *Global Culture: Nationalism, Globalisation, and Modernity*. London: Sage.

Hannerz, Ulf. 1990. "Cosmopolitans and Locals in World Cultures." In Mike Featherstone, ed., *Global Culture: Nationalism, Globalisation, and Modernity*. London: Sage.

King, Anthony. 1990. *Global Cities: Post-Imperialism and the Internationalization of London*. London: Routledge.

Lash, Scott and John Urry. 1994. *Economies of Signs and Space*. London: Sage.

Sassen, Saskia. 1998. *Globalization and Its Discontents*. New York: The Free Press.

Scott, James C. 1985. *Weapons of the Weak: Everyday Forms of Peasant Resistance*. New Haven: Yale University Press.

Wallerstein, Immanuel. 1984. *The Politics of the World Economy*. New York: Cambridge University Press.

"Glocalizing Protest: Urban Conflicts and Global Social Movements"

from *International Journal of Urban and Regional Research* (2003)

Bettina Kohler and Markus Wissen

Editors' Introduction

City streets and squares have long been sites of food riots, rent strikes, labor rallies and other political demonstrations. Activists and political organizers have benefited from the ready density and diversity of residents and the city's function as a hub of information and communication to fill the ranks at demonstrations and strengthen social movements ranging from environmental concerns to racial justice. In 1999, more than a half-million demonstrators took part in the protests against the World Trade Organization's (WTO's) ministerial meetings in Seattle, Washington. Many were affiliated with the more than 700 organizations that participated in the demonstrations, representing environmental activism, Third World debt reduction, human rights, anti-sweat-shop labor, and other interests. As many journalists and scholars would later comment, the Seattle demonstrations epitomized the overlap of global and local movements – that ostensibly local and regional interests, such as deforestation or corporate buy-outs, were simultaneously global ones. Indeed, a key purpose for many demonstrators was to highlight the fact that the global and local are deeply intertwined.

In this selection, Bettina Kohler and Markus Wissen examine what roles cities play in fomenting and hosting a variety of global social movements. It is clear that Kohler and Wissen's approach to the study of globalization is similar to that of Michael Peter Smith. The local and the global are "mutually constituted." That is, globalization is not an abstraction or external "force" that requires localities and individuals simply to respond or adjust to its pressures. Globalization is, in fact, articulated locally ("glocally") and realized in the everyday activities that occur within cities and other places. Grassroots politics is not immune to "glocalization." The shape and form local activism takes is informed by events and processes that occur elsewhere but are nonetheless intrinsically connected to problems and possibilities at "home."

Kohler and Wissen argue that the nearly universal adoption of neoliberal policies among industrialized nations and global institutions, such as the WTO, the International Monetary Fund (IMF) and the World Bank, is a key source of social problems within many cities. Neoliberalism is a political ideology that guides both domestic social and foreign economic policies. Its principles facilitate the global circulation of capital. Neoliberalism rejects bureaucratic or state-centered planning and triumphs a laissez-faire or free-market approach to major social and political decisions. Corporations, it follows, should be minimally regulated and given freedom to act across borders. In turn, social protections, such as welfare benefits (which require tax revenues) and labor unions, should be minimized so as not to hinder the free flow of money, goods, and information.

Neoliberalist policies, according to the authors, have widened the gap between wealthy and poor residents within cities, producing urban conflicts over economic inequality and increased state surveillance and regulation. Urban conflicts, therefore, become part of the global social movements against neoliberalist globalization. Kohler and Wissen call this global articulation of urban protest "scale jumping" – the intertwining of various spatial scales from the sub-local to the global which produces diverse, yet connected, social movements across the globe.

Finally, the authors offer three different types of glocalized urban collection action against neoliberalism and the contributions they can make to the strategic orientation of the global social movements. They are (1) protests against images of neoliberal restructuring, (2) the creation of alternative knowledge or possibilities, and (3) the construction of material infrastructures beyond state and market. The strategies offer a "political space" in which alternative ways of living become thinkable or provide real alternatives beyond the existing power relations.

For further information on the political ideology and social policies of neoliberalism, consult Neil Brenner and Nik Theodore's, "Cities and the Geographies of 'Actually Existing Neoliberalism,'" *Antipode*, 34, 3 (2002): 349–379; Jamie Peck and Adam Tickell's, "Neoliberalizing Space," *Antipode*, 34, 3 (2002): 380–404; and Henk Overbeek's *Restructuring Hegemony in the Global Political Economy: The Rise of Transnational Neoliberalism in the 1980s* (New York: Routledge, 1993).

Bettina Kohler is a member of the Architecture and Planning Faculty at the Technical University of Vienna, Austria. Markus Wissen is a professor in the Department of Political and Social Sciences at the Free University of Berlin, Germany. His research interests are in European integration and regional development.

Among the features which distinguish the emerging global social movements from the social movements of the last third of the twentieth century, the most obvious and most frequently cited are their international orientation, their broad range of issues and their diversity. Another feature, which is also rather striking but named less frequently, is the fact that the movements' emergence, as well as important struggles in their short history, is symbolized by the names of cities, the most prominent ones being Seattle, Genoa and Porto Alegre. This is not accidental: groups stemming from all parts of the world need 'places' to constitute themselves as a movement. Furthermore, cities represent to a large extent the global and local focal points of social movements, because a large part of the issues and institutions criticized by the movements are located in cities.

This is not to say that rural conflicts are a less heeded local focus. On the contrary, conflicts over land distribution, the exploitation of natural resources or the use of genetically modified seed form an important field of action for the movements, especially for local groups in the South, but also for NGOs and activists in the North. Like urban conflicts they are essentially linked to global neoliberal restructuring. The conflict over the Plan Puebla Panama in Mexico, for example, is shaped by interests which organize themselves on different spatial scales while claiming control over a certain place: there is a mega-infrastructure project in the service of global markets and in the interest of national competitiveness in the global economy, competing with the claims of local communities over their territory. The struggles of 'rural' actors like the Mexican Zapatistas or the movement of the landless in Brazil (MST) not only marked important starting-points of the global social movements but still form essential parts of them. Furthermore, many rural conflicts are increasingly connected to the situation in cities, through the – unevenly distributed – global flows of resources, capital and power, as well as through the migration of people.

Nevertheless, cities play a key role in neoliberal restructuring. At the same time, they have always been a favourable place where alternative practices and resistance against hegemonic projects could emerge. The role of cities in representing great parts of the movements' history and consciousness therefore can be read as a metaphor which hints at the importance of urban struggles within the global social movements. The aim of this article is – following some brief terminological remarks concerning the terms 'glocal' and 'global social movements' – is: (1) to examine to what extent and in which context urban conflicts became part of the global social movements against capitalist globalization; (2) to highlight the variety of spatial scales and forms of articulation as a central feature of these conflicts; and, finally, (3) to analyse different types of 'glocalized' protests and the contributions they can make to the strategic orientation of the global social movements.

TERMINOLOGICAL REMARKS

The term 'glocal' is used in some contexts to describe the one-directional shaping of the local in the interest of global capital, leaving space at the local level only for repairing the damages of the global. The way the term is used in this article is a more dialectical one. The local and the global, the regional and the national are deeply intertwined; their relation is one of being mutually constituted. That means that the global is not a pre-given entity external to other spatial scales. Instead it is produced, reproduced, modified and challenged in a multiplicity of actions on various spatial scales. Thus, rather than conceptualizing the relationship between the global and the local as a one-directional process, we emphasize the influence of local actors and power relations on urban living conditions as well as on the form taken by globalization. This will become obvious when looking at an important feature of urban movements from the 1990s onwards: while criticizing and acting upon very local/material urban issues, they often explicitly relate and politicize these issues in a broader context and articulate their criticism on various spatial scales – not only on a local but also on a global scale.

As far as the term 'global social movements' is concerned, we have to admit that it is rather unwieldy. Nevertheless, we consider it more appropriate than the term 'anti-globalization movement', which has gained some discursive publicity in the last couple of years – this for the following three reasons. First, the movements characterized as being 'against' globalization by the term 'anti-globalization movement' certainly constitute one of the most globalized actors of our time. Although rooted in various local and national contexts, they are able to act and to articulate themselves on a global scale. The logistics and coordination implied in such 'diffuse global networks' has much of the complexity needed for the administration of great transnational corporations. Secondly, there are indeed forces which fight globalization, but without pursuing any emancipatory aim. These groups usually belong to the extreme right. They fight for example for 'cultural homogeneity' within national borders, which they see as threatened by migrants and refugees as well as by international (financial) capital. Of course, speaking of 'global social movements' instead of speaking of the 'anti-globalization movement' alone does not prevent reactionary groups from getting a foot in the door of international protest. But at least it tries to preclude a nationalist interpretation and politicization of capitalist globalization terminologically, stressing that emancipatory movements do not fight globalization per se, but all kinds of power relations – capitalist, patriarchal, racist – which are strengthened by the globalization process. Thirdly, and regarding the plural form in 'global social movements' in contrast to the singular form in 'anti-globalization movement', we wish to stress that there is no single and homogeneous actor. Rather, the plural form 'movements' implies heterogeneity, diversity and contradictions. It hints at a variety of different actors which explicitly have no common (symbolic) representation.

Developing the last point further, it has to be emphasized that the notion of global social movements – besides showing material articulation and action in certain places – is a highly discursive phenomenon. Perceiving the variety of glocal movements as interrelated to each other and as part of a new global social movement scene is to some extent a question of the intellectual perspective, cultural context and discursive construction. Not all movements or activists of local groups referred to in this paper may perceive themselves as politicized and contributing to the global social movements. In many cases, of course, the struggles for basic needs are prior to general reflections, even though the material claims may be compatible with the aims of the global movements. Also, the question of representation is not always very clearly articulated within the movements: if one member of a local social movement participates in a Social Forum or writes an article in this context, the whole group will easily be related to the 'global social movements', sometimes without the other members being aware of this. Indeed, the emergence of global social movements from a variety of actors and differing local or urban experiences seems to rely very much on decentralized 'resonances' between sometimes very unique approaches. Against this background, terrains like the World Social Forum, which was organized for the third time in Porto Alegre in 2003, play an important role if they contribute to strengthening these resonances and avoid creating a single common representation or even new hierarchies.

NEOLIBERAL RESTRUCTURING AND URBAN CONFLICTS

Urban conflicts, of course, are not a new phenomenon. Cities and their everyday life have always been places where hegemonic projects developed. But cities have also been the milieu where conflicts, alternative practices and resistance against hegemonic projects emerged. In Europe, cities saw the development of grassroots and community groups at the end of the 1960s and in the early 1970s. The efforts of these groups were directed against the technocratic Keynesian style of urban development policy as it was manifested, for example, in large-scale renewal projects and modern housing construction. Being partially successful in their resistance to such politics, from the mid-1970s on social movements were increasingly confronted with a changing social and political environment: neoliberal politics began to shape living conditions as well as ways of thinking and horizons of expectation. Quite often they managed to integrate the social movement critique of a bureaucratic Keynesianism and to give it a regressive turn. They successfully occupied formerly progressive discursive terrains. For example, former sites of cultural and artistic resistance to the bureaucratic state and occupied spaces for autonomous ways of living were integrated into the marketing concepts of cities and became part of the cultural diversity and attractiveness of the post-Fordist city.

[. . .]

Cities played an important role in this process. As places where the 'nodal functions' of regional, national or global economies can be found, they were often starting-points and fields of experimentation for neoliberal restructuring and the corresponding image production.

With neoliberal restructuring in the 1980s and throughout the early 1990s the conditions of urban conflicts changed dramatically. Distributive politics were reduced or replaced by measures in favour of strengthening urban competitiveness. As a consequence, socio-spatial polarization increased, wealth and opportunities became more and more unevenly distributed, and the gap between the promises and the realities of neoliberalism widened. That also meant that the latter, from the early 1990s on, was increasingly confronted with its own contradictions and consequences. Neoliberalism's primary 'mission' was no longer to overcome the Fordist class compromise, but to manage its self-made disasters by methods such as the aggressive reregulation, disciplining, and containment of those marginalized or dispossessed by the neoliberalization of the 1980s.

[. . .]

In some cases the new protest generation was also able to build on the infrastructures created by earlier generations of social movements, so that these infrastructures formed an organizational starting-point and secured a degree of continuity to former struggles and experiences. In Europe, a very striking example of this is Italy. Since the early 1970s in Italian cities a lot of spaces have been created in which new forms of societalization – alternatives to capitalist production and markets, as well as to state bureaucratic intervention – could develop. The 'social centres' in many Italian cities are of particular importance in this context. Most of them were squatted in the 1980s, forming a space for various cultural and social activities like exhibitions, concerts, the production of records, CDs and video cassettes, workshops, congresses, free legal offices, playgrounds or after-school groups for children. Furthermore, they soon established international links, establishing a sort of parallel market of self produced goods. Today the social centres form an important infrastructure for a glocalized protest, which became clear when two major events of the global social movements took place in Italian cities: the demonstrations against the world economic summit 2001 in Genoa and the first European Social Forum 2002 in Florence. In the following section we wish to address the very specificity of the protests since the late 1990s: their multiscalarity or their 'glocal' character.

THE MULTISCALARITY OF PROTEST

There is a strong link between neoliberal restructuring and globalization. The latter is both a result and a medium of the former: it results (to a great extent) from capital being 'freed' of national boundaries by neoliberal politics, and it shapes the conditions which are favourable for further neoliberal restructuring. Globalization can be understood as a complex rescaling and reconfiguration of politics and economy on various spatial scales, a contested process of an uneven de- and reterritorialization. New institutions like the WTO are established on a global scale, older supranational institutions like the EU gain in importance, and on the national scale the repressive and economic state apparatuses are strengthened vis-à-vis the social ones. Thereby, terrains are formed which are highly structurally selective, so that powerful actors find favourable conditions for generalizing their own interests and marginalizing the interests of weaker actors. Constraints are created which heavily influence the terms of national and local conflicts.

This is not to say that local actors are just victims of global processes. In contrast, globalization is essentially a local product, insofar as global constraints are to a large extent created, reproduced and naturalized by local politics and everyday practices. Urban competition politics, for example, are an important transmission belt for transforming global constraints into material local realities. The logic of interurban competition turns cities into accomplices in their own subordination. Furthermore, the implementation of neoliberal politics is to a great extent carried out through its acceptance and reproduction in the local everyday life of people.

It is this complex interplay between institutions and processes on different spatial scales which influences and provokes the search for new forms and scales of resistance. With the articulation of urban protest not only on a local but also on a global scale, urban social movements confront those institutions and actors which are increasingly influencing their living conditions. Claiming the 'right to the city' today means the improvement of material living conditions in cities. At the same time, it relates to broader conceptions of dignified livelihood – including (varying and dependent on the context) aspects of democratic participation, human rights, equal access to goods and services, reclaiming a sense of a public sphere, environmental justice or solidarity within society. Thereby, material issues are politicized and linked to the various spatial scales which shape them. In this sense, housing movements, like those which have been gathering at the Social Fora, are not just fighting for affordable houses but questioning the general constitution of society and contributing to a global movement.

This means referring to realities on a global, national and regional scale, demanding simultaneously the democratization of international institutions which shape local living conditions or – in cases like the IMF or the WTO – even their abolishment and the immediate improvement of living conditions in the cities. Thereby, the intertwining of various spatial scales from the sub-local to the global is politicized and a precondition for the development of practices that go beyond the mere administering of global constraints or the management of the few spaces not penetrated by these constraints is created. The global articulation of urban protest can be interpreted as a 'scale jumping' which strengthens the symbolic representation of urban non-competition interests and essentially contributes to the formation of a global, but nevertheless diverse movement.

During recent years the global articulation of protest has taken various forms. There are, for example, the demonstrations against the symbolic representations of neoliberalism at Seattle, Genoa and Prague. Other important examples are the World Social Forum and the subsequent continental and local Social Forums. This type of meeting can be interpreted as one kind of crystallization of a glocalized protest where spaces for reflecting and exchanging specific local experiences with movements from other parts of the world are opened. Nevertheless, these types of events represent only one visible expression of the glocal movement scene. Beside them various formal or informal, short or long-lasting, widely perceived or rather hidden exchanges and networks emerge and contribute to the global movements. They differ due to local cultures, traditions and the specificity of former movement infrastructures. The variety of approaches is of special importance when conceiving and creating alternative strategies and encouraging struggles on different spatial scales.

In what follows we shall analyse three different types of glocalized urban action: a first type which focuses on attacking images of neoliberal restructuring and the imaginations associated with it, a second type which creates alternative knowledge, and a third type which aims at constructing material infrastructures beyond state and market. The typology, of course, is not exhaustive. Urban struggles are more diverse than these examples reveal. In the context of this article, however, the three types are of particular interest because they fight neoliberal hegemony, into which large parts of the movements of the 1970s and 1980s were integrated, at the different levels where it is reproduced in everyday life. They do so either by intervening in neoliberal image or knowledge production, and thereby creating the discursive space in which alternatives become thinkable again, or by setting up (or at least living and experiencing) concrete alternatives beyond the existing social power relations. Thereby, they politicize the contradictions of neoliberal globalization and redefine the city, establishing various forms of alternatives.

URBAN CONFLICTS AND THE TRANSFORMATION OF EVERYDAY LIFE

The first type to be presented here attaches special importance to the symbolic aspects of neoliberal restructuring. Going far beyond 'traditional' urban issues like housing and urban infrastructure (though these issues continue to be essential) it aims at transforming dominant narratives and attacking hegemonic images. This can, for example, take the form of a reproduction and overstretching of certain images or practices, as an experience from Hamburg shows: activists, dressed as members of a private security service and equipped with cameras, distributed flyers in which they propagated a more secure and clean central station. They underlined their claims by controlling tickets, and filming and accompanying people obtrusively 'for security reasons'. The action called 'Security now!' aimed at criticizing the control of public space by cameras and security services by means of an over-affirmation. Another example of the first type of urban action were the worldwide protests against the world economic summit on 18 June 1999 in Cologne. In more than forty cities all around the world local groups organized street parties, theatre or other actions at symbolic places like the City of London, thereby reclaiming streets and places dominated by global capital. This is a striking example of a glocalized urban protest: although the actions were directed at specific local places, they confronted global power relations as symbolized in the buildings of banks, corporations or international organizations. Furthermore, they were coordinated on a global scale in order to be perceived by a broader public.

The focal point of the first type of action are symbols and images which transport the neoliberal message of corporate power, flexibility, entrepreneurship and self-reliance, as well as security and cleanness. This approach takes into account that the quotidian reproduction of meaning systems on various spatial scales is a central means of achieving neoliberal hegemony – and therefore needs to be questioned in a context specific manner. By entering this symbolic universe, disturbing it or occupying it with alternative meaning systems, neoliberalism is fought at a point where its hegemony is essentially produced and reproduced and at which it therefore proves to be vulnerable.

Because of the interventionist and the very context-dependent character of the first type of protest, many groups only act for a short time. A second type of urban action tries to establish more organizational continuity. It also aims at undermining and questioning dominant paradigms. However, it does not primarily focus on neoliberal symbols but tries to produce alternative knowledge. In Germany, for example, many local groups of ATTAC practice this kind of intervention. ATTAC stands for 'Association for the Taxation of financial Transactions for the Aid of Citizens.' It was initiated by the editor of the French journal *Le Monde diplomatique*, Ignacio Ramonet, in 1997. Against the background of the financial crisis in Asia, Ramonet demanded the 'disarming of the markets' and proposed founding a civil society network which should strive for an international tax on currency speculation (the Tobin tax). Starting from

France ATTAC expanded into several countries and now forms an important part of the global social movements. It has also broadened its thematic focus from the financial markets to neoliberal restructuring in a more general sense. There are an increasing number of local groups gathering information about the manifestation and production of neoliberal globalization in their specific environment. For example, groups examine and campaign on the privatization of education systems, health care, the water supply and sewage disposal, and reflect upon this trend in the context of the General Agreement on Trade in Services (GATS) within the WTO. By reflecting on and questioning dominant paradigms they produce knowledge and thereby provide approaches for alternative activities and practices.

The third type of action to be discussed here is also a more long-term oriented one. It aims at establishing alternative social and economic (infra-)structures in general. This type ranges from very practical and locally rooted infrastructure concepts to rather complex and broad concepts. Some very striking examples of this type of action can be found especially in countries of the South where, as in Argentina, many people found themselves in long-term structural disintegration and poverty due to economic crisis and political failure. Out of socio-economic desperation and the absence of perspectives in the current political system people started to develop alternative approaches which provided them with material solutions but at the same time contained a fundamental system critique. Movements of unemployed workers like the *piqueteros* struggled throughout the 1990s for improvements in their working and living conditions. After the crisis of 2001 workers began to occupy factories abandoned by their former owners due to the economic crisis, and started to take over the whole production process. Another phenomenon which emerged and spread widely after the crisis of 2001 in Argentina was the alternative Local Exchange Trading Systems. In the first place this was a solution for satisfying basic needs in a situation of crisis. Though most of them collapsed after a while for various reasons, in many places they created the networks for further political action. Another example of this type are projects of cooperative housing movements like those in Uruguay or Brazil where solutions for affordable housing are developed and at the same time experience, spaces and knowledge for alternative political reflection, articulation and action are developed. Some of the examples mentioned act very locally and do not all have a clearly articulated relation to global movements. But in their way of acting and expressing system criticism it may be stated that they contribute to the glocal movement scene.

A European example is the already mentioned social centres in Italian cities, where people experiment with new forms of societalization, crossing the borders between economy and politics. They try to put together the development of productive enterprises with political action . . . The social centres are not only places for political aggregation, but also places for self-production that build a network of social cooperation outside the welfare state, and free from the intermediation of money, to produce what they need in harmony with nature. This is far from being easy. The social centres are threatened by quite familiar problems like the danger of being integrated into urban competition strategies as a special cultural attraction, a problem which is enforced by the need to finance the centres. Furthermore, within the centres there are problems concerning the division of labour, generational problems or tensions between innovation and continuity. The ambivalence between setting up alternative structures and the danger of being integrated into the system as repairers of the dismantled structures can be diagnosed for most approaches of this type – a phenomenon which is closely linked to the concept of social capital.

What the three approaches, despite their differences, have in common is that they fight the destructive influences which neoliberal globalization exerts on everyday life: the privatization of public space and urban infrastructures, the deterioration of working, housing and environmental conditions, the cuts in social services, the shrinking space for autonomous action. In doing so they make an important contribution to the strategic orientation of the global social movements: protests against neoliberal globalization do not exhaust themselves, for example, in the demand for the re-regulation of financial markets or the global economy in general. The struggles on a global scale, this is the message of the three types presented, strengthen the struggles against power relations in everyday life – in the city, in the neighbourhood, at home, at the workplace, at school or at university – and vice versa. Everyday life is like quantum reality: by going small you can begin to understand the whole structure of life. By changing everyday life you can change the world; why change the world if it doesn't change everyday life? And yet, how can you change everyday life without changing the world?

CONCLUSIONS

Urban social conflicts contribute to politicizing the contradictions of neoliberal globalization, to making them visible. They do so both on a material and on a symbolic level: on a material level insofar as actors aim at improving concrete living conditions, for example by preventing the privatization of public goods or by creating alternative infrastructures beyond state and market. On a symbolic level we have to distinguish two meanings of urban social conflicts. First, they contribute to disturbing the dominant narrative which takes globalization as given, inevitable and, 'in the long run', as for the benefit of everybody. Thus, they de-naturalize neoliberal globalization and make it understandable as a contested process, which is driven by certain social, political and economic forces. Second, urban conflicts represent an important dimension of the struggle against neoliberal globalization: the changing, or better liberating, of everyday life. We consider this a crucial issue because everyday life is the sphere where neoliberal hegemony has its roots (in shaping meaning systems, horizons of expectation and thereby corridors of action) and, consequently, where it can be challenged. Thus, urban social conflicts contain the chance, on the one hand, to be concrete by fighting for direct improvements of living conditions and, on the other hand, to be radical by opening up a perspective beyond existing power relations.

Of course, this is always a precarious balance, as the experience of urban social movements from the 1970s to the 1990s has shown. They were at the same time expressions of (antagonistic) protests and, mostly as an unintended consequence, agents of the modernization of local economies and political systems. One cannot escape this dilemma. The only way to learn from the success and failure of former movements is a permanent reflection about the chances and threats of acting within and/or against existing institutions, of fighting the 'spectacle' without becoming part of it. Besides this reflection being an important task of critical social science, the simultaneous articulation of protest on different spatial scales and the embeddedness of urban conflicts in the broader frame of the global social movements can help to deal with unavoidable tensions. It creates the space for exchanging experiences, for rethinking one's own political practice and for collectively developing strategies to challenge power relations. This is what remains crucial: the amplifying of space for a broad culture of reflection and debate for the movement itself. The multidimensional activities and strategies and the long-lasting search for, and diffusion of, alternative practices can only be tackled if alternative symbolic and material spaces are occupied and defended.

ILLUSTRATION CREDITS

PART ONE

1. A transatlantic arriving in the harbor, New York City, 1959 by Henri Cartier-Bresson. Reproduced by permission of Magnum Photos.
2. Skateboarder and biker. Reproduced by permission of Corbis.
3. Breakdancer on Pier 39, Fisherman's Wharf, San Francisco by Lindsay Hagan. Reproduced by permission of the artist.
4. Woman with tattoos, Haight Street, San Francisco by Lindsay Hagan. Reproduced by permission of the artist.

PART TWO

5. Manhattan, 1959 by Henri Cartier-Bresson. In the upper part of the city, around 103rd Street, some 61 blocks from Times Square, slums are being torn down with ruthless speed to make way for low cost housing projects such as these seen against the skyline. Reproduced by permission of Magnum Photos.

PART THREE

6. East 100th Street, New York City, 1966. Children in Spanish Harlem tenement by Bruce Davidson. Reproduced by permission of Magnum Photos.
7. Homeless in São Paulo, 2003 by Ferdinando Scianna. Reproduced by permission of Magnum Photos.

PART FOUR

8. Grandmother, Brooklyn, New York, 1993 by Eugene Richards. Reproduced by permission of Magnum Photos.
9. Gay Pride Parade, New York City, 1998 by Nikos Economopoulos. Reproduced by permission of Magnum Photos.

PART FIVE

10. Chinatown, New York City, 1998 by Chien-Chi Chang. A newly arrived immigrant eats noodles on a fire escape. Reproduced by permission of Magnum Photos.

PART SIX

11. Jazz Musician, New York City, 1958 by Dennis Stock. Reproduced by permission of Magnum Photos.
12. Times Square, New York City, 1999 by Raymond Depardon. Reproduced by permission of Magnum Photos.

PART SEVEN

13. Urban protest in New York City, January 28–30, 2002 by Larry Towell. During protests at the World Economic Forum, a protestor holds a "Make the Global Economy Work for Working Families" sign. Reproduced by permission of Magnum Photos.

COPYRIGHT INFORMATION

PART ONE

TÖNNIES Ferdinand Tönnies, "Community and Society" [1887] in *Community and Society*, C. P. Loomis (trans. and ed.) (1963). New York: Harper and Row. Public Domain.

SIMMEL Georg Simmel, "The Metropolis and Mental Life" [1903] in *The Sociology of Georg Simmel*, Hans Gerth (trans.), Kurt H. Wolff (ed.) (1950). Glencoe: Free Press: 409–424. Public Domain.

WIRTH Louis Wirth, "Urbanism as a Way of Life," *American Journal of Sociology* 44: 1–24 (1930). Reprinted by permission of University of Chicago Press.

GANS Herbert Gans, "Urbanism and Suburbanism as Ways of Life: A Reevaluation of Definitions," in *People and Plans: Essays on Urban Problems and Solutions* (1968). New York: Basic Books. Reprinted by permission of Perseus Books Group.

FISCHER Claude S. Fischer, "Theories of Urbanism," from *The Urban Experience*, second edition, Fischer (ed.). © 1984. Reprinted with permission of Wadsworth, a division of Thomson Learning: www.thomsonrights.com. Fax (800) 730–2215.

PART TWO

PARK Robert Ezra Park, "Human Ecology." *American Journal of Sociology* 42, 1: 1–15 (1936). Reprinted by permission of University of Chicago Press.

BURGESS Ernest W. Burgess, "The Growth of the City: An Introduction to a Research Project," in *The City*, Robert E. Park *et al.* (eds) (1925). Chicago: University of Chicago Press. Reprinted by permission of the publisher.

ZORBAUGH Harvey Zorbaugh, "The Natural Areas of the City," in *The Urban Community*, Ernest W. Burgess (ed.) (1926). Chicago: University of Chicago Press, pp. 219–229. Reprinted by permission of the publisher.

FIREY Walter Firey, "Sentiment and Symbolism as Ecological Variables," *American Sociological Review* 10: 140–148 (1945). Public Domain.

LOGAN and MOLOTCH John Logan and Harvey Molotch, "The City as a Growth Machine," excerpts from *Urban Fortunes: The Political Economy of Place* (1987). Berkeley, CA: University of California Press: pp. 1, 17–20, 23–25, 32, 52–62. © 1987 The Regents of the University of California.

DEAR Michael Dear, "Los Angeles and the Chicago School: Invitation to a Debate," *City and Community* 1, 1 (March 2002): 5–28. Reprinted by permission of Blackwell Publishing.

PART THREE

PART FOUR

PART FIVE

PART SIX

ZUKIN Sharon Zukin, "Whose Culture? Whose City?" in *The Cultures of Cities* (1995). Cambridge, MA: Blackwell Publishing: 1–48. Reprinted by permission of Blackwell Publishing.

FLORIDA Richard Florida, "Cities and the Creative Class," *City and Community* 2, 1 (March 2003): 3–19. Reprinted by permission of Blackwell Publishing.

GOTTDIENER Mark Gottdiener, "Looking at Themed Environments," in *The Theming of America*. © 1997 by Westview Press, Inc. Reprinted by permission of Westview Press, a member of Perseus Books, L.L.C.

MELE Christopher Mele, "Globalization, Culture, and Neighborhood Change: Reinventing the Lower East Side of New York," *Urban Affairs Review* 32, 1 (September 1996): 3–22. Reprinted by permission of Sage Publications, Inc.

PART SEVEN

DAVIS Mike Davis, "*Chinatown*, Part Two? The 'Internationalization' of Downtown Los Angeles," *New Left Review* I/164: 65–86 (1987).

CALDEIRA Teresa P. R. Caldeira, "Fortified Enclaves: The New Urban Segregation," *Public Culture* 8, 2: 303–328. © 1996, University of Chicago. All rights reserved. Reprinted by permission of Duke University Press.

HAMEL, LUSTIGER-THALER and MAYER Pierre Hamel, Henri Lustiger-Thaler and Margit Mayer, "Urban Social Movements – Local Thematics, Global Spaces," in *Urban Movements in a Globalising World*, Hamel, Lustiger-Thaler and Mayer (eds) (2000). London and New York: Routledge: 1–22. Reprinted by permission of the publisher.

KOHLER and WISSEN Bettina Kohler and Markus Wissen, "Glocalizing Protest: Urban Conflicts and Global Social Movements," *International Journal of Urban and Regional Research* 27, 4 (December 2003): 942–51. Reprinted by permission of Blackwell Publishing.

Index